T0140104

Connected Vehicles in the Internet of Things

Zaigham Mahmood
Editor

Connected Vehicles in the Internet of Things

Concepts, Technologies and Frameworks for the IoV

 Springer

Editor
Zaigham Mahmood
Debesis Education
Littleover, Derby, UK

Northampton University
Northampton, UK

Shijiazhuang Tiedao University
Hebei, China

ISBN 978-3-030-36169-3   ISBN 978-3-030-36167-9   (eBook)
https://doi.org/10.1007/978-3-030-36167-9

This Springer imprint is published by the registered company Springer Nature Switzerland AG
The registered company address is: Gewerbestrasse 11, 6330 Cham, Switzerland

My 10th and 20th publications were dedicated to my parents. This 30th book is also in memory of my late parents, Ghazi Ghulam Hussain Bahadur and Mukhtar Begum, who spent the prime of their life in fighting for the freedom and independence of their motherland. At a very young age, my father joined a para military organization, as a freedom fighter, with the mission to engage in peaceful struggle to free the country from foreign occupation. Although, the struggle for independence started many decades before and various political parties participated in many diverse ways towards it; there is one event that stands out, that took place on 19 March 1940.

On this day, the organization, that my father belonged to, decided to stage a much more decisive countrywide peaceful protest. The government at the time, fearing the shut-down of the country, had already banned the gatherings but, on this day, supporters and general pubic were out in such huge numbers that the army personnel patrolling the streets received orders to shoot to kill. Live bullets were fired; many thousands were killed or injured and many more taken as political prisoners. That day, my father was leading a group of 313 men— totally unarmed—marching on the streets to oppose the ban on political activities. According to newspaper reports, more than 200 of this group were killed and many dozens injured; majority of the remaining were captured and tried in the courts. There were thirteen freedom fighters who were sentenced to political imprisonment for life— my father was one of the 13. His organization honoured these brave men with the titles of Ghazi (survivor in the fight between the right and the wrong) and Bahadur (valiant).

This brutality by foreign occupiers and the massacre of unarmed public on 19 March 1940 proved such a turning point in the struggle for independence that only four days later, on 23 March 1940, an all-party confederation passed a unanimous resolution demanding the formation of an independent state. Soon after, a declaration was signed to transfer power to the leading political organisations of the land. The process took another seven years, and eventually the country achieved independence on 14 August 1947. On this day, all freedom fighters were released; my father also returned home ghazi and victorious. My mother, a young girl at the time, was no less courageous in her struggles: she fully supported her husband's mission and raised a young daughter totally independently while my father was away.

Now that the independence was gained and the mission accomplished, my father devoted his time to engage in the study of oriental languages and theology, bringing up his family, and serving the community as a social activist. Achieve excellence ... be bold ... side always with the truth ... make a difference: my parents would constantly remind us.

Looking back now at the life and struggles of my parents, I am proud to say that they were certainly most excellent and bold to side with the truth; and undoubtedly made a huge difference for the entire nation to remember. They are my heroes and inspiration in my life.

Zaigham Mahmood
October 2019

Preface

Overview

The Internet of Things (IoT) paradigm and device connectivity protocols are rapidly developing into a vision of pervasive and distributed computing, where *smart* objects connect to each other in a seamless manner, to establish a unified physical-virtual world. The interconnection between processor embedded objects, whether a household appliance, a vehicle on the road or a body wearable, is helping to build smart environments such as intelligent cities, automated factories and intelligent transportation systems. According to research by Gartner, connected devices across the globe, and across all technologies, will reach at least 20 billion in number, by 2020. In this context, connected vehicles are another welcome addition to the applications portfolio of the IoT.

A connected vehicle is one that is equipped with Internet access and wireless local area network that allow communication with other smart devices, whether inside the vehicle or outside in the environment. There arc a number of ways that vehicles may be connected to each other and with the environment as well as people such as drivers and those on the roads. These include: vehicle-to-vehicle, vehicle-to-infrastructure, vehicle-to-pedestrian, vehicle-to-cloud and generally as vehicle-to-everything. Various other terms relevant to the connected vehicles vision include: (1) Internet of Vehicles (IoV) that refers to convergence of the mobile Internet and the IoT; (2) Mobile Ad Hoc Networks (MANET) that are a type of ad hoc network that allow network nodes to change locations and configurations on the fly; and (3) Vehicular Ad Hoc Networks (VANET) that are a type of MANET, referring specifically to the connectivity of vehicles and created by applying the principles of MANET.

There is clear evidence that technological innovation in the field of connectivity and pervasive computing in the automotive industry is accelerating rapidly. With this progress, Advanced Driving Assistance Systems (ADAS) and Intelligent Transportation Systems (ITS) are also coming on board at a much faster pace. However, like any other technology, we are also witnessing numerous challenges and cyber-related concerns; relating to reliability, safety, security, privacy, mobility, device communication and autonomous self-drive nature of new breed of vehicles. Nevertheless, since the IoV, VANET and MANET present an attractive way forward, mainly due to the huge social, industrial and economic benefits that they offer, issues and related challenges are being researched, addressed and resolved.

It is in the above context that this book is developed. The focus of the volume is on the use of IoT for the connectivity of vehicles on the roads, in particular the relevant principles, frameworks, architectures, technologies and applications, as well as practical suggestions and solutions to the inherent barriers and challenges.

In this book, 24 researchers and practitioners of international repute have presented latest research, current trends and case studies, as well as suggestions for further understanding, development and enhancement of the much attractive IoV paradigm.

Objectives

The aim of this volume is to present and discuss the Internet of Vehicles (IoV) vision as extended to the connected vehicles paradigm. The general objectives include the following:

- Latest research on wireless connectivity, machine learning, sensor technology and autonomous operations in the IoV vision.
- Technical solutions that address the key challenges and limitations with respect to device connectivity, VANET and MANET.
- Textbook and complete reference for students, researchers and practitioners in the subject area of IoV, VANET and MANET.

Organization

There are 11 chapters in this book. These are organized in three parts, as follows:

Part I: Technologies and Architectures

This part of the book has a focus on concepts, technologies and architectures relating to connectivity of vehicles. There are five chapters in this section. The first contribution explores technologies and architectures for the connected vehicles vision. The second contribution presents the concept of special intelligence in the case of V2V connectivity and discusses relevant topologies and architectures. The third chapter focuses on V2I communication in 5G heterogeneous networks, discusses associated challenges and presents future research directions. The next chapter highlights the use and integration of vehicular technologies within the prevailing IoT environment in Egypt and addresses the emerging challenges and future advancements. The fifth contribution proposes a conceptual framework articulating an architectural configuration suitable for MANET.

Part II: Frameworks and Methodologies

This part of the book comprises three chapters. The focus is on methods and mechanisms. The first contribution presents a review of intelligent traffic management systems suitable for next generation IoV. It discusses the learning and context-aware systems for better driving experience in a smart city scenario. The next chapter reviews the existing literature on smart transportation tracking systems based on the IoT vision. Information from 30 recently published papers has been summarized to present the state-of-the-art concerning vehicle tracking systems. The final chapter in this section presents the development of a unified telemetry platform for electric vehicles and vehicle charging infrastructures. The authors promote an open-source approach to charging station software development and implementation.

Part III: Security and Privacy in the IoT

This section has three contributions that focus on security and privacy challenges relating to vehicular ad hoc networks. The first chapter presents a review of such challenges. It also addresses the related concerns and proposes solutions and countermeasures to mitigate the security and privacy-related threats. The next chapter takes a similar look at the security concerns and presents a robust solution as a need for future evolution of vehicular ad hoc Networks towards the evolution of IoV. The suggestion consists of a lightweight scheme for authentication and location of vehicles combined with a digital watermark-based revocation mechanism. The final chapter of the book proposes to integrate the cloud paradigm with VANET so that resources may be shared with the network. The chapter also proposes a novel resource management architecture.

Target Audiences

This current volume is a reference text aimed at supporting a number of potential audiences, including the following:

- *Network Specialists, Wireless Communication Engineers and Cyber Security Experts*: who wish to adopt the newer approaches to device connectivity, network security, data privacy and sensors-based devices design relevant to the connected vehicles, MANET and VANET vision.

- *Students, Academics, Researchers and Practitioners*: who have an interest in further enhancing the knowledge of technologies, mechanisms and practices relevant to IoV, MANET and VANET from a distributed computing perspective in the IoT vision.

Northampton, UK   Zaigham Mahmood
Hebei, China

Acknowledgements

The editor acknowledges the help and support of the following colleagues during the review, development and editing phases of this text:

- Prof. Zhengxu Zhao, Shijiazhuang Tiedao University, Hebei, China
- Dr. Alfredo Cuzzocrea, University of Trieste, Trieste, Italy
- Dr. Emre Erturk, Eastern Institute of Technology, New Zealand
- Prof. Jing He, Kennesaw State University, Kennesaw, GA, USA
- Josip Lorincz, FESB-Split, University of Split, Croatia
- Aleksandar Milić, University of Belgrade, Serbia
- Prof. Sulata Mitra, Indian Institute of Engineering Science and Technology, Shibpur, India
- Dr. S. Parthasarathy, Thiagarajar College of Engineering, Tamil Nadu, India
- Daniel Pop, Institute e-Austria Timisoara, West University of Timisoara, Romania
- Dr. Pethuru Raj, IBM Cloud Center of Excellence, Bangalore, India
- Dr. Muthu Ramachandran, Leeds Becket University, Leeds, UK
- Dr. Lucio Agostinho Rocha, State University of Campinas, Brazil
- Dr. Saqib Saeed, University of Dammam, Saudi Arabia
- Prof. Claudio Sartori, University of Bologna, Bologna, Italy
- Dr. Mahmood Shah, University of Central Lancashire, Preston, UK
- Dr. Fareeha Zafar, GC University, Lahore, Pakistan

I would also like to thank the contributors to this book: 24 authors and co-authors from academia as well as the industry from around the world, who collectively submitted 11 well-researched detailed chapters. Without their efforts in developing quality contributions, conforming to the guidelines and meeting often the strict deadlines, this text would not have been possible.

Grateful thanks are also due to the members of my family—Rehana, Zoya, Imran, Hanya, Arif and Ozair—for their continued support and encouragement. Every good wish, also, for the youngest and most delightful in our family: Eyaad Imran Rashid Khan and Zayb-un-Nisa Khan.

October 2019   Zaigham Mahmood

Other Books by Zaigham Mahmood

Software Engineering in the Era of Cloud Computing

This book has a focus on developing scalable complex software for distributed computing applications. It presents and discusses state of the art of Software Engineering (SE) in terms of methodologies, trends and future direction for cloud SE. Domain modelling, SE testing in the cloud, SE analytics and software process improvement as a service are also discussed. ISBN: 978-3-030-33623-3.

The Internet of Things in the Industrial Sector: Security and Device Connectivity, Smart Environments, and Industry 4.0

This reference text has a focus on the development and deployment of the Industrial Internet of Things (IIoT) paradigm, discussing frameworks, methodologies, benefits and inherent limitations of connected smart environments, as well as providing case studies of employing the IoT vision in the industrial domain. ISBN: 978-3-030-24891-8.

Security, Privacy and Trust in the IoT Environment

This book has a focus on security and privacy in the Internet of things environments. It also discusses the aspects of user trust with respect to device connectivity. Main topics covered include: principles, underlying technologies, security issues, mechanisms for trust and authentication as well as success indicators, performance metrics and future directions. ISBN: 978-3-030-18074-4.

Guide to Ambient Intelligence in the IoT Environment: Principles, Technologies and Applications

This reference text discusses the AmI element of the IoT paradigm and reviews the current developments, underlying technologies and case scenarios relating to AmI-based IoT environments. The book presents cutting-edge research, frameworks and methodologies on device connectivity, communication protocols and other aspects relating to the AmI-IoT vision. ISBN: 978-3-030-04172-4.

Fog Computing: Concepts, Frameworks and Technologies

This reference text describes the state of the art of Fog and Edge computing with a particular focus on development approaches, architectural mechanisms, related technologies and measurement metrics for building smart adaptable environments. The coverage also includes topics such as device connectivity, security, interoperability and communication methods. ISBN: 978-3-319-94889-8.

Smart Cities: Development and Governance Frameworks

This text/reference investigates the state of the art in approaches to building, monitoring, managing and governing smart city environments. A particular focus is placed on the distributed computing environments within the infrastructure of smart cities and smarter living, including issues of device connectivity, communication, security and interoperability. ISBN: 978-3-319-76668-3.

Data Science and Big Data Computing: Frameworks and Methodologies

This reference text has a focus on data science and provides practical guidance on big data analytics. Expert perspectives are provided by an authoritative collection of 36 researchers and practitioners, discussing latest developments and emerging trends; presenting frameworks and innovative methodologies; and suggesting best practices for efficient and effective data analytics. ISBN: 978-3-319-31859-2.

Connected Environments for the Internet of Things: Challenges and Solutions

This comprehensive reference presents a broad-ranging overview of device connectivity in distributed computing environments, supporting the vision of IoT. Expert perspectives are provided, covering issues of communication, security, privacy, interoperability, networking, access control and authentication. Corporate analysis is also offered via several case studies. ISBN: 978-3-319-70101-1.

Connectivity Frameworks for Smart Devices: The Internet of Things from a Distributed Computing Perspective

This is an authoritative reference that focuses on the latest developments on the Internet of things. It presents state of the art on the current advances in the connectivity of diverse devices and focuses on the communication, security, privacy, access control and authentication aspects of the device connectivity in distributed environments. ISBN: 978-3-319-33122-5.

Cloud Computing: Methods and Practical Approaches

The benefits associated with cloud computing are enormous; yet the dynamic, virtualized and multi-tenant nature of the cloud environment presents many challenges. To help tackle these, this volume provides illuminating viewpoints and case studies to present current research and best practices on approaches and technologies for the emerging cloud paradigm. ISBN: 978-1-4471-5106-7.

Cloud Computing: Challenges, Limitations and R&D Solutions

This reference text reviews the challenging issues that present barriers to greater implementation of the cloud computing paradigm, together with the latest research into developing potential solutions. This book presents case studies, and analysis of the implications of the cloud paradigm, from a diverse selection of researchers and practitioners of international repute. ISBN: 978-3-319-10529-1.

Continued Rise of the Cloud: Advances and Trends in Cloud Computing

This reference volume presents latest research and trends in cloud-related technologies, infrastructure and architecture. Contributed by expert researchers and practitioners in the field, this book presents discussions on current advances and practical approaches including guidance and case studies on the provision of cloud-based services and frameworks. ISBN: 978-1-4471-6451-7.

Software Engineering Frameworks for the Cloud Computing Paradigm

This is an authoritative reference that presents the latest research on software development approaches suitable for distributed computing environments. Contributed by researchers and practitioners of international repute, the book offers practical guidance on enterprise-wide software deployment in the cloud environment. Case studies are also presented. ISBN: 978-1-4471-5030-5.

Cloud Computing for Enterprise Architectures

This reference text, aimed at system architects and business managers, examines the cloud paradigm from the perspective of enterprise architectures. It introduces fundamental concepts, discusses principles and explores frameworks for the adoption of cloud computing. The book explores the inherent challenges and presents future directions for further research. ISBN: 978-1-4471-2235-7.

Cloud Computing: Concepts, Technology & Architecture

This is a textbook (in English but also translated in Chinese and Korean) highly recommended for adoption for university-level courses in distributed computing. It offers a detailed explanation of cloud computing concepts, architectures, frameworks, models, mechanisms and technologies—highly suitable for both newcomers and experts. ISBN: 978-0133387520.

Software Project Management for Distributed Computing: Life-Cycle Methods for Developing Scalable and Reliable Tools

This unique volume explores cutting-edge management approaches to developing complex software that is efficient, scalable, sustainable and suitable for distributed environments. Emphasis is on the use of the latest software technologies and frameworks for life-cycle methods, including design, implementation and testing stages of software development. ISBN: 978-3-319-54324-6.

Requirements Engineering for Service and Cloud Computing

This text aims to present and discuss the state of the art in terms of methodologies, trends and future directions for requirements engineering for the service and cloud computing paradigm. Majority of the contributions in the book focus on requirements elicitation; requirements specifications; requirements classification and requirements validation and evaluation. ISBN: 978-3-319-51309-6.

User Centric E-Government: Challenges & Opportunities

This text presents a citizens-focused approach to the development and implementation of electronic government. The focus is twofold: discussion on challenges of service availability, e-service operability on diverse smart devices, as well as on opportunities for the provision of open, responsive and transparent functioning of world governments. ISBN: 978-3-319-59441-5.

Cloud Computing Technologies for Connected Government

This text reports the latest research on Electronic Government for enhancing the transparency of public institutions. It covers a broad scope of topics including citizen empowerment, collaborative public services, communication through social media, cost benefits of the cloud paradigm, electronic voting systems, identity management and legal issues. ISBN: 9781466686298.

Human Factors in Software Development and Design

This reference text brings together high-quality research on the influence and impact of ordinary people on the software industry. With the goal of improving the quality and usability of computer technologies, topics include global software development, multi-agent systems, public administration platforms, socio-economic factors and user-centric design. ISBN: 9781466664852.

IT in the Public Sphere: Applications in Administration, Government, Politics, and Planning

This reference text evaluates current research and best practices in the adoption of e-government technologies in developed and developing countries, enabling governments to keep in touch with citizens and corporations in modern societies. Topics covered include citizen participation, digital technologies, globalization, strategic management and urban development. ISBN: 9781466647190.

Emerging Mobile and Web 2.0 Technologies for Connected E-Government

This reference highlights the emerging mobile and communication technologies, including social media, deployed by governments for use by citizens. It presents a reference source for researchers, practitioners, students and managers interested in the application of recent technological innovations to develop an open, transparent and more effective e-government environment. ISBN: 9781466660823.

E-Government Implementation and Practice in Developing Countries

This volume presents research on current undertakings by developing countries towards the design, development and implementation of e-government policies. It proposes frameworks and strategies for the benefits of project managers, government officials, researchers and practitioners involved in the development and implementation of e-government planning. ISBN: 9781466640900.

Developing E-Government Projects: Frameworks and Methodologies

This text presents frameworks and methodologies for strategies for the design, implementation of e-government projects. It illustrates the best practices for successful adoption of e-government and thus becomes essential for policymakers, practitioners and researchers for the successful deployment of e-government planning and projects. ISBN: 9781466642454.

Contents

About the Editor

Prof. Dr. Zaigham Mahmood is a published author/editor of 30 books on subjects including Electronic Government, Cloud Computing, Data Science, Big Data, Fog Computing, Internet of Things, Internet of Vehicles, Industrial IoT, Smart Cities, Ambient Intelligence, Project Management, and Software Engineering, including *Cloud Computing: Concepts, Technology & Architecture* which is also published in Korean and Chinese languages. Additionally, he is developing two new books to appear later in the year. He has also published more than 150 articles and book chapters and organized numerous conference tracks and workshops.

Professor Mahmood is Editor-in-Chief of *Journal of E-Government Studies and Best Practices*, Editor-in-Chief of the IGI book series on *E-Government and Digital Divide*, Technology Consultant at Debesis Education UK and Professor at the Shijiazhuang Tiedao University in Hebei, China. He further holds positions as Foreign Professor at NUST and IIU in Islamabad, Pakistan. He has also served as a Reader (Associate Professor) at the University of Derby, UK, and Professor Extraordinaire at the North West University, South Africa. Professor Mahmood is a certified cloud computing instructor and a regular speaker at international conferences devoted to Distributed Computing and E-Government. Professor Mahmood's book publications can be viewed at: https://www.amazon.co.uk/Zaigham-Mahmood/e/B00B29OIK6.

Contributors

Aya Sedky Adly Faculty of Computers and Artificial Intelligence, Helwan University, Cairo, Egypt

Mohamed Ben Zid Center for Cybercrime Studies, John Jay College of Criminal Justice of the City University of New York, New York, USA

Thomas Bräunl The Renewable Energy Vehicle Project (REV), Department of Electrical, Electronic and Computer Engineering, The University of Western Australia, Perth, Australia

Kelvin Joseph Bwalya School of Consumer Intelligence and Information Systems, University of Johannesburg, Johannesburg, South Africa

W. K. A. Upeksha K. Fernando School of Computing and Mathematics, Charles Sturt University, Melbourne, VIC, Australia

Kayhan Zrar Ghafoor School of Mathematics and Computer Science, University of Wolverhampton, Wolverhampton, UK;
Department of Software Engineering, Salahaddin University, Erbil, Iraq

Malka N. Halgamuge Department of Electrical and Electronic Engineering, The University of Melbourne, Parkville, VIC, Australia

Jennifer Holst Center for Cybercrime Studies, John Jay College of Criminal Justice of the City University of New York, New York, USA

Matluba Khodjaeva Center for Cybercrime Studies, John Jay College of Criminal Justice of the City University of New York, New York, USA

Kai Li Lim The Renewable Energy Vehicle Project (REV), Department of Electrical, Electronic and Computer Engineering, The University of Western Australia, Perth, Australia

Zaigham Mahmood Debesis Education, Derby, UK;
Northampton University, Northampton, UK;
Shijiazhuang Tiedao University, Hebei, China

Sulata Mitra Department of Computer Science and Technology, Indian Institute of Engineering, Science and Technology, Shibpur, Howrah, India

Seshadri Mohan University of Arkansas at Little Rock, Little Rock, AR, USA

Atanu Mondal Ramakrishna Mission Vidyamandira, Howrah, India

Sukumar Nandi Department of CSE, Indian Institute of Technology Guwahati, Guwahati, Assam, India

Sunit Kumar Nandi Department of CSE, Indian Institute of Technology Guwahati, Guwahati, Assam, India;
Department of CSE, NIT Arunachal Pradesh, Yupia, India

Muath Obaidat Center for Cybercrime Studies, John Jay College of Criminal Justice of the City University of New York, New York, USA

J. Rian Leevinson Infosys Limited, Chennai, India

Ruwani M. Samarakkody School of Computing and Mathematics, Charles Sturt University, Melbourne, VIC, Australia

Sachin Sharma Graphic Era Deemed to be University, Dehradun, UK, India

Pranav Kumar Singh Department of CSE, Indian Institute of Technology Guwahati, Guwahati, Assam, India;
Department of CSE, Central Institute of Technology Kokrajhar, Kokrajhar, Assam, India

Roshan Singh Department of CSE, Central Institute of Technology Kokrajhar, Kokrajhar, Assam, India

Stuart Speidel The Renewable Energy Vehicle Project (REV), Department of Electrical, Electronic and Computer Engineering, The University of Western Australia, Perth, Australia

V. Vijayaraghavan Infosys Limited, Bengaluru, India

Part I
Technologies and Architectures

Chapter 1
Connected Vehicles in the IoV: Concepts, Technologies and Architectures

Zaigham Mahmood

Abstract A connected car is an essential element of the Internet of Vehicles (IoV) vision that is a highly attractive application of the Internet of Things (IoT). The underlying technologies include Internet of Everything (IoE), artificial intelligence, machine learning, neural networks, sensor technologies, and cloud/edge computing. The connectivity between vehicles is through inter communication between sensors and smart devices inside the vehicles, as well as smart systems in the environment as part of the Intelligent Transportation Systems (ITS). In this chapter, the focus is on underlying concepts, architectures and relevant technologies. Types of connectivity and inter communication are also discussed. State of the art is articulated and, in the last sections of the chapter, future vision of vehicles' connectivity is outlined and conclusions presented. The aim is that this chapter serves as a basis and sets the scene for detailed presentations on various aspects of connected vehicles that appear later in this book.

Keywords Connected vehicles · Internet of things · Internet of vehicles · IoT · IoV · Architecture · Smart city · Smart car · Autonomous car · Automation · Intelligent transportation system · V2V · V2I · V2C · V2X

1.1 Introduction

The Internet of Things, or IoT, is a network of interrelated *things* including computing devices, mechanical and digital machines, smart sensors, and even people. These things are provided with unique identifiers and have the ability to receive, in

Z. Mahmood (✉)
Debesis Education, Derby, UK
e-mail: dr.z.mahmood@hotmail.co.uk

Z. Mahmood
Northampton University, Northampton, UK

Z. Mahmood
Shijiazhuang Tiedao University, Hebei, China

© Springer Nature Switzerland AG 2020
Z. Mahmood (ed.), *Connected Vehicles in the Internet of Things*,
https://doi.org/10.1007/978-3-030-36167-9_1

some cases process and transfer data to other smart objects, without requiring human interaction. Kevin Ashton of Auto-ID Centre at MIT [1] is attributed to having used the term IoT at a presentation in 1999.

The concept of connectivity of devices is not new. The first implementation of the idea was probably in 1982 when a coke vending machine, at Carnegie Mellon University Computer Science Department in the US, was connected to the Internet for the programming staff to status check the availability of coke in the machine [2]. The IoT paradigm promises the following benefits:

- Allows connectivity of all kind of *things* to develop *smarter* environments
- Allows automation to make peoples' lives easier and more comfortable
- Enables organisations to automate, increase efficiency and reduce costs
- Enables companies to cut down on waste and improve service delivery
- Enables companies to integrate business models and enhance productivity.

With the passing of time, new technologies and mechanisms appeared and matured, e.g. wireless and sensor technologies, machine-to-machine communication, artificial intelligence, machine learning, neural networks, cloud storage, and big data analytics. In the last two decades, the convergence of such technologies and related frameworks has resulted in a number of highly attractive applications of IoT, including the following:

- Smart Homes: where status of heating, lighting and security, etc. can be remotely monitored, switched on/off and their operations modified
- Smart Roads: where sensory systems embedded in roads warn vehicles of road incidents ahead, and provide information regarding traffic situations
- Smart Healthcare: where technology and smart wearables are used as more efficient diagnostic tools, for better treatment of patients
- Smart Manufacturing: where computer-integrated manufacturing results in better adoptability, rapid design changes and intelligent automation
- Industrial IoT: which refers to the implementation of IoT technologies and capabilities in the industrial and manufacturing space
- Intelligent Transportation System: where innovative services relating to traffic management enable safe and smarter use of transport networks
- Internet of Vehicles: where the moving network of IoT enabled vehicles share information collected via in-car smart devices to help decision-making.

In general, IoT touches every industry including manufacturing, agriculture, transportation, logistics, education, healthcare, and retail. Nevertheless, connectivity of devices, especially when these are heterogeneous and their number in the network is increasing with time, imposes numerous new challenges. For example:

- as number of devices increase, amount of information flow increases, which in turn has implications with respect to data safety, security, storage, and data analytics, as well as, decision-making in real-time

- as the variety of devices increase, it poses challenges with respect to device compatibility, device-to-device (D2D) communication, and information exchange protocols
- as users become more knowledgeable, their expectations grow exponentially, which in turn raises issues of availability, reliability and quality of service of physio-cyber systems.

Having said the above, the future is bright. Communication protocols are getting better, new security mechanisms are being developed, standards organisations are working towards enhancing safety of systems and people, and compatibility of devices.

The remainder of this chapter is organised as follows:

- The next section introduces the concept of connected vehicles and Internet of Vehicles (IoV); and explores connectivity architectures.
- The following section presents the underlying technologies and discusses related benefits and inherent issues
- The final section concludes our contribution and presents some future directions.

1.2 Connected Vehicles: Concepts and Architectures

A connected vehicle is one that has access to the Internet for the sharing of data with smart devices inside the car as well as with smart objects outside the car including other cars, traffic lights, and road sensors, etc. A connected car is an essential element of the Internet of Vehicles (IoV) vision that is a highly attractive application of the IoT. This, in turn, becomes part of a smart city vision that includes smart living in smart homes, and smart working in the more supportive Industrial IoT (IIoT) scenario. In 2019, some of the leading manufacturers of cars, who are advancing the IoV vision, include: General Motors, Google, BMW, Audi, Mercedes Benz, Tesla, VW, Jaguar, Porche, and Nissan [3]. CMSWire [4] suggest that, by 2020, we are expected to see more than 380 million cars on the roads. Business Insider [5] expect 94 million connected cars to be shipped in 2021 and 82% of these cars will be connected.

In the following sub-sections, we discuss different types of vehicle connectivity and the varieties of vehicle applications.

1.2.1 Types of Vehicle (or Car) Connectivity

There are generally five major approaches to vehicle communication and connectivity viz: Vehicle-to-Vehicle (V2V), Vehicle-to-Infrastructure (V2I), Vehicle-to-Cloud (V2C), Vehicle-to-Pedestrian (V2P) and Vehicle-to-Everything (V2E). These are further explored below. Here the term 'vehicle' is used more generically to refer to all types of road vehicles including cars.

V2V Communication

This refers to a vehicle (e.g. car, bus, lorry, etc.) being connected to other vehicles of any type. Communication is necessarily wireless for purposes such as determining location, position, direction of travel, and speed of moving vehicles. The technology behind V2V communication allows vehicles to send and receive omni-directional messages (normally every 5–10 s), creating a 360-degree 'awareness' of other vehicles in proximity. Through such messages, vehicles can determine potential crash threats as they develop. V2V technology uses dedicated short-range communication (DSRC), sometimes described as being a WiFi network because one of the possible frequencies is 5.9 GHz, normally used by WiFi. The technology can then employ visual, tactile, and audible alerts, or a combination of these, to warn drivers to take appropriate action to avoid crashes, e.g. change speed or change direction. All this helps to provide better and safer experience for drivers and passengers; and reduce accidents and fatalities.

V2I Communication

This refers to a vehicle being connected to highway infrastructure that includes traffic lights, traffic signal sensors, road sensors, speed cameras, communication satellite, parking metres, bus stops, etc. Communication is wireless and bi-directional. Hardware technologies and components include sensors such as RFID, radars, and cameras. Communication takes place using dedicated DSRC frequencies which are similar to V2V connectivity. The purpose is to capture infrastructure data to provide travellers with real-time advisory information to change speed, apply brake, follow diverted roots or avoid certain situations. Like V2V, this also provides better and safer experience for drivers and passengers; and reduces accidents and injuries. Difference between V2V and V2I is that, whereas V2V requires other vehicles in the environment to be also connected, V2I needs connectivity only between the infrastructure and the vehicle itself.

V2C Communication

This refers to a vehicle being wirelessly linked up to a cloud environment mainly to: (1) access, transfer, share, and analyse information; and make decisions in real time; (2) to store and retrieve data; and (3) to provision and consume other cloud-based services of all varieties. The idea of using these services in vehicles is attractive for many reasons, including that: (1) cloud technologies can help to reduce the size of central consoles in vehicles; (2) these technologies require less hardware under the dash board; and (3) many data-intensive tasks can be outsourced to remote cloud-based servers [6]. This, in turn, means reduction in expensive data storage elements in the vehicles and thus possible reduced prices of increasingly computerised cars. Cloud related capabilities can also enhance automotive safety as well as augmented reality in vehicles.

V2P Communication

This involves direct communications between a vehicle and pedestrians, within close proximity of the vehicle. Pedestrians encompass a broad set of road users including people walking or crossing the road, children being pushed in strollers, people using wheelchairs or other mobility devices, passengers embarking and disembarking buses, and people riding bicycles. The aim is to increase pedestrian safety and to avoid accidents involving road users, as well as to help improve safety of vehicle occupants. One important aspect of V2P is the pedestrian detection system [7] that can be implemented in a number of ways, viz: (1) within vehicles (for blind spot warning, incident ahead warning, etc.); (2) embedded within the roadside infrastructure (e.g. lane closure warning); and (3) carried by pedestrians (e.g. smart devices to warn the drivers). Car manufacturers are already well-advanced regarding testing of V2P technology. One example is Telstra in partnership with Cohda Wireless, who have successfully conducted V2P technology experiments over a mobile network in South Australia [8]. The technology was tested using some common every day scenarios such as a car and a cyclist approaching a blind corner.

V2X (Vehicle-to-Everything) Communication

This is a relatively nascent market. It refers to the technology and processes that allow vehicles to communicate with different parts of the traffic system around them. In this respect, V2X is a context aware system that includes V2V, V2I and V2P communication [9]. In addition to the safety benefits, V2X technology offers a range of other convenient benefits, e.g. payments at toll stations and for parking metres, etc. However, in order for the V2X to become more pervasive, other vehicles and objects in the traffic system in the immediate environment (e.g. other vehicles, road infrastructure elements such as traffic lights, etc.), must also have V2X technology embedded in them. Even the pedestrians and cyclists must also really carry the V2X devices. It is for this reason that V2X needs to advance much more and very quickly. Nevertheless, many V2X suppliers such as Delphi, Denso, Qualcomm and Continental are working hard towards V2X integration plans [9].

1.2.2   Mobile Ad Hoc Network (MANET)

MANET is used for communication between vehicles and roadside smart devices. In this network, also known as Wireless Ad Hoc Network (WANET) or Ad hoc Wireless Network, nodes (i.e. vehicles) are autonomous and mobile in nature and free to move independently in any direction. Nodes are, therefore, interfaced dynamically in an arbitrary fashion. Hence, the network uses ad hoc nature of routing protocols that are either table-driven or of on-demand variety.

MANETs have a self-configuring infrastructure-less network. These have the following features: dynamic topology; variable capacity links; energy constrained operation; and limited physical security. Some MANET are generally restricted to a local area of wireless devices, while others may be connected to the Internet. In MANET, every node is a potential router. Here, the routing protocols that are often used include: ADOV (Ad Hoc On-demand Distance Vector), DSR (Dynamic

Source Routing), OLSR (Optimised Link State Routing) and TORA (Temporally Ordered Routing Algorithm).

MANETs are decentralised and, therefore, typically more robust than centralised networks due to the multi-hop nature of data transmission. Since the MANET architecture evolves with time, it has the potential to resolve issues such as isolation or disconnection of nodes or the slowness of Internet data flows. Also, due to the fact that MANET topology changes, these networks are more flexible and scalable, with lower administration costs. However, with the time evolving nature of the network and changing mobility patterns of vehicles and devices, there can be variations in the network performance and quality of service. There are also concerns relating to signal protection, reliability of nodes, limited processing power, and even the adequacy of power supply. Still, as said before, flexibility of MANET is a huge positive attribute.

1.2.3   Vehicular Ad Hoc Network (VANET)

VANET is a subclass of MANET type of connectivity, applicable specifically to vehicles. It consists of groups of moving (and stationary) vehicles, connected via a wireless network. Until recently, VANETs were used mainly to provide safety and comfort to drivers in vehicular environments. Now, however, these networks are seen as infrastructure for an intelligent transportation system (ITS) for an increasing number of autonomous vehicles on the roads. One of the distinguishing characteristics of VANET is the content-centric distribution—the content being important, rather than the source. This is in marked contrast to the Internet where an agent demands information from a specific source.

VANET communication protocols are similar to those used in wired networks, where each host has an IP address. However, here, assigning such addresses to moving vehicles is far from trivial and often requires a Dynamic Host Configuration Protocol (DHCP) server, a heresy for Ad Hoc networks that operate without any infrastructure, using self-organisation protocols. One difference between VANET and MANET is that the routing protocols of MANET are not feasible to be used in VANET architecture.

The VANET architecture generally consists of three types of categories viz: (1) cellular and WLAN network where fixed gateways and WiMaX/WiFi Aps are used for connection to the Internet for routing and getting traffic information; (2) Pure ad hoc i.e. between vehicles and fixed gateways; and (3) hybrid i.e. a combination of both infrastructure and ad hoc networks. Various popular architectures of VANET include: WAVE (Wireless Access in Vehicular Environment) by IEEE, CALIM (Continuous Air Interface for Long to Medium range) by ISO, and C2CNet (Car-to-Car Network) by C2C Consortium.

VANETs can be seen as systems consisting of entities that can be divided into three domains as mentioned below:

- Mobile domain: that consists of two sub-parts: the vehicle domain (i.e. cars, buses, etc.) and mobile device domain (i.e. handheld devices, smart phones, navigation system, etc.)
- Infrastructure domain: that also comprises two sub-parts: the roadside infrastructure domain (i.e. roadside units and traffic lights) and the central infrastructure domain (i.e. traffic and vehicle management centres)
- generic domain: that, in turn, consists of Internet infrastructure domain and the private infrastructure domain.

1.2.4 Internet of Vehicles (IoV)

The new era of the IoT is driving the evolution of conventional VANET into the new IoV vision. As such, IoV integrates VANET, IoT and the mobile Cloud Computing. IoT and VANET have been discussed above; the Cloud paradigm provides data storage and data analytics facilities, and provision of all kind of services, whether software related or virtualised hardware. In Vehicle-to-Cloud (V2C) communication, any vehicle in VANET can directly communicate with the cloud environment, and provision cloud-based services.

IoV refers to the data interaction, in real time, between vehicles and roadside units using mobile-communication technology, vehicle navigation systems, smart-terminal devices, and information platforms. The objective is to enable information exchange and interaction and a driving–instruction–controlling network system.

IoV can be regarded as an extension of Vehicle-to-Vehicle (V2V) connectivity that enhances driving aids for full autonomous driving by furthering the vehicles' artificial intelligence awareness of the surrounding environment (including other vehicles and driving-related smart devices). The components of IoV include the following [10]: Vehicles (as nodes of the ad hoc network), Roadside Units (RSU) and infrastructure (e.g. road and traffic related sensors, signals and smart objects), personal devices (e.g. smart phones and PDAs), and humans (e.g. drivers). Therefore, there are interactions such as the following:

- Vehicle-to-Vehicle interaction
- Vehicle and Infrastructure interaction
- Vehicle and sensors interaction
- Vehicle and RSU interaction
- Vehicle and human interaction
- Personal devices and vehicle interaction
- Cloud and vehicle interaction
- Device-to-device (D2D) interaction.

To organise the various interactions within the IoV, several architectures have been proposed by various researchers, e.g. [11–15]. These are all layered architectures to incorporate various kind of interactions, as suggested above. Well-organised architectures promise benefits such as better connectivity, fast response, and increased efficiency of communication.

IoV is especially useful for autonomous self-driving vehicles. However, it is highly complex integrated network system mainly because: (1) number of nodes in the network (i.e. cars) can increase and decrease in real time; (2) nodes are mobile in nature that have varying speeds and changing directions in real time; and (3) distances between nodes change in real-time. As in any network, there are also issues relating to trust, privacy and confidentiality of data, as well as cyber threats to comprise the network in various ways.

1.2.5 Vehicle-Embedded Software Applications

According to one estimate [16], the sale of connected cars services will reach $155 trillion by 2022. Of these, services that relate to driver assistance, safety applications, information and entertainment (infotainment) capture the greatest market share. Some of the necessarily required application domains are briefly discussed below:

Human Machine Interface (HMI)

This is probably the most important domain to ensure safety and smooth operation of machines (in this case, vehicles). However, the dilemma is to determine how much of control should be retained with the human operators, e.g. drivers and passengers. Vision of having driverless autonomous vehicles on the road suggests that human interaction should be minimal, tending to zero, except possibly in emergency situations. Another issue refers to determining how much of automated services should be embedded within the vehicles (e.g. close to dashboard), and how many kept on, say, drivers' mobile phone for remote handling of vehicles. Related issue is that, since the life cycle of vehicles is much longer than that of mobile phones, embedded systems will need to be updated regularly.

Infotainment

There is no doubt that smart phones have become the de facto way people communicate with each other and with other smart devices, e.g. listening to music through speaker systems embedded in the cars via Bluetooth connectivity, while controlling the music channels and volume through the smart phones. In this context, connected vehicles designers need to find innovative ways to encourage third-party developers to help build connected car ecosystems. This is attractive, as this can help smart phone app designers to penetrate the car market, while at the same time, helping car manufacturers to keep their core systems isolated from non-OEM software [16]. Numerous relevant platforms already exist (e.g. Android Auto, Apple CarPlay, MirrorLink, Baidu CarLife, etc.) that provide infotainment on smart phones.

Onboard Diagnostics (OBD)

Such a system, in the form of OBD-II, already exists but it is a one-way data port, rather rudimentary in the current smart cars scenario. Engineers are progressing with the development of systems that provide at least the following: (1) much extended information for drivers on vehicle performance and in graphical form; and (2) better management mechanisms for fleet managers with enhanced efficiency, better safety, and auto scheduling of maintenance services.

Smart Device Link (SDL)

Smart vehicles in smart environments require better connectivity with smart objects, not just within the vehicles but also in the environment. Toyota and Ford are already adopting and utilising SDL devices for their next generation of vehicles [16]. Through the use of such devices, all stake holders (whether vehicle manufacturer, vehicle users, or software designers) will benefit and also further the car manufacturing activity into other related domains.

1.3 Connected Vehicles: Enabling Technologies

Technologies that support the connected vehicles paradigm already exist. These include IoT, Artificial Intelligence (AI), Machine Learning (ML), Dedicated Short-Range Communication (DSRC) protocols, 5G mobile technology, and Cloud/Edge Computing platforms. In this section, we discuss some of these technologies and mechanisms that underpin the recent developments in the connected vehicles vision.

1.3.1 Internet of Things

This is the main underlying paradigm, without which, concepts and developments such as connected vehicles, smart environments, Industry 4.0, etc. are not possible. IoT is a cyber-physical system, which relies on processor-embedded devices, sensors and actuators to collect, process and transmit data from the physical world. It allows the objects connected to the Internet, to be detected and remotely controlled via the Internet-connected infrastructure. This, then, provides direct integration between physical and digital worlds. The outcome is in the form of smart cities, Industrial IoT, smart grids, intelligent transport system, connected healthcare, etc.

At the basic level, the IoT technology stack comprises three layers, viz: sensors, microcontrollers and Internet connectivity, and service platforms [17], as briefly outlined below.

- Layer 1—Sensors and electronic measuring devices are embedded in objects in the physical environment to capture and transmit information. Sensors do not necessarily need to be connected directly to the Internet—they can synchronise with smartphones and other devices via, e.g. Bluetooth LE.
- Layer 2—Microcontrollers and Internet connectivity share information captured at Level 1 and analyse for Layer 2 to take further action. The main and most important capability of this layer is networking, which can be either wireless or wired.
- Layer 3—After the analysis of data, service platforms take over to take actions to adjust, modify, maintain and monitor the physical environment. Generally, service platforms operate on cloud infrastructure and utilise a multi-tenant software architecture to deliver seamless software-as-a-service (SaaS).

1.3.2 Cloud Computing

This refers to a distributed computing environment that provides online availability of virtualised computer systems including software, hardware and platforms; as well as storage facilities and other technology related services. Environment is multi-tenant where services can be provisioned by consumers as and when required, on a pay-as-you-go basis. Cloud environments can be publicly organised or privately developed or a combination of both.

Cloud services are, generally, of three varieties: Software-as-a-Service (SaaS), Platform-as-a-Service (PaaS), and Infrastructure-as-a-Service (IaaS). Underlying technologies include: virtualisation, replication, service-oriented architecture (SOA), autonomic and utility computing. Cloud Computing shares characteristics with the following:

- Client-Server Model: for distributed computing applications that differentiate between service providers (servers) and service consumers (clients)
- Grid Computing: for distributed and parallel computing where a virtual machine (e.g. computer) is composed of clusters of networked computers
- Fog Computing: where initial storage and processing services reside much closer to the client (i.e. the end users) for later processing/storage in the cloud
- Utility Computing: which refers to packaged computing resources and services, provisioned like a utility such as water, power and gas, etc.
- Peer-to-Peer organisation: for a distributed network architecture where nodes are both clients and servers, where central coordination is not required
- Green Computing: which refers to environmentally responsible use of computing resources e.g. by using energy efficient CPUs and servers etc.

Core characteristics of the Cloud paradigm include: device and location independence, scalability of applications, virtualisation of software and hardware platforms, advanced quality of services, pooling of resources that can then be shared by multiple consumers.

1.3.3   AI and Machine Learning

Artificial Intelligence (AI), sometimes called Machine Intelligence, refers to intelligence demonstrated by machines. The term is often used to describe machines (e.g. computers and cars) that mimic *cognitive* functions that humans associate with *learning* and problem solving. Machine intelligence is gained by performing statistical operations on data to discover patterns therein, rather than intelligence being explicitly programmed within the systems. AI can be categorised into three different types, as mentioned below:

- Analytical: this has characteristics consistent with cognitive intelligence, including learning based on prior experience
- Human-inspired: this has elements from cognitive and emotional intelligence that contribute to the decision-making processes
- Humanised: this has characteristics of several different types of competencies (e.g. cognitive, emotional, social, etc.) and has the ability to be self-aware.

Machine Learning (ML), sometimes called Deep Learning, is the application of AI. It enables a system to automatically learn and improve with experience, without human intervention, beyond the initial programming that sets up the learning process. ML methods include: supervised ML algorithms, unsupervised ML algorithms, semi-supervised ML algorithms, and reinforcement ML algorithms. Some well known classes of algorithms include: genetic algorithms, rule-based ML, learning classifier systems, similarity and metric learning, clustering algorithms, Bayesian algorithms, neural networks, decision tree learning, etc. ML is currently the most promising path to strong AI [18].

In the context of connected vehicles, some useful features of AI-based systems include speech recognition, driver monitoring, virtual driver assistance, camera-based vision systems, and radar-based detection (of other vehicle and roadside units) systems.

1.3.4   5G and DSRC Technologies

DSRC (Dedicated Short-Range Communication) is a two-way short-to-medium range wireless communication technology that enables vehicles to communicate with each other and other road users directly, without involving cellular or other infrastructure. It is specifically designed for automotive industry, and the Intelligent Transportation Systems (ITS), specifically for safety applications. It operates on 5.9 GHz band of radio frequency spectrum.

DSRC has low latency and high reliability; it is secure and supports interoperability. It provides a key foundation for V2V and V2I safety by enabling connectivity between vehicles and road users. This protocol has been thoroughly tested for V2X applications.

5G is the 5th generation digital cellular network technology for mobile Internet connectivity, offering faster speeds and better reliability. The new 5G wireless devices also have 4G LTE capability. However, 5G can support up to a million devices per square kilometre, as opposed to 4G supporting only up to 100,000 devices per sq. km. There are generally three main uses of 5G, as follows:

- Enhanced mobile broadband: this uses 5G as a progression from 4G LTE mobile broadband services, with faster connections, higher throughput, and more capacity
- Ultra-reliable low latency communication: this refers to using the network for mission critical applications that require uninterrupted and robust data analysis and exchange
- Massive machine type communication: this is used to connect large number of low power devices that have high scalability and increased battery lifetime, in a wider geographical area.

Discussing DSRC versus C-V2V (cellular V2V), one notes that C-V2V has several key advantages over DSRC including longer range, better reliability, and enhanced performance. Longer range capability allows vehicles to travel at higher speeds while still being able to stop in time to avoid hazardous conditions [19]. While 5G is still developing, there are exciting new opportunities, especially with respect to vehicles connectivity, waiting to be further explored.

1.3.5 Other Related Technologies

There are numerous other technologies and methodologies that are relevant to the connectivity of vehicles in the IoV vision including: Blockchain, Voice Over Internet Protocol (VOIP), Telematics, Big Data analytics, and Cloud SaaS platforms. These are briefly explained in [20].

1.4 Connected Vehicles: Issues and Challenges

The opportunities for connected vehicles vision are huge. However, a realistic look at the new technologically based automotive industry reveals multiple challenges. There are also numerous obstacles that will need to be addressed. Some of these are outlined below:

- **Currency of embedded software**: In the connected vehicles scenario, a vehicle's life cycle is about 5 years, mainly due to constant updating of operating systems and software applications. To gain full usefulness, it is important that such systems are always fully up to date. And, this is a challenge. Ideal solution lies in developing automotive-related software systems that can automatically adopt to changing needs, without human intervention.

- **Diversity of smart devices**: Huge diversity of inter-connected smart devices poses another challenge, due to different, and sometimes, incompatible communication protocols. Solutions is for the mobile connection providers and device manufacturers to have a long-term partnership based on trust, and establishment of relevant communication standards, as well as agreeing upon certain legislation to ensure smooth, reliable and wider connectivity.

- **Bandwidth consumption issues**: Consumption of power and battery life is an obstacle that connected vehicles manufacturers face. Bandwidth consumption is another major challenge. Building connected vehicles on a 'data messaging' backend with low charges, vehicle manufacturers can manage to keep socket connections open with a limited amount of bandwidth usage. Using open socket can allow streaming of bi-directionally data without hammering the server several times per second [21]. Inherent problems are well known and solutions are being worked out by manufacturers.

- **Data protection issues**: Data generated through the use of connected vehicles is considered to be 'personal'. Since, many drivers may be permitted to use the same vehicle, it is not clear how consent may be given or obtained for the use of such data by other drivers. Additionally, in case of vehicle manufacturers in the European Union, there might be additional data transfer issues as jurisdiction changes as a vehicle moves from one geographical region to another [22]. Solution is to revise data protection regulations to also consider the various connectivity-of-vehicles scenarios.

- **Network integration issues**: There is much work in progress on different standards of communication (in the V2V and V2I scenarios, especially) between WAN, and between local networks. Even if the macro network is not available at some point, the autonomous driverless vehicles must still work reliably and safely. Solution is to have end-to-end architecture that spans from the chip level of the CPU to the network of the vehicles to the gateway of the vehicular networks, and additionally up to the cloud environment especially against cyber threats [23].

- **Cyber-crime risks**: Given the huge amount of data generated and transmitted through connected vehicles, the issue is how such data may be protected against cyber-attacks. Additionally, there are some relevant questions, e.g.: (1) where and how much of this data should be stored in the car and how much to be transferred to a cloud or edge environment? and (2) which security measures should be implemented to protect such vehicles from hackers? Answers do not necessarily exist in the current legislation, at least at the present time. However, the relevant issues need to be addressed.

- **Liability of accidents**: Currently, vehicle owners or drivers are generally liable for accidents caused by their vehicles. In the case of autonomous self-driving cars, clarity is lacking. Solution is to have contractually regulated mechanisms agreed between the vehicle manufacturers and buyers of cars (and fleet managers) as well as new regulation in the law.

There are numerous other social, ethical, regulatory and technological challenges that need to be addressed by not just the vehicle manufacturers, but also for smart device designers, communication engineers, security experts, and law makers: some at local level, others at international level spanning different jurisdictions.

1.5 Future Directions

There were nearly 20 million connected cars on the roads in 2017. Gartner [24] estimate that the number will increase to around 250 million by 2020. These cars and other connected vehicles generate, gather, and transmit huge amounts of data that is currently being used to develop better technologies that are resulting in increased efficiency, improved safety, better driving experience, and even better connectivity.

In the near future, technologies and tools such as AI, machine learning, and GPS will work even better to detect obstacles on the roads, or bad weather, to reduce risks of traffic incidents and avoid accidents. Stored driver profiles will create personalised driving experiences, e.g. personalised navigation. Smarter infotainment may result in playing the drivers' favourite music when, for example, it detects the drivers' presence near the vehicle. Car radios and CD players have already become things of the past, as these are now getting replaced by 'infotainment units'. In-vehicle sensors will provide even better information on fuel consumption, engine performance, emission data, and driving experience.

In the distant future, as the data mining techniques become more advanced, new revenue opportunities will emerge in the form of hyper targeted marketing adopted to individual drivers' driving habits, visited locations, in-car entertainment, and content choice [25]. Smart devices and machine learning tools, as they get smarter, will enhance the drivers' digital lifestyle. Smart cars scenario and Intelligent Traffic Systems (ITS) will become pivotal elements of the smart cities vision. The possibilities are endless.

Currently, much of the publicity on future of smart vehicles is focusing on automated 'autonomous' vehicles that are self-driving, implying complete independence. In-built cameras and sensors detect and interpret road signs, lane control, location of other vehicles in the proximity, and general traffic situations. The Google self-driving car project is probably the most advanced. They have already conducted over 5 million miles of road testing. Concerns of cyber-security will be well addressed.

1.6   Conclusions

Internet of Things (IoT) technologies are helping to build smart cites to further improve resources utilisation, enhance safety on the roads, and manage traffic and transportation systems. In this respect, Internet of Vehicles (IoV) and Intelligent Transportation Systems (ITS) are important aspects of a successful smart city. This further entails development of connected vehicles technologies in terms of V2V, V2I, V2X, and development of smart road and intelligent traffic environments.

Much work has already taken place and the connected vehicles vision has moved from hype to reality. Smart and autonomous vehicles already exist. Although there are still issues, the progress is continuing at a very fast pace. Google has already conducted over 5 million miles of road tests on their self-drive cars. Tesla's Model S rolled out in 2012 was already leading the connected vehicles market at that time. Its current models have, amongst other systems, the infotainment system that allows users to monitor vehicle's car performance; has a full browser, full navigation, music media, and voice control features. Its operating, safety and road condition monitoring systems are updated remotely and drivers notified automatically. In Pittsburgh, Pennsylvania, 50 intersections are already equipped with smart traffic signals that are helping to reduce traffic congestion. Austin transportation department in the US has already tested smart parking metres as part of its smart mobility project. In connected vehicles vision, there is much more to come. The future of connectivity vision is, undoubtly, highly attractive.

References

1. PostScapes (2019) Internet of things (IoT) history. https://www.postscapes.com/internet-of-things-history/
2. Engineers Rule (2006) How a coke machine and the industrial internet of things can give birth to a planetary computer. https://www.engineersrule.com/how-a-coke-machine-and-the-industrial-internet-of-things-can-give-birth-to-a-planetary-computer/, https://internetofbusiness.com/coca-cola-drinks-to-the-future-of-the-internet-of-things/
3. Technavio (2019) Top 10 connected car companies leading the global connected car market] in 2019. https://blog.technavio.com/blog/top-10-connected-car-companies-leading-global-connected-car-market
4. CMSWire (2019) Connected car experiences in 2019. https://www.cmswire.com/digital-experience/connected-car-experiences-in-2019-exploring-the-possibilities/
5. Business Insider (2019) Automotive industry trends: IoT connected smart cars and vehicles. https://www.businessinsider.com/internet-of-things-connected-smart-cars-2016-10?r=US&IR=T
6. Techopedia (2012) Cloud computing for vehicles: tomorrow's high-tech car. https://www.techopedia.com/2/28137/trends/cloud-computing/cloud-computing-for-vehicles-tomorrows-high-tech-car
7. ITS (2019) Connected vehicles: vehicles to pedestrian communication. https://www.its.dot.gov/factsheets/pdf/CV_V2Pcomms.pdf
8. Auto Connected Car News (2017) Vehicle to pedestrian (V2P) and bicycle demoed. https://www.autoconnectedcar.com/2017/07/vehicle-to-pedestrian-v2p-and-bicycle-demoed/

9. Investopedia (2017) V2X (vehicle-to-vehicle or vehicle-to-infrastructure). https://www.investopedia.com/terms/v/v2x-vehicletovehicle-or-vehicletoinfrastructure.asp
10. Indu SK (2019) Internet of vehicles (IoV): evolution, architecture, security issues and trust aspects. Int J Recent Technol Eng 7(6):2019
11. Bonomi F (2013) The smart and connected vehicle and the internet of things, synchronization. In: Telecommunication systems, pp 1–53
12. Nanjie L (2011) Internet of vehicles: your next connection. Huawei WinWin, pp 23–28
13. Wan J, Zhang D, Zhao S, Yang L, Lloret J (2014) Context-aware vehicular cyber-physical systems with cloud support: architecture, challenges, and solutions. IEEE Commun Mag 52 (8):106–113
14. Raj D et al (2016) Fog. In: IEEE 8th international conference on cloud computing technology and science, pp 90–93
15. Gandotra P, Kumar Jha R, Jain S (2017) A survey on device-to-device (D2D) communication: architecture and security issues. J Netw Comput Appl 78:9–29
16. Ignite (2018) 5 approaches to connected car app development. https://igniteoutsourcing.com/automotive/connected-car-app-development/
17. Ashwini A (2017) What is the technology stack used in IOT product (Internet of things)? https://www.quora.com/What-is-the-technology-stack-used-in-IOT-product-Internet-of-Things
18. Quora (2019) What are the main differences between artificial intelligence and machine learning? Is machine learning a part of artificial intelligence? https://www.quora.com/What-are-the-main-differences-between-artificial-intelligence-and-machine-learning-Is-machine-learning-a-part-of-artificial-intelligence
19. FierceWireless (2018) 5G Americas delves into cellular V2X/5G versus DSRC. https://www.fiercewireless.com/wireless/5g-americas-delves-into-cellular-v2x-and-5g-versus-dsrc
20. Intellias (2019) 7 technologies empowering connected cars. https://www.intellias.com/7-technologies-empowering-connected-cars/
21. TeckInAsia (2015) 8 challenges faced by connected cars. https://www.techinasia.com/talk/top-8-challenges-roadblocks-faced-connected-cars
22. Coraggio G (2014) Connected cars—legal issues and hurdles. https://www.gamingtechlaw.com/2014/10/connected-cars-legal-issues.html
23. Carter J (2015) Challenges aplenty: what are the roadblocks facing the connected car? https://www.techradar.com/news/world-of-tech/challenges-aplenty-what-are-the-roadblocks-facing-the-connected-car-1287028
24. Ismail N (2017) The present and future pf connected car data. https://www.information-age.com/present-future-connected-car-data-123466309/
25. Epam (2019) Open source, cloud computing and the connected car. https://www.epam.com/about/newsroom/in-the-news/2019/open-source-cloud-computing-and-the-connected-car

Chapter 2
Spatial Intelligence and Vehicle-to-Vehicle Communication: Topologies and Architectures

Kelvin Joseph Bwalya

Abstract As the developing world positions itself towards implementing Smart Cities, concepts such as intelligent transport systems and spatial intelligence come to the fore. Smart Cities require contemporary pervasive and dynamic topologies and architectures to achieve spatial intelligence which is supported by intelligent transport systems. In such systems, vehicles can communicate with one another using Vehicle-to-Vehicle (V2V) communication models. V2V requires the availability of information on demand and anytime; also, that this information must be accessible in real time by the vehicles as they traverse through the city. Advanced information provision in Smart City environments enable vehicles to exchange information and make intelligent decisions on the roads. A whole array of both functional and non-functional requirements such as usability, aesthetics, security (access, availability and reliability), topology and information architecture, etc. need to be considered to achieve the desired level of spatial intelligence. Putting in place a network to handle the different network dimensions to achieve ubiquity can be significantly costly and beyond the reach of many of the developing world countries. Although, there have been some pockets of research on different aspects of vehicular networks, there is no significant research that brings a great deal of spatial intelligence together. This chapter aims to comprehensively explore the concept of spatial intelligence in the realm of V2V communication. Without carefully thought topologies and architecture, given the context, spatial intelligence in V2V communication cannot be realised. This chapter contributes to knowledge by exploring the different topologies and architectures in mobile agents (vehicles) where cost is one of the key inhibiting factors influencing the actual design.

Keywords V2V · V2V communication · Issues · Spatial intelligence · IoV · MANET · VANET · FANET · Topologies · Architecture · Big data · Network · Markov decision process · Q-learning · Fuzzy logic

K. J. Bwalya (✉)
School of Consumer Intelligence and Information Systems, University of Johannesburg, Johannesburg, South Africa
e-mail: kbwalya@uj.ac.za

© Springer Nature Switzerland AG 2020
Z. Mahmood (ed.), *Connected Vehicles in the Internet of Things*,
https://doi.org/10.1007/978-3-030-36167-9_2

2.1 Introduction

There is an increased interest in the design of networks and information systems to ensure pervasive access to information and allow for on-the-go decisions [2, 4, 9, 13]. As a result, there is a constant need for information to facilitate the making of instantaneous decisions. In vehicular networks, there is a huge demand for information to make decisions with respect to vehicle routing, avoidance of accidents and autonomous driving, etc.

Vehicular networks are characterised with constantly changing information which can be difficult to manage. Handling of dynamic information is complex and it calls for advanced techniques with higher efficiency and accuracy levels. In vehicular networks, ever-changing network topologies and dynamic information regarding the nodes (vehicles) calls for real-time management of huge sets of information from different objects in the environment as well as from within the vehicles.

An ideal environment for vehicular networks is where vehicles can communicate with the objects in its vicinity and capture the current information to make decisions. For this to happen, sensors are embedded into the environment through Road Side Units (RSUs) and basic nodes which in turn communicate with in-vehicle sensors and devices. Internet of Vehicles (IoV) is the starting point to having efficient Vehicle to Vehicle (V2V) networks. Communication is complete in vehicular networks using IoV, that senses information from outside the vehicle using Vehicle Information Systems (VIS) to alert the driver of the circumstances in the environment in which he or she is driving. The VIS allows one vehicle to communicate to another. In this context, a vehicle is considered as a mobile terminal which uses a given communication network. In most cases, this network is a wireless integrated communication system. Because of the dynamic nature of vehicular networks, there are many issues that need to be explored bordering on design aspects of the networks.

This chapter aims to investigate the already-published topologies and architectures articulating spatial intelligence in V2V communication and highlight the key issues for contemporary and future designs. The issues are discussed from the developing world perspective. The expected contribution is that the contents of the chapter can be used by resource-constrained countries to jump onto the bandwagon for putting in place intelligent transport systems and networks that can support effervescent V2V communication.

The organisation of this chapter is as follows. The next section presents the background, setting the tone of this research. Thereafter, the concept of spatial intelligence is introduced together with concepts articulating data management in Mobile Ad Hoc Networks (MANET). Subsequently, communication principles are discussed before exploring the network topologies and architectures in MANET. Furthermore, the different challenges are looked into in realising appropriate communication in vehicular networks. Before the conclusion, the chapter explores the different design innovations for MANET and discusses future research directions in this domain.

2.2   Background

Just less than a decade ago, the vehicle was just an ambulatory system for people, providing limited information to the driver, such as inside or outside temperature, fuel range, etc. and relying so much on the driver's intuition. This is now changing, given the rapid advances in wireless technology, presenting the vehicle as a node in an Intelligent Vehicle Grid (IVG). The IVG contains both the vehicles on the ground (supported by MANET) and in the air (supported by Flying objects Ad Hoc networks: FANET).

There is a growing interest in IoV from researchers and practitioners from all over the world. This is because of the emergence of the Fourth Industrial Revolution (4IR) which demands that the value chains are automated and significantly shortened. In the transport sector, autonomous vehicle communication, intelligent road transport decisions, and need for increased safety, entails that many designers and innovators are pursuing automation. Given the emergence of the 4IR, artificial intelligence has kept penetrating into different socio-economic setups. Specifically, machine learning is being used in vehicular networks especially in situations where automation is pronounced [13]. Machine learning can be used in autonomous vehicles to automate many things such as route navigation, V2V communication, etc. Ye et al. [13] posits that there are many instances in vehicular networks where machine learning has been utilised especially with a bid to provide a data-driven approach to solving problems in VANET. As a major component of artificial intelligence (AI), machine learning (ML) is a good candidate for intelligent processing of huge amounts of data as it can quickly help in finding patterns and structures in data [13].

In real vehicular environments, information is highly dynamic/pervasive and is constantly evolving in both its meaning (integrity) and scale (towards big data which is partly structured, partly unstructured). Managing such kind of information involves a complex system of coordination for the interaction between vehicles, people, and the environment. Coordination is important because it allows information to be exchanged among the participating agents in the information space. As a result, contemporary vehicles are installed with necessary gadgetry and smart devices that enable them to communicate with the environment and other objects in road networks so that there is appropriate coordination and data aggregation for sensing and further processing.

Vehicular communication is advancing due to rapid advances in wireless and mobile technologies [4]. To achieve anticipated contemporary intelligent road transport infrastructure, both the vehicles and the environment are installed with sensor nodes and intelligent devices. Sensor-enabled intelligent devices, properly connected in different communication systems, can monitor their environments and make intelligent decisions. This ensures that there is information everywhere to achieve spatial intelligence where information is available pervasively to achieve intelligent decision-making anywhere and anytime in all situations.

In vehicular ad hoc networks, one of the key components is the propagation model which is needed to ensure that signal encapsulation and propagation happen as quickly as possible given the highly mobile nature of participating units. For desired on-demand information communication to take place, there is a need to take into consideration the highly dynamic nature of vehicular networks where vehicles travel at varying speeds, information changes every second due to mobility, higher density of vehicles within a confined distance, etc. Given the huge dynamic nature of vehicular networks, it is important to have intelligent networks that can quickly change their topology and structure to accommodate the highly dynamic vehicular networks, where nodes are nearly always mobile.

2.3 Spatial Intelligence

Given the availability of many objects (nodes, devices, etc.) found in the environment in which vehicles operate, it is not a secret that there is a lot of spatial diversity in vehicular networks. Spatial diversity points to the fact that in different geographical points in an environment, diverse information can be easily accessed. When vehicles traverse through the environment, they make decisions on mobility based on the information they are able to access. Ubiquitous access to dynamic information is cardinal for achieving ambience intelligence. Ambience intelligence entails pervasive intelligence where information is available anywhere and anytime. To achieve ambience intelligence, there is need for an information infrastructure where adaptive technologies such as sensors and actuators can seamlessly be integrated into information superhighways.

In mobile environments, where information dynamically changes every second, there is need to maintain a special version of ambience intelligence called spatial intelligence—which entails that information digital agents are able to harness information resources instantaneously while they are in a given environment. This information is used to make intelligent decisions.

Many of the challenges for achieving spatial intelligence in vehicular networks emanate from sparse connectivity, mobility (high and varying speeds of vehicles), heterogeneity of smart devices, and high density of vehicles especially in cities [10]. Since vehicles are seen as network nodes able to collect information, vehicular networks usually have too much dynamic information from the many vehicles in the network.

In many cases, the exponential growth of available information in vehicular networks makes it hard for vehicles to process information instantly to make intelligent decisions. In order to address these challenges, there is need to understand the principles that define architectural configurations and network topologies that can handle dynamic information architectures and topologies. This understanding enables technology innovation relevant to the given context.

2.4   Managing Data in Vehicular Networks

In contemporary vehicular networks, there are huge sets of dynamic information that guide mobility and intelligent decision-making. Vehicular information networks are usually highly nested and designed with sensors and actuators scanning information from the environment. These sensors are linked to the nodes which relay this information to the vehicles. With the ability of the different nodes to share information, each object in such an environment obtains pervasive access to information enabling it to make decisions anytime and anywhere thereby achieving spatial intelligence. In autonomous vehicles, there are often more than 300 sensors with a capacity of generating over 5000 GB of data on an everyday basis [11]. This data, with multiple dimensions, needs to be stored so it can be used as historical data for making future decisions such as prediction of traffic conditions on certain days and time of the week and the calculation of optimal navigation directions [11].

Due to increased connectivity, Internet of Vehicles (IoV), as discussed below, presents an environment where data is rapidly shared in connected vehicular networks. As a result, large quantities of structured and unstructured data are generated culminating into big data. This data is gathered from nearby vehicles (on-road and on-board), passengers, road side units (RSUs), cloud storage facilities, and the Internet. On-board and on-road data can be vehicle speed, status of brakes, rear camera data, distance from barrier(s), fuel gage, engine parameters, temperature, road map, etc.

Data in a VANET is highly dynamic, given the ever-changing position, speed, and density of vehicles on the road calling for efficient big data analysis for vehicle traffic. Not only the ability to analyse captured data, but also to identify trends and patterns in the data to achieve predictive scenarios on the roads a few minutes from the moment the data is captured. As data is captured, it needs to be stored in the cloud and accessed by all vehicles in the network in order for them to make ad hoc intelligent decisions thereby increasing traffic safety [1] and better traffic flow.

Due to the nature of highly dynamic big data in vehicular networks, some innovative data processing techniques have been designed to handle data processing. For example, one of the widely used technique is machine learning [13]. This and natural language processing (NLP) provide a lot of potential in handling multi-dimensional, unstructured and structured data, heterogeneous data and huge quantities of data to quickly identify underlying structures and patterns for evidence decision-making. Machine learning makes available tools and techniques for mining data, combing through huge sets of data, and exploiting multiple sources of data [13]. Machine learning tools can be embedded into the design of the topology routing mechanisms and protocols to enable communication to go on given the change in the topology of the networks. These tools can comb through real-time and historical traffic data, traffic flow, etc. to make routing decisions [13].

To ensure that there is intermittent dynamic learning of the clusters to enable efficient communication, Wu et al. [10] propose a Q-learning model based on advanced deep learning algorithm. Rapid application of reinforcement learning, i.e. Markov decision processes (MDPs), with the help of Q-learning, enables machine learning to be used in contexts where there is need for quick rapid interaction and scanning of the environment to determine optimal routes in the network [13]. Using Q-Learning, deep learning and fuzzy logic, efficient routing protocols for transferring data in congested networks. It is desired that this learning takes place as fast as possible so that there are no significant delays in making desired decision in highly dynamic network environments [13].

2.5   Vehicle Communication Models

In the contemporary world, a safer vehicle is one which is able to access as much information as possible from the environment and is able to communicate with every device it is connected to, in real time. In order to achieve this, there are different modules through which the vehicles communicate with the environment.

2.5.1   Internet of Vehicles (IoV)

As a result of continuous expectations of vehicles to process information on large quantities of data, traditional VANET are transcending into Internet of Vehicles (IoV) which have brought intelligence in vehicular networks. Because of these happenings, it is important to explore the design of topologies and infrastructure that can handle big data storage, access and distribution among the heterogeneous distributed agents.

IoV is critical in the exploration of advanced opportunities of ICTs to come up with a converged system where vehicles are adequately networked. In this chapter, IoV is defined as a converged multi-modal system which enables pervasive connectivity for data and information sharing, to enable vehicles and humans to make intelligent decisions as they traverse the environment. The IoV presents opportunities where vehicles, people and the environment exist in cohesion to the general advancement of humanity [12].

Together with the internal telecommunication systems integrated into the IoT, the IoV enables a vehicle to be considered as a mobile intelligent sensor, actuators and controllers as it scans the environment as it traverses through. Several concepts come together to enable a true V2V network where dynamic information is constantly shared among vehicles in the vicinity. There is a need for distributed data and information system, intelligent sensors that are embedded into the environment, wireless communication and properly designed road infrastructure. This infrastructure is needed in intelligent transport systems [12].

It is worth noting that the realisation of the IoV vision is a key milestone towards realisation of commercial autonomous vehicles. In future, IoV will enable vehicles to be more of open platforms which integrate data and information intelligence,

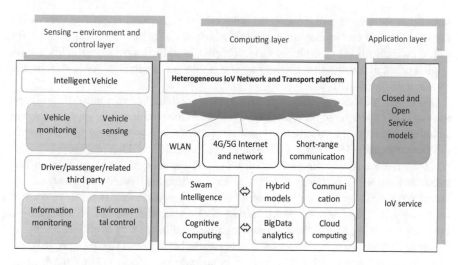

Fig. 2.1 Basic components for IoV architecture

increased Internet connectivity and information agent for intelligent decision-making [12]. A requirement of IoV is the need for an open and integrated system which enable easier and quicker sharing of dynamic data and information. For this to be achieved, it is important to have dynamic network topologies which are able to handle huge volumes of structured or unstructured data [9].

Because of similarities in the characteristics, there is no way that V2V can be separated from IoT, except that 'things' in the IoT are vehicles. Therefore, understanding of V2V architecture begins from the understanding of IoT which is the conceptual basis of internet of vehicles (IoV), and correspondingly, the V2V communication.

A simplified version of the IoV architecture comprises application tier, communication tier articulating different protocols and technologies and the perception tire which is a layer for different sensors deployed to capture the information and the actuators. Figure 2.1 shows a simplified IoV architecture, based upon the idea put forward in [9].

Implementation of innovations based on IoT provides opportunities for vehicles to communicate with each other (V2V) and vehicles with various networks in the environment, e.g. Vehicle to Infrastructure (V2I) and Vehicle to Everything (V2X).

2.5.2   Vehicle to Vehicle (V2V) Communication

Vehicle to Vehicle (V2V) vision is the overall basic conceptual underpinning upon which vehicles exchange information to make intelligent decisions. Efficient V2V communication happens when pervasive networks are in existence which allow ubiquitous access to information as and when it becomes available. Pervasive networks are all-encompassing networks which allow networking and support for different applications in different environments in a dynamic manner. This allows application and information processing and network connectivity for both terminal and mobile agents [5].

To realise V2V communication, the physical medium access control (MAC) and the network layers have seen a lot of novel techniques tailored to allow rapid and seamless exchange of information through wireless networks, along the cellular 3/4G networks, WIMAX IEEE 802.16 and Wi-Fi IEEE 802.11.

One of the critical components of V2V communication is the mobile ad hoc network (MANET) which is essentially an autonomous system, made up of different mobile stations at any one point in time. The Ad Hoc networks are formed intermittently. The sensors (vehicles) in V2V networks are not considered as ordinary nodes because these can have higher capabilities such as collecting highly dynamic information to influence the network behaviour. The idea is that the information collected will most certainly contribute towards increasing the level of automation in vehicle behaviour on the roads. In a relatively developed V2V network, the nodes are able to relay information at certain intervals, e.g. on the state of the road, level of congestion, accidents reports, etc. The timely availability of such information may help motorists make intelligent decisions to reduce road carnage.

Different communication media can be considered in the realm of V2V communication. Some of the basic media types include the following: multimode fibre embedded onto the nodes on the road infrastructure, power line communication (PLC) that uses electric transmission lines to transmit digital communication signal, creating Wi-Fi zones (IEEE802.11) used in conjunction with the Multiple-Input Multiple-Output (MIMO) system to obtain a stronger signal, and the Worldwide Interoperability for Microwave Access (WiMAX, IEEE802.16) for wireless metropolitan area networks, etc. It is worth noting that wireless ad hoc networks have a huge potential for facilitating V2V communication in real life. This is because contemporary vehicles are embedded with networking capabilities which can enable relatively good communication using wireless media [5]. To sum it all, V2V communication can be defined by the following Lemma.

Lemma 1 *A hybrid multimodal communication system may be represented as $C = [a_1, ..., a_n]$ where a_i represents different wavelengths of signal which can be multiplexed through multimode fibre system able to support simultaneous heterogeneous communication protocols. These wavelengths with different bit rates and covering different distances may be assigned to $a_1 = 4G/LTE$, $a_2 = IEEE\ 802.11ah$, ..., $a_n = IEEE\ 802.11p$.*

There are other variants of V2V communication. For example, e.g.:

- Vehicle to Infrastructure (V2I) mode allows vehicles to communicate with different nodes and devices such as roadside units (RSUs) in the environment.
- Vehicle to Everything (V2X) which assume that a vehicle should be tagged to all the objects in the environment and literary communicate with everything.

This is made possible by rapid advances in the Internet of Things (IoT). With advances in vehicular communications, new communication models such as Vehicle to Road-Signs (V2RS), Vehicle to Cloud (V2C) and Vehicle to Pedestrians (V2P) are also emerging.

2.6   Node Information Dissemination Behaviour

One of the key challenges in understanding information dissemination behaviour in vehicular networks is the determination of node behaviour and therefore designing network and routing mechanisms based on that. Thus, modelling of node mobility behaviour is an important research area. To understand node behaviour, numerous simulations have been carried out by researchers throughout the world in computational sciences. Many of the simulations for model node mobility behaviour have been done using NS-3 network simulator. It is important to consider developing node mobility models using alternative development platforms and applications such as C++ for Linux platforms. The state of the nodes can also be modelled using Bayesian-based probabilistic models such as the Hidden Markov Chains.

To model the randomness of the nodes in dynamic networks, the random access MAC protocol for Ad Hoc networks, and random access code division multiple access (RA CDMA) can be used. This allows multiple nodes to access the ad hoc network simultaneously. Alouache [2] has proposed reliable Reservation ALOHA (RR-ALOHA) for use as a distributed channel location in highly information dynamic networks. These MAC protocols offer a great promise for VANET however these are not currently commercially utilised in practice. Currently, it is protocols based on IEEE 802.11b that are mostly utilised to model mobility in VANET [2].

One of the modelling decisions that need to be carefully considered is the heterogeneity of a MANET network in terms of mobility of the different nodes (vehicles) due to varying speeds. The mobility speed may complicate the capturing of the network in the speedy nodes and diminishes their ability to integrate to the existing Ad Hoc networks in their vicinity. This translates into network fragmentation problem due to dynamic spatial-temporal conditions [2].

2.7   Network Topologies and Information Architecture

Wide-range dynamic information including text, images and videos is shared in vehicular networks. Topologies and information architectures need to be designed to handle any type of information in any format [1]. It is not a secret that VANETs handle huge sets of dynamic information which demand for a well-designed information architecture to manage it without significant delays [13]. From the design perspective, topologies and architecture geared towards achieving V2V networks occupy at least four layers of the ISO protocol stark: Physical, Data Link, network, and the Transport layer. These layers are needed to provide wireless communication and information exchange in real-time between vehicles and environment (cities), vehicles and roads, vehicle and vehicle, and vehicle-human. By so doing, there is a clear integration between vehicles, people and the environment.

In information intensive environments, it is important to have intelligent topologies and information architectures which change their form and structure given the dynamic nature of the ad hoc vehicular networks. In real environments, vehicles communicate by belonging to one of the clusters in the network. Therefore, the network topology in VANET is determined by vehicle speed and the position of the vehicle. Topology design is important in information networks because it gives a conceptual outlay of the information interchange routes and possibilities that can be explored in any given situation.

The most simplistic topology desired in video streaming is a generic peer to peer (P2P) video streaming model. The P2P network is easy to design and deploy in real environments and is less costly; therefore, highly suitable in the context of developing nations. In information-intensive environments, mesh networks are usually utilised. Here, two or more network communication types are integrated. It is a key concept for increasing the capacity of a network.

In a typical V2V network, the communication tier is realised by the utilisation of traditional information capturing and routing protocols such as the global positioning systems (GPSs). The network topology is mostly influenced by dynamic structures due to the highly pervasive clusters of vehicles. Exact topologies often use graph algorithms which may use dynamic network or data routing mechanisms such as the Open Shortest Path First (OSPF). Another candidate for routing protocol is the SDN-based geo-routing protocol (SDGR) which uses a combination of advanced version of path routing algorithm and packet forwarding algorithm. Use of such a system allows the capturing of intermittent network states of the vehicles [2]. Alouache [2] has developed a multi-criteria routing protocol conceptualised upon a hybrid SDN architecture. This routing protocol aims to minimise link breaks so that there is continuous link connection between the different vehicles.

A key requirement for information and network distribution in VANET is a reliably designed middleware. Middleware is a key component of distributed networks. It is a software layer that masks the heterogeneity and provides the distributed systems complex abstraction to different applications and multiple distributed resources [4]. It ensures that the communication between distributed agents (nodes) is simplified as much as possible by providing needed support such as real-time communication support, security enhancements, resource utilisation and management, etc. The conceptual arrangement of a middleware in vehicular networks is shown in Fig. 2.2.

Commercial software does exist to act as middleware in different contexts. For example, a real-time event-based middleware, the RT-STEAM, is used for message propagation in VANET. Another middleware, SRAware, acts as a publish-subscribe middleware in heterogeneous vehicular environments. To address the

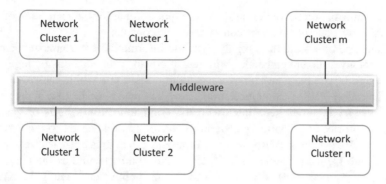

Fig. 2.2 Conceptual outlay of middleware in V2V networks

heterogeneity and dynamic nature of VANET, MARCHES provides a context-aware middleware which uses self-reflection to provide a software layer for real-time applications.

With the use of middleware, distributed systems are used as key network configuration principle for the design and deployment of V2V communication networks. The use of distributed information systems in V2V communication allows applications and systems to share information autonomously anywhere and at any time. The distributed system allows information to be easily shared among the participating mobile stations or parties in the system, mostly using a wireless communication medium [2].

Pervasive communication where vehicles are integrated with communication and networking technologies of unmanned aerial vehicles (UAVs) or drones can enable efficient V2V communications. The drones form what is known as flying Ad Hoc networks (FANETs) as they communicate with ground nodes of the different vehicles on the ground. When this happens, a Drone-Assisted Vehicular Networks (DAVN) is formed [8]. Because of the dynamic nature of the DAVN, it is a highly heterogeneous network which comes with many networking issues [8]. In order to effectively participate in VANET, vehicles have to be installed with IEEE 802.11p based Dedicated Short Range Communication (DSRC) On-board Unit (OBU) coupled with additional sensors to be fully aware of the environment in which the vehicle finds itself in [11]. For vehicular networks, the IEEE 802.11ah is long range Wi-Fi which can be usefully used especially when vehicles are dispersed to distances of over 1 km in road networks. In environments where vehicles are running at a higher speed of up to 160 km/h, use of WiMAX may be better advisable [7].

Realisation of efficient V2V communication is not only about the vehicles' ability to communicate with one another using in-built communication systems; it also needs cognizance of all non-negligible physical objects; and thus all these objects will need to have some form of digital identity. Therefore, it can be posited that the space of IoT cannot be ignored in the race for ubiquitous communication of dynamic information by vehicles [5]. A core requirement for the realisation of V2V and IoV is an effective efficient communication console.

Efficient communication in highly dynamic information environments demands a 5G topology structure embedded with recent advances in edge computing (or fog computing). Given the many innovations around the globe within the space of 5G communication, V2X communication can be made easily possible. Many researchers have shown keen interest in the communication aspects of vehicular networks. Storck and Duarte-Figueiredo [9] have investigated the vehicular video services in V2V networks especially in rural and urban areas. Their research uses the Software-Defined Internet-of-Vehicles (SDIoV) architecture as the main communication protocol [14]. Figure 2.3 shows the architectural configuration of a 5G-enabled V2V communication architecture.

To handle the different challenges and issues that come with the need for efficient routing mechanisms in VANETs, recent research has explored probabilistic routing (e.g. random Hidden Markov Chains), geo-routing and social relationship-aware routing. The main limitation of such routing approaches is that they are not able to locate nodes in the environment in which the current node is. Therefore, it is important that the routing mechanisms and protocols are intelligent. Consider the following Lemma.

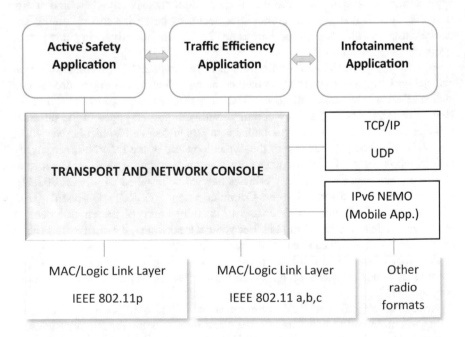

Fig. 2.3 Architectural configuration of a 5G-enabled V2V communication

Lemma 2 *A protocol is intelligent, if and only if, it can make correct decisions in complex, uncertain environments endowed with inaccurate information AND self evolves, given a change in the communication environment.*

Furthermore, to address communication challenges in VANETs, it is advisable to employ hybrid architecture such as utilisation of both WLAN or cellular network and ad hoc network configuration [4].

2.8 Design of Vehicular Communication Models

Given the importance of appropriate communication models in vehicular networks, many researchers have attempted the design of optimal communication models. Because of increased complication in vehicular networks, new dissemination and communication models are constantly being invented. For example, the Inter-Vehicle Geocast (IVG) intends to overcome contemporary pronounced challenges such as reliability bad neighbour determination, and network fragmentation by generalising previous dissemination mechanisms such as TRADE and DDT. To overcome the deteriorating quality of communication, Alouache [2] has proposed the hybrid Software-Defined Networking architecture. Yang et al. [12] have proposed a layered architecture which can be used to guide the implementation of vehicular networks as follows:

- Vehicular Network Sensing and Control Layer: For instant capturing of dynamic information from the environment and guiding of vehicle navigation. This may be cardinal for autonomous vehicle systems. Spatial sensors are deployed in the environment to capture information as events occur. The information from the captured events are dispatched to a control layer which allows vehicles to access. In other network configuration types, the sensing function is made possible using Global Positioning Systems (GPS)
- Network Access and Transport Layer: This provides heterogeneous handling of cases to the network by different vehicles and ensures that there is accurate information interchange between different modes (such as vehicle to vehicle, vehicle to environment, vehicle to humans, etc.). This layer ensures that there is a reliable information and communication transmission media. Depending on the needs for information sharing, a given network configuration can be chosen. In many cases, a mesh virtual network configuration allows sensed information to be shared by all the vehicles in the network. Standard transport protocols such as TCP/IP and UDP are used at the transport layer
- Coordination Computing Control Layer: This layer is an important layer for coordinating a series of computing requirements in the IoV environment. In vehicular networks, coordination involves segregating information with regards to what vehicles receive what information. Specific hardware and software is designed to achieve such kind of functionalities. To achieve desired coordinative capabilities, this layer also coordinates coordinated communication management

- Application layer: This provides the different requirements regarding the computing coordination endeavours. In the exchange of information, there is need for clear coordination in the intermittent computing schedules vehicle-human-environment model. This layer also provides an open interface to enable third-party information providers to easily dovetail to the IoV network [12].

Lathrop [5] investigated the concept of cognitive structures exploring knowledge structures and representation paving way for architectural design for deployment of intelligent systems. Understanding cognitive architectures entails understanding of how humans understand the environments in which they live by understanding the visual depictive representations in the environment. Such understanding is cardinal to the design of intelligent system design and representation in the real environments. Sherazi [7] has proposed a cost-effective heterogeneous multi-interface IoV architecture that uses inexpensive but effective RAU transmission mechanism. In this case, the overall cost of the architecture may become high if installed in environments where the fibre is installed newly. It is thus suitable for environments where fibre networks have already been installed. Further, because of the ability to switch from one fibre channel and interface to the other, the proposed architecture is able to enforce congestion control. Because of the already designed interfaces in the multi-channel infrastructure, the proposed architecture is able to support future generations by its ability to support a wide range of frequency bands. Hajj et al. [3] proposed the automotive open source architecture (AUTOSAR), a standard for supporting applications. This standard is based on IEEE 802.11p. Since it is open source, developers can explore this model further and come up with newer applications. Sherazi [7] proposed a heterogeneous VANET network architecture which aims to incorporate different sets of wireless interfaces to explore the radio-over-fibre approach. This ensures context-aware network connections at all times despite the changing topologies evident in the vehicular networks.

Bespoke network and routing algorithms have already been developed by researchers and practitioners and are embedded onto vehicular networks based on the varying contexts [2]. Some of these are the following:

- MDDV (Mobility-Centric Data Dissemination Algorithm for Vehicular Networks): Here, vehicles are modelled using a series of intersections and connections as weights in a directed graph. The MDDV uses the sum of weights as forwarding paths
- Urban Multi-Hope Broadcast Protocol: This is based upon a modified version of the access layer of the 802.11 specifically to achieve reduced collisions and efficient use of the bandwidth
- Role-Based Multicast (RBM): In this case, each node is going to maintain two pieces of information: a list of neighbour nodes and a list of V2V communications especially with a focus on transmitting nodes. The nodes rebroadcast a message intermittently with a view that the new neighbours will access this information to their benefit.

2.9 Inherent Challenges and Issues

There have been many attempts to realise and exploit the potential of vehicular networks for intelligent transport systems. However, there have been a plethora of challenges and issues that need to be addressed. Since, research and practice in V2V communication is relatively new, the existing issues need to be addressed to achieve agile, adaptive and intelligent V2V communication networks. For example, in vehicular networks, issues such as security, seamless connectivity, data rates, scalability, data forwarding through the dynamic networks, etc. are still of concern. The following are some of the main challenges that need to be addressed in vehicular networks:

(1) One of the key challenges in realising efficient V2V communication is the ever-changing network topologies resulting in less effective information networks for a sustainable period of time. Situation gets worse when this is coupled with a high dynamic scale and network density. MANETs are momentous and highly dynamic, so maintaining relevance of a system that keeps changing may be very difficult in a given context.

(2) Although there are rapid advances in IoV design and implementation with the Japanese Advanced Telematics Information System, the European Eureka Plan, Google's Android Auto Platform, etc. there are still prevalent challenges. Some of these include: (a) integration and coordination of people, vehicles and environments in IoV; and (b) lack of global understanding of what IoV entails. Overcoming the challenges may involve developing a conceptual IoV architecture [12].

(3) The topology of a pervasive network that can allow V2V ubiquitous interaction is near field communication (NFC). In such a network configuration, some of the key issues include location and service discovery, security, delay tolerance, seamless integration, agent identification and authentication, etc. [5].

(4) In highly mobile environments, handling intermittent connectivity, given the embedded delays in propagation or connection encapsulation, is problematic. In such environments, there is a requirement of delay tolerance mechanisms embedded into the communication networks using, for example Delay-Tolerant Networking (DTN). Network and communication requirements cannot be solved only by relying on IEEE 802.11p technology. The 5G network is a good option for enhancing the VANETs in as far as data latency, coverage and QoS are concerned [9].

(5) There are challenges regarding video streaming in V2X networks. One of these in case of VANETs has been attributed to limited channel capacity or throughput to allow huge video streams through communication channels. Many videos slow the system due to their huge size; therefore, distorting the bandwidth capacity [1]. To avert this problem, communication systems need to be designed with encoding and decoding layers. To allow for seamless flows of videos, they are usually partitioned into sub-streams with a view of reducing

size. The videos are mostly compressed using the H.263 scalable video coding standard, H.264/AVC standard and MPEG1-2-4-7.

(6) There are several challenges attributed to software-defined vehicular networks (SDVN) enshrined on volume of connected vehicles at any time instant, ever-changing network topologies, and heterogeneous network environments. Of these, the central problem is the frequent topology changes. These challenges have a significant negative impact on QoS [14].

The articulated aforementioned challenges are not mutually exhaustive as there are still other challenges in vehicular networks especially V2X network environments. Because of the higher efficiency expectations in VANET, QoS levels are uncompromisingly high. A properly designed VANET should handle the different issues surrounding the dynamic nature of vehicular networks.

To handle these challenges, Wu et al. [10] proposed multi-tier multiple access edge clustering and decentralised moving edge. Software-defined vehicular network (SDVN) presents another good option for the design of vehicular networks [14]. VANETs suffer from high density of vehicles, heterogeneous network clusters, privacy and security issues, delays, and constantly changing network topologies given the dynamic nature of the network. SDN separates the network traffic control from data transfer increasing flexibility by overcoming limitations in intelligence and facilitating programmability in vehicular networks [2]. This decoupling is important to ensure that the slowness in computation that can be introduced in the system by the data processing and analysis are separated from the network traffic management model.

Reliable vehicular networks are mainly intelligent network infrastructures which are highly dynamic to handle the high mobility of nodes, heterogeneity and high density of devices, and can handle uncertainty [10].

2.10   Future Directions in Vehicular Networks

Many researchers have been working in this domain. Although a great deal has been achieved, there is still much that needs to be done in different aspects of vehicular networks [4]. An analysis of the body of knowledge on the topic of MANET, the following are some more key issues that need to be investigated to achieve spatial intelligence in these networks:

- Many vehicular networks provide communication based on fixed mobile nodes and infrastructure. As mobility aspects get better in VANETs, it is important that mobile agents are able to communicate using mobile infrastructure embedded in GPS or the vehicles themselves. A relevant question is: how we can have nodes or sensors embedded in the vehicles themselves which are able to read information off the roadside sensors using GPS to get information about the environment?

- In advanced VANETs, links exist for a short lifetime presenting themselves as highly dynamic aspects of the communication system. The nodes in the V2V communication network scenarios are highly mobile presenting a future challenge of signal propagation. Future research needs to investigate the software nuances to design dynamic protocols that can synchronously change with the changing network states; and thus maintaining reliable communication.
- With the advancement of autonomous vehicles, many developing nations are left behind, in as far as jumping onto the bandwagon of driverless vehicles is concerned due to the high costs involved. It is important for future research to focus on developing networking standards and protocols on open platforms so that they can easily adopt and deploy.
- Although many delay-tolerant mechanisms have been developed (as already discussed in this chapter), MANETs usually demand communication applications with strict QoS expectations such as reduced delay jitter, higher bandwidth and reduced packet loss. In real mobile environments, it is difficult to have systems with real-time multimedia applications that can handle stringent QoS requirements.

With the emergence of fifth generation (5G) cellular networks, the realisation of efficient ubiquitous Internet and a truly pervasive paradigm is not far-fetched. For a reliable V2V communication, there is need for a reliable network and 5G makes a promising technology. Given the fact that there are many objects with Internet identity, the need for vehicles to connect with everything in the Vehicle-to-Everything (V2X) vision, cannot be overemphasised [9]. Of late, a lot of research is being done to explore the different opportunities provided by the IEEE 802.11p for VANETs. There is need to explore Mobile Edge Computing (MEC) in the realm of IEEE 802.11p as well, to come up with innovative solutions that can be used in contemporary vehicular networks.

2.11   Conclusion

The chapter explored different principles that are at the centre of design of V2V communication in contemporary vehicular networks. The chapter relies on a synthesis of work already in the public domain to come up with positions on the different issues affecting V2V paradigm. Specifically, the chapter explored spatial intelligence in the realm of information management in relation to vehicular networks. The different principles guiding the topological and architectural configurations of MANET are explored to an appreciable extent. Because of the dynamic and ad hoc nature of MANET, topologies are discussed that are required to change with the changing speeds, location and density of vehicles [6]. For effective communication, network clustering is mostly used in real environments. In such scenarios, these networks need to have agile and intelligent structures to accommodate the change given the highly dynamic nature of contemporary MANETs.

Although a lot has been achieved in this domain, there are so many things that need to be done to achieve true network topology agility and dynamic architectures in MANET. This chapter has articulated the key areas that need exploration. A key generic architecture is also proposed with a focus of bringing out the key entities in contemporary vehicular networks.

References

1. Aliyu A, Abdullah AH, Kaiwartya O, Cao Y, Lloret J, Aslam N, Joda U (2018) Towards video streaming in IoT environments: vehicular communication perspective. Comput Commun 118:93–119
2. Alouache L, Nguyeny N, Aliouatz M, Chelouah R (2018) Toward a hybrid SDN architecture for V2V communication in IoV environment. In: 2018 fifth international conference on software defined systems (SDS), Barcelona, Spain. https://doi.org/10.1109/sds.2018.8370428, 23–26 Apr 2018
3. Hajj HM, El-Hajj W, Kabalan KY, El Dana MM, Dakroub M, Fawaz F (2009) An efficient vehicle communication network topology with an extensible framework. https://pdfs.semanticscholar.org/8985/a4ee452f25bcd2d22451fbff3417f0870c53.pdf?_ga=2.92291477.942827893.1558614756-2097276758.1558614756. Accessed 3 July 2019
4. Jawhar I, Mohamed N, Usmani H (2013) An overview of inter-vehicular communication systems, protocols and middleware. J Netw 8(12):2749–2761
5. Lathrop S (2008) Extending cognitive architectures with spatial and visual imagery mechanisms. Unpublished PhD dissertation, University of Michigan, USA
6. Raza S, Wang S, Ahmed M, Anwar MR (2019) A survey on vehicular edge computing: architecture, applications, technical issues, and future directions. Wirel Commun Mob Comput 1–19. https://doi.org/10.1155/2019/3159762. https://www.hindawi.com/journals/wcmc/2019/3159762/. Accessed 3 July 2019
7. Sherazi HHR, Khan ZA, Iqbal R, Rizwan S, Imran MA, Awan K (2019) A heterogeneous IoV architecture for data forwarding in vehicle to infrastructure communication. https://doi.org/10.1155/2019/3101276. Accessed 3 July 2019
8. Shi W, Zhou H, Li J, Xu W, Zhang N, Shen XSN (2018) Drone assisted vehicular networks: architecture, challenges and opportunities. IEEE Netw 99:1–8. https://doi.org/10.1109/mnet.2017.1700206. Accessed 3 July 2019
9. Storck CR, Duarte-Figueiredo F (2019) 5G V2X ecosystem providing internet of vehicles. Special issue "Recent advances in software-defined internet of vehicles (SDIoV)". Sensors 19(3):1–20. https://doi.org/10.3390/s19030550. Accessed 3 July 2019
10. Wu C, Liu Z, Zhang D, Yoshinaga T, Ji Y (2018) Spatial intelligence towards trustworthy vehicular IoT. IEEE Commun Mag 56(10):22–27. https://doi.org/10.1109/mcom.2018.1800089. Accessed 2 July 2019
11. Xu WC, Zhou HB, Cheng N, Lyu F, Shi WS, Chen JY, Shen XM (2018) Internet of vehicles in big data era. IEEE/CAA J Autom Sin 5(1):19–35
12. Yang FH, Li JL, Lei T, Wang S (2017) Architecture and key technologies for internet of vehicles: a survey. J Commun Inf Netw 2(2):1–17
13. Ye H, Liang L, Li GY, Kim JB, Lu L, Wu M (2018) Machine learning for vehicular networks. IEEE Veh Technol Mag. https://arxiv.org/pdf/1712.07143.pdf. Accessed 3 July 2019
14. Yi F, Zhang N (2017) A survey on software-defined vehicular networks. J Comput 28 (4):236–244. https://doi.org/10.3966/199115592017062803025

Chapter 3
Seamless V2I Communication in HetNet: State-of-the-Art and Future Research Directions

Pranav Kumar Singh, Roshan Singh, Sunit Kumar Nandi, Kayhan Zrar Ghafoor and Sukumar Nandi

Abstract Vehicle-to-infrastructure (V2I) communication enables a variety of applications and services, including safety, infotainment, mobility, payment, and so on, to be accessed and consumed. However, V2I requires seamless connectivity without having to worry about transitions between and across heterogeneous networks. In the next generation 5G heterogeneous networks (HetNet), which is a combination of multi-tier and multi-radio access technologies (RAT), the main challenges for V2I communication are having better network discovery, selection, and implementation of fast, seamless, and reliable vertical handover; maintaining QoS; and providing better quality of experience (QoE). To meet these challenges, considerable research contributions exist, and various generic solutions have also been proposed. In this chapter, the authors discuss the state-of-the-art of such technologies and V2I communication that consists of available radio access

P. K. Singh (✉) · S. K. Nandi · S. Nandi
Department of CSE, Indian Institute of Technology Guwahati,
Guwahati 781039, Assam, India
e-mail: snghpranav@gmail.com

S. Nandi
e-mail: sukumar@iitg.ac.in

P. K. Singh · R. Singh
Department of CSE, Central Institute of Technology Kokrajhar,
Kokrajhar 783370, Assam, India
e-mail: roshansingh3000@gmail.com

S. K. Nandi
Department of CSE, NIT Arunachal Pradesh, Yupia 791112, India
e-mail: sunitnandi834@gmail.com

K. Z. Ghafoor
School of Mathematics and Computer Science, University of Wolverhampton,
Wulfruna Street, Wolverhampton WV1 1LY, UK
e-mail: kayhan@ieee.org

Department of Software Engineering, Salahaddin University, Erbil, Iraq

© Springer Nature Switzerland AG 2020
Z. Mahmood (ed.), *Connected Vehicles in the Internet of Things*,
https://doi.org/10.1007/978-3-030-36167-9_3

technologies, handover management, access network discovery and selection function (ANDSF), and media-independent handover (MIH)-based standard solutions for vertical handover. Besides, the chapter presents associated challenges with these technologies and highlights the possible research directions on multi-path technologies, SDN-based solutions, hybrid solutions, and 5G-enabled internet of vehicles (IoV). Primarily, the chapter discusses seamless V2I connectivity in HetNet and presents the state-of-the-art and future research directions in this domain.

Keywords IoV · IoT · ANDSF · MIH · DSRC · HetNet · Radio access network · RAT · Handover · 5G · V2X · MPTCP · MPIP · WAVE · WLAN

3.1 Introduction

Vehicle-to-infrastructure (V2I) communication takes place between a vehicle and the roadside communication/network infrastructure to access various services available in the vehicular network [1]. Figure 3.1 illustrates a generic vehicular network architecture that has three essential components: vehicular plane, radio access network (RAN) plane, and services plane. In the vehicular plane, we have vehicles equipped with on-board units (OBUs) to facilitate computation, storage, navigation, and communication facilities. Like the smartphones, vehicle's OBUs also have multiple network interfaces to provide connectivity in heterogeneous networks (HetNet) [2]. RAN facilitates communication between the vehicular plane and the services plane via the core network. The last mile connectivity from vehicle to RAN and vice versa is wireless communication. The V2I connectivity via RAN can be of multi-tier and heterogeneous type, for example, Wi-Fi, cellular, or dedicated short-range communications (DSRC) [3].

It is well recognized that better and seamless V2I connectivity can serve consumer demands for safety, infotainment, and other services related to mobility, environment, infotainment, vehicular social networking, maintenance, payment, and so on. For example, the driver and occupants can access many services from their user interfaces such as audio–video on demand, live TV, software updates, route guidance, navigation, messaging, voice-over inter protocol (VoIP), and many more.

In a wireless network, there have always been trade-offs between capacity, coverage, latency, data rate, mobility, spectral efficiency, and reliability. Improving them simultaneously in a single radio access technology is not that simple, and it may not deliver everything. For example, the IEEE 802.11p/wireless access in vehicular environments (WAVE) [4] have been designed explicitly for vehicular communications and provides seamless handovers. However, it has low market penetration and suffers in terms of data rate support (capacity). Wi-Fi has the largest market penetration and supports better speed and capacity; however, it suffers in

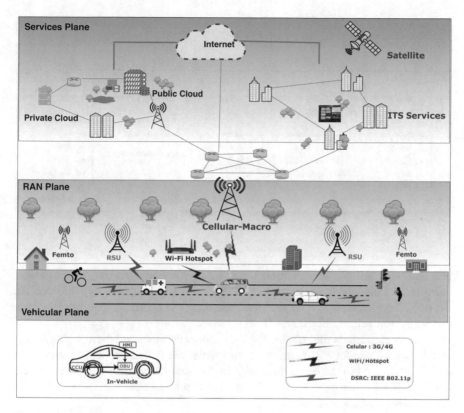

Fig. 3.1 Vehicular network architecture

terms of coverage, delay spread, high handover latency, and fading in the context of vehicular communication. The cellular networks such as long-term evolution (LTE), LTE-advanced, and LTE-pro also have its advantages and limitations. Heterogeneous connectivity based on multi-RAT is now essential to provide the best connectivity at all times. The next-generation network 5G IMT-2020 [5] is envisioned to provide seamless V2I connectivity across multiple RATs such as LTE/LTE-A, WLAN, DSRC or worldwide interoperability for microwave access (WiMAX), and so on.

The seamless vehicular communication in HetNet implies switching or roaming between radio access technologies (RATs) which should hardly be perceived by the driver and occupants of the vehicle while accessing various services. Therefore, the biggest challenge is how to switch seamlessly from one radio access technology to another. Switching across different RATs disrupts any ongoing communication, and it is one of the biggest challenges of making it transparent to transport and application layer protocols. The emergence of intelligent transportation systems (ITS) [6], smart grids, cloud computing [7], and manufacturer services; the explosion of social networks, online gaming, multimedia content, VoIP; and the

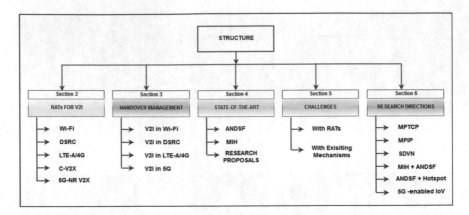

Fig. 3.2 Chapter organization

growing popularity of live streaming are among the trends driving the academia, industry, and manufacturers to re-examine traditional V2I connectivity options. The list of V2I applications is endless. Thus, seamless V2I communication in HetNet is one of the top priority objectives in this context.

This chapter details the suitable RATs, connectivity between these technologies, solutions available for V2I seamless connectivity, associated challenges; and at the end, discusses future research directions in this domain. Figure 3.2 illustrates the structure of this chapter. A list of relevant acronyms and abbreviations used in this contribution appears in Table 3.1.

3.2 Radio Access Technologies for V2I Communication

Vehicle-to-infrastructure connectivity is a core feature in smart cars today for internet access, infotainment, safe and eco-friendly drive, and access to ITS and cloud-based services. The industry and academia focus on the development of better RATs that can enable connectivity between vehicles and roadside communication infrastructures, which in turn enables various services. The four leading radio access technologies (RATs) for V2I communication are: Wi-Fi, DSRC, 4G/long-term evolution advanced (LTE-A), and 5G. Automotive industries are already experimenting with the available RATs to enable both safety and non-safety (mainly internet access) applications. For example, Audi and General Motors have the most vehicles that have 3G or 4G LTE-powered Wi-Fi connectivity for internet access. In this section, we provide details of Wi-Fi, DSRC, LTE-A, cellular-vehicle-to-everything (C-V2X), and 5G, as well as available RATs for V2I communication.

Table 3.1 List of acronyms and abbreviations used in the chapter

ANDSF	Access Network Discovery and Selection Function
CAM	Cooperative Awareness Message
CoA	Care of Address
C-V2X	Cellular-V2X
DENM	Decentralized Environmental Notification Message
DHCP	Dynamic Host Configuration Protocol
DSRC	Dedicated short-range communication
EPC	Evolved Packet Core
FMIPv6Fast MIPV6	Fast MIPV6
HetNet	Heterogeneous Network
HSPA	High-speed Packet Access
IETF	Internet Engineering Task Force
LTE	Long Term Evolution
LWA	LTE Wi-Fi Aggregation
LWIP	LTE Wi-Fi Radio Level Integration with IPSec
MAPCON	Multiple-Access PDN Connectivity
MIH	Media-Independent Handover
MICS	Media-Independent Command Service
MIES	Media-Independent Event Service
MIIS	Media-Independent Information Service
MPIP	Multi-path IP
MPTCP	Multi-path TCP
MPQUIC	Multi-path-enabled QUIC
OBU	On-Board Unit
OMA-DM	Open Mobile Alliance Device Management
RAN	Radio Access Network
RLI	Radio Level Integration
RAT	Radio Access Technology
RSSI	Received Signal Strength Indicator
RSU	Road Side Unit
SDR	Software Defined Radio
SRC	Single Radio Controller
VHO	Vertical Handover
WAVE	Wireless Access in Vehicular Environment
WiMAX	Worldwide Interoperability for Microwave Access

3.2.1 Wi-Fi

Wi-Fi has undergone a significant transformation since its conception. All the existing Wi-Fi variants are tagged with IEEE 802.11 followed by a letter or two, which signifies properties such as the speed, range, security features, and roaming. For example, IEEE 802.11a/b/g/n/ac/ad are incremental standards that define enhancements mostly in terms of speed [8]. Frequency bands, carrier aggregation mechanisms, spread spectrum techniques, modulation and coding schemes, medium access mechanisms, antenna technologies (such as multiple-input and multiple-output (MIMO)), interference cancellation techniques, and so on are the key elements of enhancements made.

Apart from communication enhancements, Wi-Fi has also seen several security standards enhancements mainly for authentication and encryption. It starts from wired equivalent privacy (WEP) and then enhanced in an incremented way to Wi-Fi protected access (WPA), WPA2, WPA2-pre-shared key (PSK), WPA2-enterprise (802.1X) [9], and IEEE 802.11i [10]. Now the management frames can be protected using IEEE 802.11w.

In addition to speed and security enhancements, the Wi-Fi is also evolving for fast transition or handover/roaming. For example, IEEE 802.11r for fast transition (FT) roaming via the over-the-air or over-the-distribution system [11, 12], IEEE 802.11 k for assisted roaming via prediction, and IEEE802.11v for network-assisted roaming and power saving. The IEEE 802.11u specification ratified in 2011 [13] also defines various enhancements related primarily to security, access, roaming, QoS integration, and so on. These enhancements make Wi-Fi one of the strong contenders for V2I communication.

3.2.2 DSRC/IEEE 802.11

In the USA, Europe, and Japan, the dedicated short-range communication (DSRC) is the key enabler of vehicular communication to ensure road safety and traffic efficiency through V2X connectivity. It uses the IEEE 802.11p standard defined at the physical (PHY) and lower medium access control (MAC) layers [14]. The IEEE 802.11p is an enhanced version of the Wi-Fi standard family to support the vehicular communication requirements of better and reliable connectivity in different mobility scenarios. In the USA, the Federal Communications Commission (FCC) allocated 75 MHz of licensed spectrum for DSRC in the 5.9 GHz band. In Europe and Japan, it has been allocated by the Electronic Communications Committee of the European Conference of Postal and Telecommunications Administrations (CEPT) and the Ministry of Internal Affairs and Communications (MIC), respectively. However, these allocations vary considerably. In all these three regions, the dedicated protocol stacks have also been developed to support V2V and V2I connectivity for this allocated DSRC spectrum.

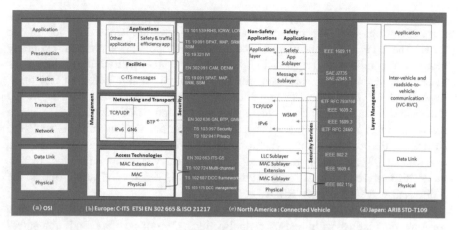

Fig. 3.3 Protocol stacks of Europe, USA, and Japan for vehicular communication

As shown in Fig. 3.3, the WAVE protocol stacks in the USA, the Cooperative-ITS (C-ITS) protocol in Europe, and the ARIB STD-T109 in Japan have been developed for vehicular communication over DSRC [14, 15]. The protocol stacks in those regions have considered the unicast, multicast, and broadcast communication options. The V2I connectivity for infotainment-related services can be made via the layer-3 and layer-4 options of IPv6 and TCP/UDP, respectively. The lower layers of these protocol stacks differ from each other. The connected vehicle and C-ITS standards are based on WLAN extension in outside the context of a BSS (OCB) mode operating in the 5.9 GHz frequency band (IEEE 802.11p), which uses CSMA/CA to coordinate multiple access. However, the ARIB STD-T109 uses the same physical layer, but in the frequency band of 700 MHz (center frequency of 760 MHz) it employs an adapted MAC layer that mixes CSMA/CA with TDMA. The lower layer of C-ITS protocol stack is referred to as an access layer that merges the physical and the data link layer corresponding to OSI model, and the details of the technology are specified in ETSI ES 202 663. It constitutes of three parts: IEEE 802.11p, IEEE 802.2 LLC, and DCC. The lower layers of connected vehicle follow a similar approach with MAC sublayer extension functionalities of multi-channel operation specified in IEEE 1609.4.

Network and transport layers (in Fig. 3.3) specify the required functionalities of layers 3, 4, 5, and 6 of the OSI reference model [16]. In connected vehicles and C-ITS protocol stacks, at these layers, we have two separate data planes; one with TCP/UDP over IPv6 for V2I communication, and other using WSMP (specified in IEEE 1609.3) and BTP with geo networking protocols (GN6: defined in ETSI EN 302 636-4-1) for V2V communication, respectively. The WAVE short message protocol (WSMP) is a highly efficient messaging protocol and designed primarily for optimized operation in a vehicular mobility environment. WSMP is utilized mainly for safety and fee collection scenarios. It is well suited for message-based applications and in intermittent connectivity. GN6 is used as a network protocol that facilitates both single-hop as well as multi-hop communication via geographical addressing. Thus, for multi-hop, it uses routing, which is based on geographic areas. The BTP is a connectionless, best effort transport layer protocol explicitly designed for traffic safety applications, which are delay-sensitive. It also facilitates means for distinguishing between different protocols at the layer on top of it. In ARIB STD-T109, the functionalities of these two layers are defined in the inter-vehicle and roadside-to-vehicle communication (IVC-RVC) layer, which maintains channel access parameters, handles communication control, and synchronizes the clock. Time synchronization among vehicles is achieved via over-the-air (OTA) synchronization. IVC deals with V2V communication, that is, the traffic between vehicles, whereas RVC defines I2V communication, that is, the traffic sent to vehicles from RSUs.

The application layer (Fig. 3.3) of these protocol stacks represents an interface for communicating with end-user applications. Various safety and non-safety applications and protocols have been developed in this respect, such as basic safety messages (BSM), cooperative awareness messages (CAM), and decentralized environmental notification messages (DENM), which use underlying layers to offer road safety, infotainment, payment, and commercial services in vehicular networks.

IEEE 802.11bd

Formation of a new study group has been announced by the IEEE and the IEEE Standards Association in May 2018, with the aim to enhance the existing 802.11 technology for next-generation V2X communications. It has been named as IEEE 802.11 next generation V2X (NGV), and amendments are being prepared as IEEE 802.11bd [17, 18]. The study group is in its initial stage and investigating the usage of more advanced physical layer (PHY) technologies in amendments after 802.11p. The objective is to enhance the throughput and transmission range. The core idea is adapting the state-of-the-art IEEE 802.11ac Wi-Fi technologies and making it suitable to high mobility scenario by enabling OCB mode. It will also guarantee backward compatibility with legacy devices. As of now, the main emphasis is at the PHY layer and the candidates being investigated at this layer are OFDM numerology re-design, definition of new guard interval and optimized tone spacing, MIMO diversity through STBC codes or cyclic shift diversity (CSD), LDPC codes, addition of midambles, dual-carrier modulation (DCM), 20 MHz channel widths, and so on.

3.2.3   LTE-A/4G

The cellular technologies prior to 4G, such as 3G, HSPA, UMTS, EDGE, and GSM, had two paths for communication: one dedicated for voice and the other for data. Voice calls were circuit switched while internet connectivity was packet switched. The first release of LTE, Release-8 (LTE Rel-8), was launched in 2008 by third-generation partnership project (3GPP). The LTE technology uses all-in-one approach, that is, everything over IP, including voice. In the next 3GPP release, Release-9 of LTE/4G, it offered substantial enhancements with features like multi-input multi-output (MIMO) to allow spatial multiplexing (use of multiple channels in parallel), inter-cell interference coordination (ICIC) for better coordination between base stations, use of orthogonal frequency division multiplexing access (OFDMA) in the downlink, use of single-carrier OFDMA (SC-FDMA) in the uplink, support for both FDD and TDD, enhancement in spectrum efficiency, support of small cell deployments such as picocell and femtocell support of high mobility (terminals speed up to 350 km/h), and so on [19].

LTE-advanced (LTE-A) was specified in 3GPP Release-10 and later revised and enhanced in Releases 11 and 12 to extend capabilities in terms of higher bit rate, higher spectral efficiency, increased number of simultaneous active users, minimized handover latency, and improved performance at cell edges. The key features of the LTE-A are carrier aggregation (CA), advanced interference management coordinated multi-point (CoMP), use of MIMO, support of relay nodes, and so on. All these features make it a highly suitable RAT for V2I communication [20].

3.2.4   Cellular-V2X (C-V2X)

LTE-V2X, which is based on the device-to-device (D2D) communication of 3GPP Release-12, was defined in Release-14 [21]. It is also called cellular V2X (C-V2X), a unique initiative by the 3GPP for the automotive industry. C-V2X specifications are based on LTE technology that supports V2X communication, including V2V as well as V2I/N communication. C-V2X being a variant of 4G, the most significant advantage is that it can be rapidly deployed leveraging the existing network infrastructures. It has better coverage and can help with the continuity of services. C-V2X supports relative speeds of approx. 500 km/h and the coverage is greater than 450 m. It also leverages the robust security features built into cellular networks. The backhaul linking to the roadside infrastructure can be utilized for various internet and cloud services enabling V2I/N communication. C-V2X enables drivers to benefit from a wide range of applications and services, including safety, connected infotainment, eCall, and so on [22]. C-V2X employs two complementary communication modes: direct and network. Direct communications mode, also called sidelink, which is independent of cellular network, operates in ITS bands of 5.9 GHz and enables V2V, V2I, and V2P communication. All these direct communications happen using sidelink interface named PC5. In the network mode, the connectivity from vehicle-to-network employs traditional mobile licensed spectrum. It is implemented over "Uu interface" and supports longer range than direct mode (>1 km). C-V2X needs high clock synchronization and takes the help of global navigation satellite system (GNSS) to establish synchronization and accurate positioning. 3GPP Release-14 for C-V2X and its enhancement in Release-15 was frozen in March 2017 and March 2019, respectively.

3.2.5   5G New Radio (NR)

C-V2X of 3GPP Release-14 established the foundation for V2X communication, and in Release-15 it has been refined for enhanced safety by improving the range and reliability. It continues to evolve as part of the 5G roadmap to fulfill diverse V2X requirements of lower latency, ultra-high reliability, higher throughput, better coverage, high mobility, the connection in high density, and so on. 5G NR, which is also known as NR-V2X (in the context of vehicular communication), is 3GPP Release-16 [23] that complements and co-exists with C-V2X. The study of NR-V2X started in June 2018, and the subsequent work item is expected by December 2019. The study item for NR-V2X targets enhancements of the sidelink and Uu interface and identifies enhancements to the NR Uu interface to offer advanced V2X applications. Some other objectives are Uu interface-based allocation and configuration of sidelink resources, a mechanism for better RAT/interface selection among sidelink and Uu of LTE and NR, QoS management and co-existence of NR-V2X and C-V2X [24].

It is expected that in the 5G era, NR-V2X will be able to enhance both the co-operative and automated driving in a much better way. High throughput (approx. 20 Gbps) and ultra-low-latency (approx. 1 ms) NR-V2X connectivity will enable the exchange of both raw and processed data collected via local sensors and live video frames. The higher peak data rate will enable to build local, dynamic maps based on sensor data, and cameras. It will provide a bird's eye view and allow vehicles to see-through even if any vehicles behind. It will also support a large number of simultaneous connections in dense scenarios. By enhancing and refining the radio access network and core network infrastructure, 5G NR will be able to bring new capabilities and tremendous changes to the existing C-V2X connectivity. C-V2X and its evolution to the NR-V2X will, therefore, revolutionize the automotive industry in the future.

3.3   Handover Process in Radio Access Technologies

Seamless V2I connectivity in a homogeneous or heterogeneous network can only be assured if a better handover mechanism has been designed and implemented. Vehicles when move across Wi-Fi, base stations and roadside units may need to change their associations that may include layers 2 and 3 addresses, need to reauthorize themselves and exchange keys, and may also require additional resources. These activities fall under the handover process, which is one of the crucial factors for seamless V2I connectivity. Improvements to the existing ones can only be made if we have information about the base handover process.

This section is dedicated to discuss the details of the handover process and describe how V2I connectivity may be achieved in radio access technologies (RATs). As shown in Fig. 3.4, based on the mobility of a vehicle and the radio

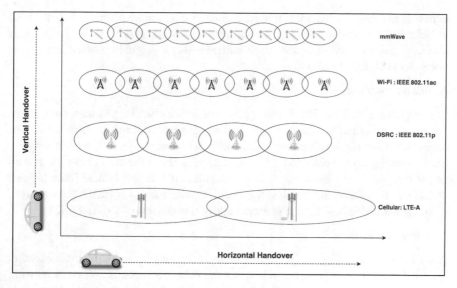

Fig. 3.4 Illustration of horizontal and vertical handover

access technology, it re-associates with the handover, which can be of two types: horizontal and vertical [25].

Horizontal Handover

Handoffs in a homogeneous network are referred to as horizontal in the sense that the next point of the attachment of a vehicle belongs to the same RAT. For example, the handover between base stations of the cellular network due to the mobility of the vehicle.

Vertical Handover

Handoffs in heterogeneous networks are referred to as vertical. Here, the next point of the attachment of a vehicle belongs to different RATs; for example, handover between cellular network base station and Wi-Fi access points.

3.3.1   Handover Process

Many research works and studies [26, 27] have divided the handover process into three phases: handover initiation, decision, and execution. These are now briefly elaborated below.

Handover Initiation

In this phase, a vehicle checks its connectivity with the associated network (roadside unit (RSU)/access point (AP)/evolved NodeB (eNB)). If the connectivity with the currently associated network is not good enough to continue a connection, then the vehicle initiates the handover process. The traditional approach for initiation is checking the received signal strength indicator (RSSI)/signal-to-interference-plus-noise ratio (SINR) threshold.

Handover Decision

In this phase, a vehicle first discovers the available networks and then collects the information required for the selection of the target network for its association. The decision criteria can be based on RSSI, SINR, bit error rate (BER), location, and so on. Depending on the selection strategy, it decides about the target network as well as the time when the handover must be executed. If it is the vehicle (mobile host) that makes this decision, then it is a mobile-initiated handover (MIHO); otherwise, if the network decides, then it is a network-initiated handover (NIHO) [28].

Handover Execution

After the selection of a suitable target network for handover, the vehicle executes the handover for its re-association to it. The vehicle gets resources in a new network and releases the resources of the old network. In this phase, either the vehicle (mobile) or the network executes and controls the actual handover process. If the vehicle does, then it is a mobile-controlled handover (MCHO), otherwise a network-controlled handover (NCHO) [28]. Figure 3.5 illustrates the concept of handover management process [27], which includes different entities related to it.

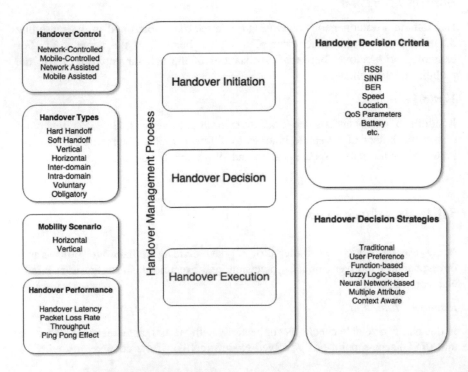

Fig. 3.5 Concepts of handover management

3.3.2   Handover in Wi-Fi

As illustrated in Fig. 3.6, when a vehicle is moving across Wi-Fi access points, the handover can be of two types: intra-domain and inter-domain handover.

Intra-Domain Handover in Wi-Fi

Handoff due to vehicular mobility in Wi-Fi is the process of re-association to the new AP, when a vehicle is reached at the edge of the currently associated AP where the received signal strength has dropped down to the defined threshold.

The intra-domain handover is a basic service set (BSS) transition within the same distribution system (DS). All access points (APs) in the DS are configured with the same IP subnet mask. This helps the vehicle to avoid the changes in IP address if it is roaming within the same DS. The intra-domain handover is also referred to as layer-2 handover or link layer handover.

This handover process in Wi-Fi starts with the discovery process of the new AP that involves initiation and scanning. As the device moves away from the connected

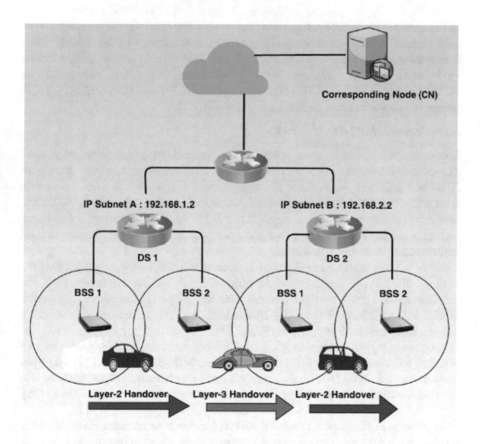

Fig. 3.6 Intra-domain and inter-domain handover in Wi-Fi

AP, the RSSI begins to drop and force the mobile node to discover new accessible APs.

There are two types of mechanism that allow the mobile node to discover target AP: passive scanning and active scanning. In the passive mode, the mobile nodes listen for the beacon management frames broadcasted by the APs. Beacon frames are transmitted periodically by the AP to announce its presence. In the active mode, the mobile node switches to a new channel and broadcasts probe request frame (probing) on it and waits for the probe response frames from APs operating on that channel.

If no response is received on that channel, it is assumed empty, and the mobile node switches to a new channel. This process repeats for all operating channels. The set of the channel depends on mode and country. Finally, the received probe response frames are processed by the mobile node to obtain information about the candidate access point.

After the discovery phase of the APs, the mobile node selects one for its association based on some specific criteria such as RSSI, location, speed, and heading.

The layer-2 handover process based on the 802.1X/EAP security framework could consist of six phases: initiation, discovery, 802.11 open authentications, re-association, re-authentication, and key exchange [29, 30]. Figure 3.7 shows management frames which are exchanged during handover in IEEE 802.11i. The re-authentication phase shown in Fig. 3.7 may differ based on the authentication mode used, that is, whether WPA/WPA2 using pre-shared key (PSK) or WPA2-enterprise with RADIUS server using 802.1X/EAP and protocol [31]. These phases at layer-2 incur high handover delay [30, 31].

Inter-Domain Handover in Wi-Fi

The inter-domain handover is referred to as layer-3 handover or network layer handover or inter-subnet handover [32]. This is a DS transition where the change of IP address will take place. Once the layer-2 handover process is over, the mobile node starts the layer-3 handover process. The vehicle tries to get a new IP address to establish a connection with the target AP. All the applications using transmission control protocol (TCP) and user datagram protocol (UDP) will get interrupted until vehicle acquires a new IP address.

When a vehicle moves to different IP subnet and retains the origin address the routing fails, and if changes the origin address, the IP session will break. Thus, the solution is to keep its origin address and try to get a temporary address, a care-of-address (CoA). The mobile IP mechanism provides such a solution. IETF has proposed mobile IPv6 (MIPv6) (RFC 3775), a network layer protocol [33], to support mobility in such scenarios of IPv6-based network. Moreover, MIPv6 protocol has been enhanced to fast handover in MIPv6 (FMIPv6) [34] and hierarchical mobile IPv6 [35] to reduce layer-3 handover latency. The researchers also proposed variants of MIPv6 such as inter-domain proxy mobile IPv6 [36]. Figure 3.8 illustrates the MIPv6-based inter-domain handover in Wi-Fi.

As shown in Fig. 3.8, a complete MIPv6 handover mechanism consists of the layer-2 and layer-3 handover. Layer-3 handover cannot start unless the mobile node

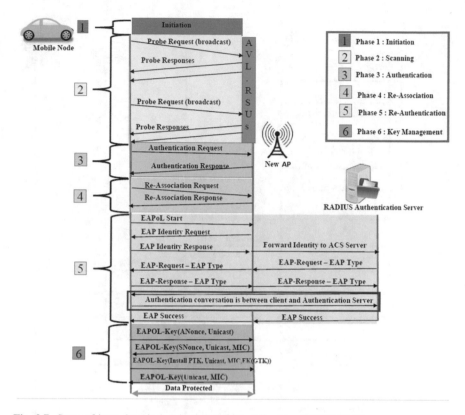

Fig. 3.7 Steps of intra-domain handover in Wi-Fi while using IEEE 802.11i

finishes the layer-2 handover. Later, in this section, the working principle of the MIPv6 is discussed because it is the base for all other enhancements.

The layer-3 handover process includes discovering the new routers, detection of movement, configuration of the address (auto-configuration), duplication of address detection, and then IP address registration. The mechanism that supports the mobility of mobile nodes with IPv6 is called mobile IPv6. In the IPv6 address, the leftmost bits are called prefix. The prefix bits are of variable length and help to distinguish the network. A mobile node attached to a subnet with IPv6 needs to know the subnet prefix of that network.

The IPv6 address has three parts of global routing prefix: subnet id, interface id, and link local scopes. The mobile node is serviced by an home agent (HA) when it is present in its home network area. However, when a mobile node enters into a target network, it needs to have a new CoA of that subnet zone. The layer-3 attachment process with the other subnets is initiated via the router solicitation, through which the mobile node (MN) tries to determine who is the router for the target network (AP in different subnet: nAP). In response to that advertisement, messages are advertised by the router in the target network (NAR), which contains

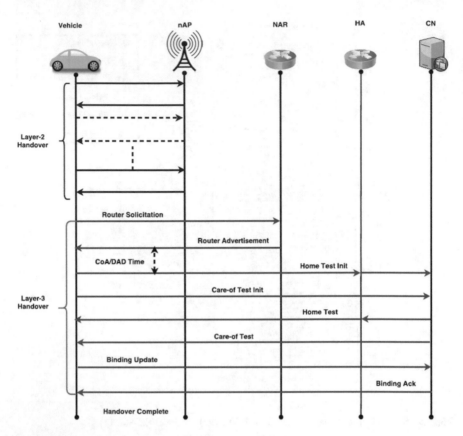

Fig. 3.8 MIPv6-based inter-domain handover in Wi-Fi

the subnet ID. Since IPv6 supports auto-configuring feature of node's IP addresses, a mobile node can generate its new CoA comprising of subnet ID advertised by NAR and its own MAC address. However, this new care of address (CoA) is required to be unique and for that duplication address detection (DAD) check must be performed followed by some new signaling message for binding authentication such as: Home Test Init (HoTI), Care-of Test Init (CoTI), Home Test (HoT), and Care-of Test (CoT).

The mobile node (MN) registers this address with home agent (HA) via binding update (BU) and, in turn, receives a binding acknowledgment (BA) from the HA. This step confirms that the BU has been received and processed successfully. The BU is issued over IPSec tunnel to protect integrity and authenticity. Now, a packet can reach to the MN from the CN via the HA using the tunnel. Afterward, the return routability (RR) procedure is performed by the MN with the CN. The RR process ensures the CN that the BU is received from a correct MN, and MN owns both HoA and CoA. Then, MN sends BU to the CN, and after receiving the BA, it sends further packets in the optimized route. In this way, layer-3 handover completes using MIPv6 mechanism.

The layer-3 handover process based on MIPv6 consists of three key latency components:

- Movement detection (MD) latency (TMD): The time which is taken by the mobile node to detect its mobility to a new network via IPv6 router advertisements and neighbor discovery.
- Duplicate address detection (DAD) latency (TDAD): The time taken by the mobile node to form care of address (CoA) and perform DAD to confirm the uniqueness of the IPv6 address.
- Binding update (BU) latency (TBU): The time which is taken by the mobile node to get binding acknowledgment (BA) from the corresponding node (CN) and home network router after sending the binding update (BU) signals.

3.3.3   Handover in DSRC/IEEE 802.11p

IEEE 802.11p is an amendment of IEEE 802.11, in which various enhancements and changes have been made to adopt it for V2X communication, primarily for V2V and V2I communications. In these enhancements, some of the important changes are also made at layer-2 of the 802.11 communication protocol stack by eliminating or shortening phases of layer-2 handover. The mobile nodes using IEEE 802.11p over 5.9 GHz DSRC band in the USA or EU send data frames outside the context of a BSS (OCB), and thus, avoid the authentication, association, or data confidentiality phases of layer-2. Since the OCB communication of V2V and V2I occurs in a dedicated frequency band for safety and non-safety, there is no need to scan the channel. Therefore, IEEE 802.11p eliminates the layer-2 handover delay due to discovery, authentication, association, and key establishment phases. However, there is a provision to secure the V2I communication outside the MAC layer. These overall changes make the IEEE 802.11p standard fast and suitable for V2I communication [37].

As shown in Fig. 3.9, the RSUs deployed alongside the road also act as an infrastructure router. In IEEE 802.11p, deployment and movement detection between two adjacent RSUs are harder for the mobile nodes because it does not rely on layer-2 triggers (unlike IEEE 820.11 association and dissociation phase). Thus, in OCB mode it is difficult to detect when they are about to leave the associated RSU and move to another RSU. In such a context, there is a need for movement detection mechanism that can utilize other triggers and detect handovers. One such mechanism can be taking the advantages of positioning data (latitude and longitude) broadcasted in various messages.

MIPv6 specifies a new option for router advertisement (RA) called the "advertisement interval option (AI)." Using this option, an RSU can indicate the maximum interval between two consecutive unsolicited router advertisement

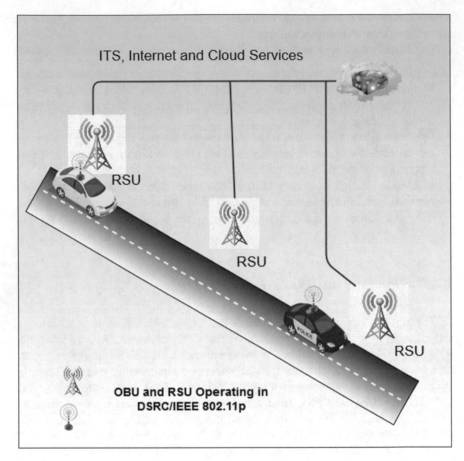

Fig. 3.9 V2I in DSRC/IEEE 802.11p

messages broadcasted by this RSU. Thus, this allows a mobile node to learn when it supposed to get the next RA from the same RSU and in turn can help in detecting the movement. When specified time interval elapses without the mobile node receiving any RA from that RSU means that the RA has been lost. Now, it is up to the mobile node to decide how many lost RAs from that RSU can be considered as a handover trigger.

In addition to the "advertisement interval option," there is another option called the neighbor unreachability detection (NUD), which can also be used to determine the reachability of the RSU. However, these mechanisms do not provide information about the availability of other potential RSUs that might be in the range of the mobile node. Thus, if RA frequency is increased, it reduces the discovery delay of the target RSU and enables faster detection of potential RSUs. When multiple RSUs are detected, it is recommended that the mobile node should consider better signal quality for a handover decision. Once target RSU has been selected, the optimistic DAD is recommended to speed up the auto-configuration to reduce layer-3 handover delay. Thus, a combination of all these mechanisms for IEEE 802.11p ensures faster layer-2 and layer-3 handover for V2I communication.

3.3.4 Handover in LTE-A/4G

Although LTE does not support soft handover (make before break), it still provides seamless mobility with the hard handover. There are three different types of handover possible in LTE [38], depending on whether it is one-tier or multi-tier.

- Intra-LTE Handover: Here, the source and target eNBs belong to the same LTE network. Intra-domain handover in LTE can be performed in two modes: S1 and X2-handover [39].
- Inter-LTE Handover: The handover takes place in two different LTE network.
- Multi-tier LTE Handover: This approach is similar to the umbrella concept. In this approach, small cells (picocell and femtocell (HeNB)) are deployed next to the larger cells (macrocell (eNB)) [40].

In a two-tier (macro and femtocell) LTE-A system, the possible handover scenarios are given as follows:

- Handover from a macrocell and femtocell (HOM2F).
- Handover from a femtocell to a macrocell (HOF2M).
- Handover from a femtocell to another one (HOF2F).

In a multi-tier LTE-A system (macro, pico, and femto), the possible handover scenarios are even more complex and are given as follows [41]: Handovers from macro to pico, pico to femto, femto to macro, or vice versa and other possible combinations.

Multiple small cells deployed under the coverage area of a macrocell can boost the communication link quality (due to the proximity of vehicle with a base station) and traffic capacity (spectral reuse). Vehicles which are located at the macrocell edges can connect to available small cells for better connectivity. Small cells can also be utilized by the vehicle for delay-tolerant applications such as downloading multimedia contents, whereas macrocells for delay-sensitive applications such as VoIP, and ITS-safety-related applications.

In this section, we discuss the intra-LTE handover over the X2 interface, which is popular and found suitable for V2I connectivity in LTE-A/4G. Various components of the access networks and core network (EPC) are shown in Fig. 3.10.

Handover means relinquishing the control and context of a mobile device or the user equipment (UE) from one base station to another. In LTE, the base stations are intelligent with decision-making capabilities and are called eNodeB (eNB). Handovers are required in cases such as mobility of the UE where the QoS received from the initial eNB degrades and for efficient load distribution among the eNBs.

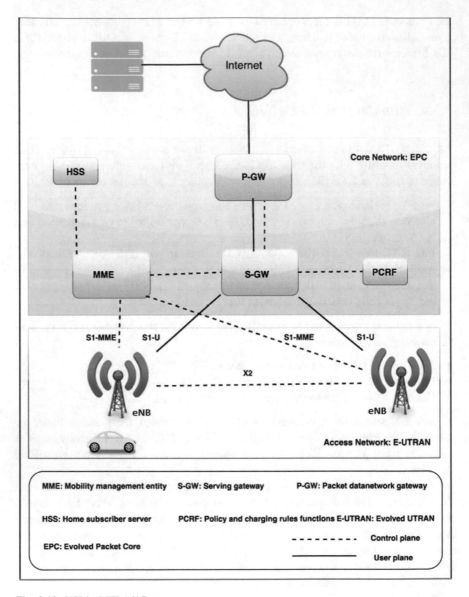

Fig. 3.10 V2I in LTE-A/4G

In LTE, handovers can be performed in two ways using the X2 interface and the S1 interface. Studies have shown that the handovers performed with the X2 interface can lower the EPC signaling in comparison with the S1 interface. Also, a decreased handover triggering time can be achieved with an increased eNB transmission power.

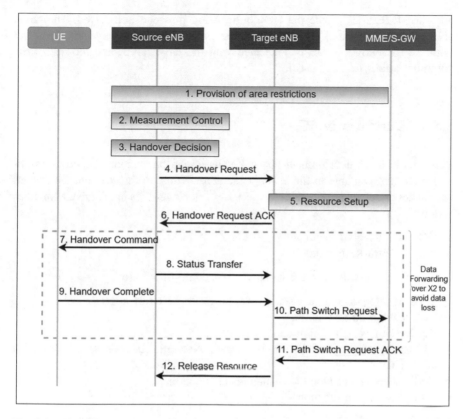

Fig. 3.11 LTE-A/4G handover with X2 interface

The handover procedure with the X2 interface in LTE [42] is shown in Fig. 3.11 and can be described as follows:

A UE is connected to the initial eNB with proper uplink/downlink channel established for communications and the resources reserved for it. UE periodically updates the initial eNB with information such as neighboring eNBs, received signal strength. Based on the collected information, the initial eNB makes the decision of relinquishing the control and context of the UE to another neighboring eNB. In case when a handover is required, the initial eNB establishes an X2 interface with the target eNB. The X2 interface facilitates both the eNBs communicate with each other directly. The initial eNB requests for a handover to the target eNB and in response, the target eNB acknowledges the request. After getting the acknowledgment from the target eNB, the initial eNB sends an HO command to the UE with reconfiguration details for synchronization with the target eNB.

Upon receiving the reconfiguration details from the initial eNB, the UE triggers its detachment from the initial eNB. When the UE completes synchronization with the target eNB, it sends an HO confirm message to the target eNB indicating a successful handover. The path switch procedure is initiated between the target eNB and the MME/S-GW once the HO confirmation message is received. The target eNB instructs the initial eNB to release the established context and the S1 bearer with the UE. Now the UE is connected with the target eNB, and the rest of the communications takes as usual.

3.3.5 Handover in 5G

Since 5G is in its development phase, details about the standard handover mechanism are not available in the existing literature. Thus, in this section, we discuss the handover details provided in Sect. 3.9 of [43]. Handovers in 5G are classified as follows:

- 5G NR inter-gNB handover
- 5G NR intra-RAN handover

The entities involved in the handover procedure in 5G are:

- gNB: It refers to the base station similar to the eNodeB in LTE.
- NR: It refers to new radio.
- Ng NGC: next generation core.
- NG-RAN: Next generation radio access Network. It consists of gNB and ng-ENB.
- AMF: Access and mobility management function.
- UPF: User plane function.

Inter-gNB Handover

In this case, the source gNB initiates the handover procedure by executing a handover request (HO request) to the target gNB. The target gNB after performing the admission control acknowledges the request by sending HO ACK. The source gNB provides the UE radio resource control (RRC) configuration details via HO command. The command contains the information that the UE utilizes to access the target gNB. Upon receiving the HO command, the UE moves the connection to the target gNB and marks the completion of the handover. Figure 3.12 presents the steps involved in the inter-gNB handover mechanism in 5G.

Intra-NR RAN Handover

This handover is also called control-plane handover. Figure 3.13 depicts the handover where neither the AMF nor the UPF changes. The exchange of messages directly takes place among the gNBs, and the resource release at the source gNB is triggered by the target gNB. The user-plane-based handover management is still in its development phase.

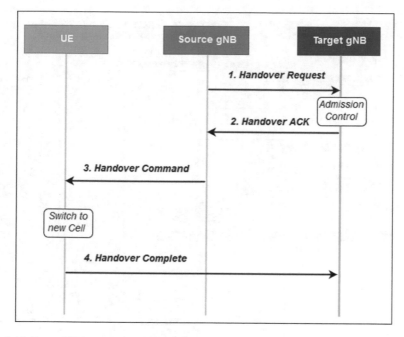

Fig. 3.12 Inter-gNB handover in 5G

3.4 Connectivity in HetNet: State-of-the-Art

A considerable number of research works have been carried out in HetNet context. The two well-known standard mechanisms, which are most promising for V2I connectivity in HetNet, are: (1) access network discovery and selection function (ANDSF) by 3GPP [44]; and (2) media independent handover (MIH) [45] by IEEE (IEEE 802.21).

- The MIH protocol facilitates network discovery, network selection, and handover execution between heterogeneous networks. This protocol has been explored to support handovers between UMTS, WiMAX, and Wi-Fi, but not with the LTE, which is now the main candidate network for V2I communication.
- ANDSF enables a mobile node to discover and select the most appropriate underlying RATs based on certain policies predefined by the network operators. It supports the discovery, selection, and connection to both 3GPP and non-3GPP access networks. ANDSF provides the mobile node, a list of radio access networks that may be available in its vicinity and mobility domain, along with their deployment coordinates. This may help the mobile node to preselect and store the target network for its handover.

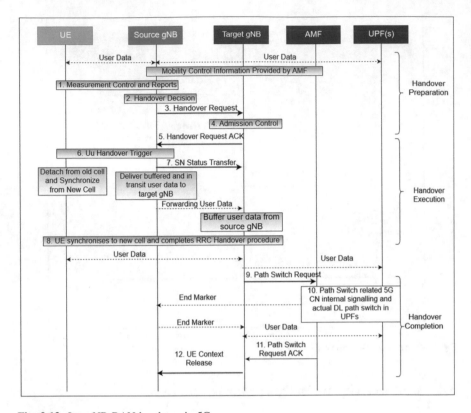

Fig. 3.13 Intra-NR RAN handover in 5G

Apart from these two key technologies, there exist several other research pro-posals for HetNet solutions. In this section, we discuss the various HetNet solutions proposed by the 3GPP, including ANDSF and MIH solution of IEEE.

3.4.1   Solutions by 3GPP

3GPP from its Release-8 started defining various 3GPP and non-3GPP inter-working architectures. These are as follows:

Solution in Release-8

The user mobility from 3GPP to non-3GPP access with IP address preservation for all kind of traffics has been standardized in this release. It has been further enhanced to incorporate WLAN accessibility via 3G-core. The two standard interfaces have been introduced as S2-a and S2-b for trusted and non-trusted non-3GPP access, respectively. These two interfaces exist between cellular and Wi-Fi networks. S2-a implements trusted access to cellular data through Wi-Fi. Wi-Fi APs, which are

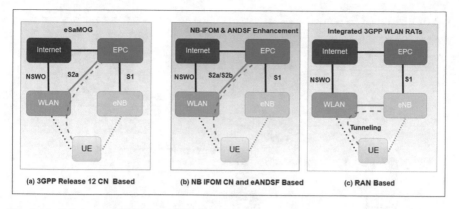

Fig. 3.14 Cellular-Wi-Fi inter-networking evolution

connected via the S2-a interface mostly include Wi-Fi hotspots deployed by the operator. S2-b interface is a proxy mobile IPv6 (PMIP)-based interface. It is an interface between P-GW of EPC and non-trusted non-3GPP access, which provides the user-plane-related control and mobility support between P-GW and the evolved packet data gateway (ePDG). Since S2-b is for non-trusted non-3GPP access, an IPSec tunnel must be established between UE and e-PDG so that mobile operator need not trust the Wi-Fi access network. In case of both solutions of S2-a and S2-b based inter-working, the offloading decision is taken at the EPC in P-GW, which costs high signaling overhead and incurs high latency. Also, with this approach, the UE can be attached to either LTE or Wi-Fi, not to both at any given time.

Figure 3.14 shows 3GPP and non-3GPP inter-networking architectures and their evolutions in latter 3GPP releases [45].

Access Network Discovery and Selection Function (ANDSF)

ANDSF is an entity within the EPC (evolved packet core) introduced in 3GPP Release-8. It aims to assist UE in the discovery and selection of non-3GPP access networks such as Wi-Fi and provide them with rules and policies for connection to these networks. This non-3GPP access then can also be used for data communications in addition to 3GPP access networks. UE may then employ IP flow mobility (IFOM), multiple-access PDN connectivity (MAPCON), or non-seamless Wi-Fi offload according to operator policy and user preferences. Figure 3.15 illustrates the high-level interaction between the UE and ANDSF server.

The ANDSF client–server interaction happens using open mobile alliance device management (OMA-DM) protocol over the S14 interface. This interface is an IP-based interface that supports pull and push communication mechanisms. Based on the operator configuration, the following category of information is provided to UE by the ANDSF server: discovery information, inter-system mobility policy (ISMP), inter-system routing policy (ISRP), inter-APN routing policy (IARP) rule selection information, WLAN selection policy (WLANSP), home network

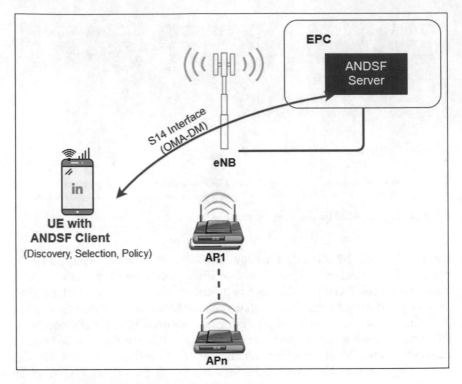

Fig. 3.15 UE and ANDSF server interaction

preferences, visited network preferences, and so on. The ANDSF-based architecture employs various ANDSF objects for smooth communication between the ANDSF client and the ANDSF server. Some of these objects are:

- UE location management object (UEL-MO): This object monitors and reports the location information of the UE to the ANDSF server. Having the information in advance allows the ANDSF server to manage the resources well.
- Network discovery information (NDI) management object (NDI-MO): It is responsible for accessing location and connection information of the desired network.
- Inter-service mobility policy (ISMP) rules management object (ISMP-MO): It defines the policy for the device for network discovery, selection, and establishment of a connection using some access priority. ISMP is valid for those UEs which cannot handle multiple (more than one) active access network connection at a time. For example, a UE can support 3GPP (LTE) as well as non-3GPP (Wi-Fi) interface; however, only one of them can be active at a time.

The ANDSF also features inter-system routing policy (ISRP), which is valid for those UEs which can have multiple (more than one) active access network connection simultaneously, and they can also route IP traffic concurrently over multiple

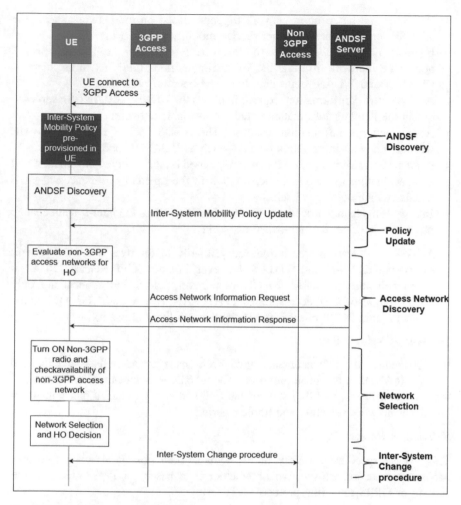

Fig. 3.16 ANDSF for connectivity in HetNet

radio access interfaces. It defines IFOM policies, and UEs can use it to implement seamless mobility protocols such as DSMIPv6.

Figure 3.16 shows the ANDSF mechanism for connectivity in HetNet. The steps involved in performing handover using ANDSF (3GPP to non-3GPP) are as follows:

- Initial connection: The UE is connected with a 3GPP access network and is utilizing the network resources.
- Pre-provisioned policies: The network selection policies are pre-provisioned on the UE. Based on these policies, UE decides the preference of selecting a non-3GPP network over various available non-3GPP networks.
- ANDSF discovery: The UE discovers ANDSF server using DHCP queries.

- Policy update based upon network triggers: Based on the network status, the ANDSF server updates the inter-service mobility policy at the UE.
- Evaluate non-3GPP networks to discover: It specifies the UE to select a non-3GPP network from a list of available non-3GPP networks such as (WiMAX and WLAN) depending upon the operator preferences.
- Access network information request: It allows the UE to query ANDSF server to obtain the network information details of available networks.
- Access network information response: The ANDSF server replies to the UE with the network information details—such as PLMN ID and network ID.
- Evaluate and select non-3GPP networks: Based on the received information and after evaluating the available networks with the operator preferences, the UE decides which network to select.
- Inter-system change procedure: Finally, the UE initiates the inter-system change procedure to select the non-3GPP network.

ANDSF in Release-8 was introduced primarily to discover non-3GPP access networks such as Wi-Fi and WiMAX; however, in latter 3GPP Releases 10–12, it has been enhanced to enable the UE to discover, select, and connect not only non-3GPP but also 3GPP access networks (UTRAN, HSPA, and LTE) [46]. Other solutions in latter 3GPP releases for HetNet connectivity are as follows [47]:

Solution in Release-9

In this release, the enhancements have been made to ANDSF, and enhanced ANDSF (eANDSF) has been proposed. The eANDSF includes the information of the cellular network, mobility state of the UE, and also provides a mechanism for intelligent network selection and traffic steering.

Solution in Release-10

This release specifies a variety of deployment scenarios. It allows a universal network connection irrespective of whether it is based on PMIP with the UE support or GPRS tunneling protocol (GTP) [48].

Solution in Release-11

In this release, S2-a mobility over GTP (SaMOG-I) [49] has been introduced, which has an S2-a interface using GTP through trusted WLAN. Figure 3.14a shows core network-based enhanced SaMOG architecture. This release also discusses the location-based selection of gateways for WLAN.

Solution in Release-12

In this release, multiple IP connectivities through trusted WLAN using GTP and IP flow mobility were introduced. Figure 3.14b shows the architecture that involves network-based IP flow mobility (NB-IFOM) [50].

Solution in Release-13

This release introduces the radio-level integration (RLI) of LTE and Wi-Fi, which enhances the capability of inter-working between LTE and Wi-Fi. Figure 3.14c shows the evolved RLI architecture. RLI architectures include:

- LTE Wi-Fi radio-level integration with IPSec tunnel (LWIP)
- LTE Wi-Fi aggregation (LWA).

Solution in Release-14

In this release, the RLI architectures have been further enhanced to support mobility, uplink aggregation, and enable Wi-Fi inter-working in high-frequency bands (60 GHz).

3.4.2 Solutions by IEEE

With advances in user equipment, radio access, and core networks, applications and services and various usage models multi-RAT operation is likely to become the norm. In this direction, one of the promising and well-known solutions initiated by the IEEE is media-independent handover (MIH), which was standardized as IEEE 802.21. The aim of MIH is to provide solutions for handover initiation (network discovery, selection, and handover negotiation), handover preparation (layer-2 and layer-3 connectivity), and interface activation in HetNet. In this section, we discuss the details of the MIH protocol [45, 51].

Media-Independent Handover (MIH)

MIH is required in the existence of a heterogeneous network (HetNet). It provides the mobile terminal the capability of discovering wireless networks of diverse technology in its vicinity and allows them to select the best network based upon the information gathered. Thus, the MIH specification enables handovers between heterogeneous technologies such as IEEE 802.x and cellular technologies, which are main V2I communication candidates. Concerning V2I in HetNet, MIH can provide not only the list of available RATs but also the relevant information, which will allow the vehicle to select suitable target RAT.

Figure 3.17 illustrates the general architecture of IEEE 802.21 [26, 45]. This framework allows higher levels to interact with lower layers to enable session continuity independent of underlying media or RAT. As shown in Fig. 3.17, the IEEE 802.21 specifies a middleware protocol called media-independent handover function (MIHF) that encapsulates different underlying RATs (e.g., 802.x, 3GPP, and 3GPP2) to the upper layers. This allows the handover management process to operate independently of the datalink layer and physical layer. MIHF provides services to both the lower and upper layers. These layers communicate with the MIHF via the service access points (SAPs). MIHF communicates with the link layer (L2) via a technology-dependent interface called MIHF_LINK_SAP. However,

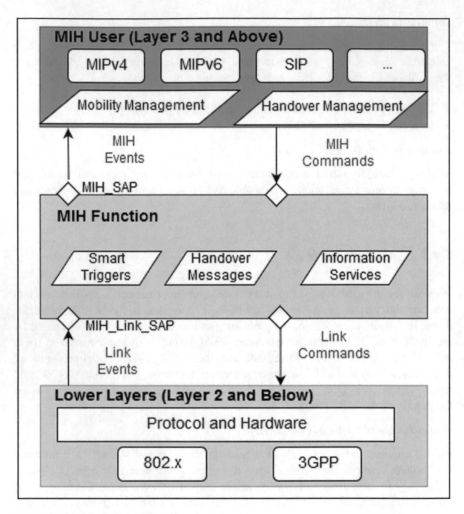

Fig. 3.17 IEEE 802.21 MIH architecture

communication between the MIHF and the media-independent handover user (MIHU) takes place via an independent technology interface called MIHF_SAP. These two SAPs aim to collect information and control behavior of the link during handovers. To implement the vertical handover, the MIHU of the mobile node queries its MIHF for further information.

The specification provides three major services to the users, namely media-independent event service (MIES), media-independent command services (MICS), and media-independent information services (MIIS). These services that enable the MIHUs to access handover-related information and to deliver commands to the link layer are briefly articulated below.

- Media-independent event services (MIES): These provide information on events related to the handover. MIES provide services on two types of events:

  - Link events: These are the events that are generated by the link layer and are received by the upper layers. Link events can be further classified into link parameter events, link synchronous events, and link transmission events. Link parameter events are generated due to the change in link layer parameters. Link synchronous events report information about link layer activities that are relevant to higher layers. Link transmission events report about the transmission status such as losses in ongoing handover.
  - MIH events: These are the events that are generated by the MIHF and are received by the MIHU. These events state the changes in MAC or PHY layers such as link up and link down. This information is crucial while deciding when to initiate a handover.

- Media-independent command services (MICS): These services provide the upper layer flexibility to control and configure the lower layers with a set of commands. MICS enable reconfiguration of the lower layers for optimal performance. MICS provide two kinds of commands.

  - MIH commands: Here, upper layers execute commands addressing a remote MIHF using MIH commands. The MIH commands are first received by the local MIHF and are then delivered to the destination using the MIHF transport protocol by the local MIHF.
  - Link commands: These commands are local to the MIHU and are used for controlling and reconfiguring the lower layers. Link commands are executed by the MIHF on behalf of the MIHU.

- Media-independent information services (MIIS): Here, a mobile node (MN) obtains the information pertaining to the heterogeneous networks present around itself from the MIIS provided by the MIHF. MIIS facilitates the handover decision process by providing relevant information to the network selection algorithm. The information tells about the details of the available networks, such as the network type and service provider ID.

More details for MIH architecture, framework, reference models, service flow mode can be found in [45].

Handover in HetNet using MIH: A Use Case from 3G to WLAN

Figure 3.18 shows the handover procedure (handover from 3G to WLAN) of MIH in HetNet. The summary of the handover process is as follows [45].

The handover procedure starts by the MIHU of the MN (operating on a 3G network) by querying its MIHF for surrounding networks (message1). The MIHF forwards this query to the information server (message2), which is located in the operator network. This generated initial query of the MN gets back the response in message4 via the MIHF. Through these messages the MN gets the required information of the target networks to which handovers can be performed. In this

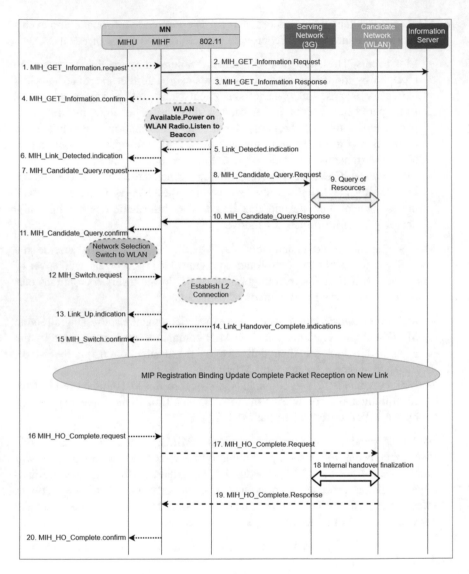

Fig. 3.18 Handover from 3G to WLAN using MIH

example, the target network is WLAN. Thus, the response contains information regarding possible WLAN. Now, MN switches on its WLAN interface and passively listens for beacons, which are broadcasted by the access points.

Once the MN receives the beacon, the Link_Detected.indication (message5) event is generated by the link layer of the WLAN interface. This 802.11 defined primitive indicates detection of new link, which is mapped into the event through MIH_LINK_SAP. This indication is forwarded to the MIHU by the MIHF in message6.

After receiving the indication from MIHF, the MIHU triggers the mobile initi-ated handover (MIHO) by sending the gathered information for potential candidate networks discovered to its service point (located on the 3G network). This infor-mation is sent to the MIHF in message7, which in turn forwards the query to the serving service point (message8).

After receiving the query via message8, the service point of 3G starts querying the available candidate networks asking for the list of available resources and user's QoS requirements (exchange 9). This step is performed with one or several candidate networks. The response of the queries is sent to the MIHF (message10), which in turn forwarded to MIHU (message11). At this point, the mobile node has enough information about the surrounding networks, which can help it to take a handover decision on the network, that is, to which network handover can be performed.

When MIHU has decided the target network to handover, it sends the switch command to MIHF (message12), which in turn triggers layer-2 connection of WLAN. Once the connection is established, the WLAN MAC layer issues an event reporting the end of L2 handover to the MIHF. After that, the higher layer handover procedures can start. At last, MIHF sends a MIH_HO_Complete.confirm (mes-sage20) to the MIHU informing that the target network (WLAN) has now become the new serving network.

IEEE 802.21, after being standardized in 2008, had various amendments made to it by the IEEE, which are as follows:

- IEEE 802.21a-2012: This is the first amendment and an extension of the initial IEEE Std 802.21-2008. The security mechanisms have been introduced to protect MIH services. Another mechanism has been introduced, which uses MIH in assisting proactive authentication so that latency in media access authentication and key establishment with the target network can be reduced.
- IEEE 802.21b-2012: This amendment discusses handovers with downlink only technologies.
- IEEE 802.21c-2014: Additional IEEE 802(R) media access independent mechanisms have been specified to optimize handovers between heterogeneous IEEE 802 systems and cellular systems and enable improved handover perfor-mance for single-radio devices.
- IEEE 802.21d-2015: This standard specifies mechanisms to enable multi-cast group management for MIH services. It defines management primitives and set of messages that can enable a user to join, leave, or update group membership. It also specified security mechanisms to protect multi-cast communication.
- IEEE 802.21-2017: Now, in this series, the standard IEEE 802.21-2017 is presently active, which defines media access independent services framework including its function and protocol that enables the optimization of services including handover and other key services when performed between heterogeneous networks.

3.4.3 Other Solutions

In addition to the standard solutions for HetNet connectivity, various other popular solutions are also available in the literature. V2I connectivity remains a hot topic due to the continued evolution of radio access and core network technologies and the growing demand for infotainment applications and services. Thus, it also aroused a keen interest among researchers. In this subsection, we present some of the popular solutions proposed by the research community for V2I connectivity in HetNet.

Single Radio Controller

In 2013, single radio controller (SRC) entity was introduced by Huawei for unified radio resource and traffic management in a HetNet environment. The SRC consists of an integrated radio network, base station, and Wi-Fi controller to facilitate handover in HetNet and reduce VHO latency [52]. This unified entity minimizes the signaling overhead of handover in multi-RAT because it directly changes the radio interface without involving the core network entities in the handover procedure. Thus, it helps in implementing fast handover, which makes it one of the solutions for V2I in HetNet.

Software Defined Radio

In 2013, the Haziza et al. [53] proposed integration of software-defined radio (SDR) in vehicle for seamless V2V and V2I connectivity across DSRC/IEEE 802.11p and LTE. SDR is a wireless system driven by software routines to support various wireless radios by the same hardware based on software changes. SDR-enabled hardware has the potential to switch across different RATs depending on its requirements and context. Thus, the use of SDR is another potential solution, which is being explored by the research community to support V2I connectivity in HetNet.

Software-Defined Networking

Software-defined network (SDN) is a new network paradigm that was introduced for the first time in 2014 for the vehicle network [54]. There are several studies, which have explored SDN for the heterogeneous vehicular network. In [55], the authors proposed an architecture for soft-defined heterogeneous vehicular network, which is based on cloud-RAN architecture. They investigated its feasibility and also proposed a new hierarchical control layer to enable a unified and flexible network. In [56], the authors studied the opportunities and challenges of applying SDN to a vehicular network. The proposed architecture abstracts all the vehicular networks components as SDN switches in a unified way, which mitigates the heterogeneity of the vehicular network. The proposed system utilizes vehicle trajectory predictions to reduce the overhead in SDN management caused by the highly dynamic mobility of vehicles.

Transport Layer Solutions

Three best transport layer protocol candidates for multi-RAT connectivity are stream control transport protocol (SCTP) [57], multi-path transport control protocol

(MPTCP) [58], and multi-path quick user datagram protocol internet connection (MPQUIC) [59]. These protocols can simultaneously utilize multiple interfaces of the mobile node such as Wi-Fi and cellular to achieve better throughput and reliability and also provide seamless connectivity across HetNet. Thus, these could become promising solutions for V2I connectivity in HetNet. These protocols have their pros and cons, and very few studies have investigated them for V2I connectivity. To the best of the authors' knowledge, MPTCP has been explored for V2I in three studies [60–62]. Similarly, SCTP has been investigated and extended in only two [63, 64] and MPQUIC in one [65]. We believe that these protocols require more research for their adoption to V2I connectivity.

Tight Coupling and Loose Coupling-based Solutions

Another possible solution for the seamless connectivity in HetNet could be the implementation of fast vertical handoff between 3GPP and non-3GPP networks. However, we need to have some mechanism for their inter-networking. Different inter-networking strategies can be found in the literature [e.g. 66–68]. The two popular strategies for inter-networking of 3GPP and non-3GPP networks are tight coupling and loose coupling.

- In tight coupling-based solution [69], the non-3GPP (WiMAX/WLAN) is integrated as part of 3GPP network (UMTS/3G/4G) and the entire traffic (data and signal) are transferred through 3GPP networks. In this approach, both the networks follow common mechanism for the authentication. This approach also solves the problem of centralized control. However, for V2I communication, this could not be an effective solution in dense conditions. In dense conditions, the high traffic generated from vehicles will be received by high-speed non-3GPP networks (WLAN) that it passes through the 3GPP network (cellular), which can be a bottleneck due to congestion in 3GPP network.
- In loose coupling-based solution [70, 71], the integration between 3GPP and non-3GPP is done via gateway nodes. The underlying RATs functions (protocols, QoS, etc.) autonomously, and the traffic of non-3GPP does not pass through 3GPP network. This solution is cost-effective and simple but increases the handover delay due to signaling between gateway nodes. Thus, this is not suitable for delay-sensitive applications via V2I connectivity such as VoIP.

3.5   V2I Connectivity: Inherent Challenges

In this section, we present the associated challenges of V2I connectivity in available RATs and state-of-the-art HetNet solutions.

3.5.1 Challenges Associated with RATs

In this subsection, we discuss the challenges associated with the radio access technologies (discussed in Sect. 3.2) when used for V2I connectivity.

V2I Connectivity in Wi-Fi

Using massively deployed APs can save considerable deployment costs of any new technology. It should be noted that the IEEE 802.11 protocols are not designed to support high mobility network services. Using Wi-Fi to support seamless connectivity for different services introduces several challenges. Major challenges are as follows:

- **Dealing with mobility**: High mobility of a vehicle in urban scenario (50 km/h) and on the highways (120 km/h) is a major challenge. Since Wi-Fi has small coverage, thus dwell-time is very short. The residence time is minimal, and the discovery of appropriate access point in minimum time is a challenging task.
- **Dealing with high traffic**: At peak times, there are a large number of vehicles on the roads, which will put a heavy load on the access points. Heavily loaded access points severely degrade the overall performance of the network.
- **Different QoS requirements**: Providing QoS requirements (delay, jitter, throughput) to diverse applications (real-time, VoIP, audio, safety-related, etc.) is one of the biggest challenges of Wi-Fi.
- **Frequent handover**: Dense deployment of APs in the urban scenario (due to small coverage) will cause frequent handover and may interrupt the ongoing communications.
- **Interference**: Dense deployment of APs will increase the interference, and the increased interference might reduce the handover decision quality.
- **Handover latency**: Layer-2 and layer-3 handover latency in Wi-Fi may disrupt various delay-sensitive applications. Minimizing the handover latency to the extent that it allows uninterrupted communication is a major challenge.
- **Delay spread and multi-path fading**: Doppler shift, spread, and multi-path effects in Wi-Fi deployments of urban scenarios are major challenges for V2I connectivity in Wi-Fi.

Possible solutions for these challenges are to adopt some of the changes which have been made at lower layers for IEEE 802.11p; however, without compromising with the peak data rate.

V2I Connectivity in DSRC/IEEE 802.11p

Although the IEEE 802.11p has been enhanced at PHY, MAC and network layer to improve performance and minimize handover latency at layer-2 and layer-3 for V2I connectivity, it still faces several challenges, which are as follows:

- **Low market penetration**: Unlike cellular and Wi-Fi deployment, DSRC-based deployment has low market penetration. DSRC/IEEE 802.11p is widespread across the USA, Europe, and Japan only and still in its pilot deployment phase.

The global acceptance of DSRC/IEEE 802.11p, which can provide a solution not only for V2V but also for V2I is one of the significant challenges.

- **High deployment cost**: The deployment of RSUs to provide full coverage across highways and urban is estimated to be very costly. This is another biggest hurdle for seamless V2I connectivity.
- **Low peak data rate**: The DSRC/IEEE 802.11p can provide a maximum of 54 Mbps when two adjacent channels of 10 MHz are combined. These days most of the applications demand high data rate and are bandwidth-hungry; thus, the provided speed can be a bottleneck for various V2I applications.
- **Performance in high density**: The delay spread and providing QoE in dense conditions are some of the biggest challenges for DSRC/IEEE 802.11p solutions for V2I connectivity.

Possible solutions for these challenges are to minimize the cost and have better RSU deployment policies. The standard is currently being amended as IEEE 802.11bd standard. Thus, it will be interesting to see if the new standard can solve the other associated challenges such as low peak data rate and delay spread of DSRC/IEEE 802.11p or not.

V2I Connectivity in LTE-A

As a vehicle moves in single-tier LTE-A, the current eNodeB must transfer the vehicle context (all information related to a mobile node), together with any buffered data to target eNodeB. Better techniques are therefore needed to avoid packet loss during the handover. In multi-tier LTE-A, a vehicle moving at high speed in multiple small cells (low coverage) will lead to unavoidable frequent handovers. This may increase the chances of ping-pong effects. However, as the dwell-time under picocell and femtocell is relatively short, a vehicle will frequently perform unwanted handovers and may lead to higher packet losses. These are the potential challenge for ITS and real-time applications. The handover process of HOM2F is quite challenging because macro and femtocells have different backhaul routes. The decision of whether to remain connected to the macrocell or to switch to an available femtocell is a key challenge. Whenever a vehicle is moving across multi-tier LTE-A, the handover initiation, decision, and execution remain open research problem.

V2I Connectivity in 5G

5G is widely unexplored for V2I connectivity. It would be interesting to see whether the 5G will be able to fulfill their claims of high throughput, ultra-low latency, high reliability, availability in different traffic conditions or not. Some of the key challenges which are lying ahead for 5G could be as follows: ensuring fairness, availability, reliability and seamless connectivity, resource provisioning, interference cancellation, and fulfilling diverse QoS and QoE requirements in V2I connectivity.

Given the positives and negatives of each RAT for V2I communication, a combination of RATs and HetNet approach might ultimately be an ideal solution for V2I connectivity. A HetNet approach can combine the pros of each RAT to result in a more robust solution. The promising solution in the future will likely be:

combining the benefits of the LTE-A for better coverage and connectivity between vehicles and the internet and ITS clouds; gradual RSU deployment is needed for DSRC communication for shorter-range, high-speed connectivity for content fetching via Wi-Fi in urban scenario and evolution of 5G to support emerging applications of connected and automated vehicle of higher level.

3.5.2   Challenges Associated with Current HetNet Solutions

The state-of-the-art solutions for vertical handover proposed by IEEE and 3GPP have laid down the stones for seamless V2I connectivity in HetNet, and it is continuously being enhanced and amended for better performance. However, with the evolution of more applications and integration of various features by the automotive industry, a lot of efforts need to be put in, and still, there is a long way to go. In this section, we discuss some of the significant challenges associated with ANDSF and MIH solutions for HetNet.

Concerning V2I connectivity over HetNet using MIH, it can provide not only the list of available RATs but also other relevant information which can enable the vehicle to select the best and appropriate target network. However, the MIH functions and features in its current state will not be able to assist the vehicle in getting the prioritized list of candidates RATs which it may encounter toward its moving direction. Consequently, a vehicle needs to spend much time in the discovery of target RATs by scanning all available candidate RATs. Thus, during handover, finding a potential candidate RAT to which the vehicle should handoff is time-consuming and one of the biggest challenges. This may cause higher handover latency and due to which there will be packet loss that can affect the QoS severely. Thus, the current MIH standard needs major modifications to fit in for V2I communications requirements.

In contrast to the MIH protocol, ANDSF provides the mobile node with a list of target networks that may possibly be present in its service area. As already mentioned, it also provides the geographical coordinates of those target networks. These characteristics of ANDSF may enable the mobile node to pre-select and store potential target RATs, to which it may likely handover to in the near future. However, most of the solutions proposed by 3GPP support non-seamless WLAN offload (NSWO) and has the ability for a mobile node to offload traffic directly to the Wi-Fi access network. ANDSF can be considered as more promising than the MIH. Having said this, a considerable amount of work is still needed to design and develop mechanisms for network discovery, association, authentication for seamless connectivity that would support high-speed mobile nodes moving in a multi-tier (mix of macro, micro, pico, and femtocells), multi-RAT (mix of DSRC, Wi-Fi, LTE) HetNet environment.

These standard solutions and other HetNet solutions such as SRC, SDR, SDN, transport layer, and coupled architectures are not widely explored with the latest combinations of RATs such as LTE-A, DSRC/IEEE 802.11p, IEEE 802.11ac, 5G. These standard solutions and other HetNet solutions such as SRC, SDR, SDN, transport layer, and coupled architectures are not widely explored with the latest RATs combinations such as LTE-A, DSRC/IEEE 802.11p, IEEE 802.11ac, and 5G. Radio access and core networks are being enhanced; therefore, existing solutions need to be reviewed and modified accordingly before adopting them for seamless V2I connectivity.

3.6   Future Research Directions

Providing seamless V2I connectivity in a multi-tier, multi-RAT network (with rapid topology changes, varying vehicle speed, and node density and in the presence of obstacles (buildings, trees, etc.)) is a challenging task. V2I communication suffers from connectivity issues, either in terms of high packet loss and delay or with low throughput. Now the question is how to provide seamless V2I connectivity (always-up connectivity with QoS support)?

Despite various standard solutions and research proposals, seamless V2I connectivity in HetNet remains an open problem for research and is currently a hot topic in the automotive industries, standards developing organizations, and universities. In this section, we present some of the critical areas that can help ensure "seamless V2I connectivity in HetNet" in the future.

3.6.1   Multi-path TCP (MPTCP)

MPTCP is an ongoing effort of the internet engineering task force (IETF) working group [72]. The V2I communication mechanisms discussed so far can only use one interface for connectivity. To seamlessly connect to different networks or RAT, V2I communication relies on the fast and seamless handover mechanisms. Now, two multi-path technologies MPTCP [73] and multi-path IP (MPIP) [74], which can allow a vehicle to use multiple network interfaces and IP paths simultaneously, are gaining momentum. Most of the end-user devices are now equipped with multiple network interfaces such as cellular and Wi-Fi. The legacy mechanism uses only one path even when there is more than one path exist between vehicle and corresponding node. MPTCP and MPIP can help in capitalize all available network interfaces and benefit in terms of better performance and reliability. The most significant advantage of MPTCP and MPIP is that unlike stream control transmission protocol (SCTP) technology of IETF, there is no need to change anything in the legacy backbone infrastructure (LTE network, Wi-Fi network, and IP infrastructure). They only require the end devices to be MPTCP and MPIP capable. The MPTCP abstract view and protocol stack is shown in Fig. 3.19 a, b, respectively.

MPTCP is an extension of TCP and a new transport layer protocol. It aggregates several TCP connections, using "subflows." A subflow is similar to the TCP connection and characterized by the same tuple (Source_IP, Source_TCP − port,

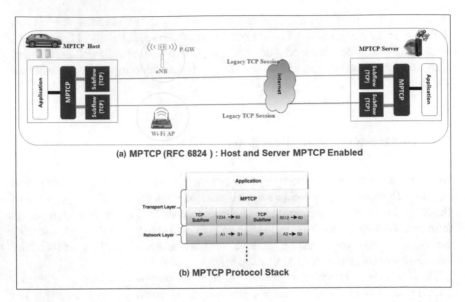

(a) MPTCP (RFC 6824) : Host and Server MPTCP Enabled

(b) MPTCP Protocol Stack

Fig. 3.19 Multi-path TCP

Destination_IP, Destination_TCP − port). Each subflow is also assigned a unique identifier called subflow-id, which is generated by the MPTCP stack. The key areas in which one can contribute are better congestion control, packet scheduling, synchronizing asymmetric paths, path management, subflow management, and dealing with lower layer handovers (e.g., layer-2 of WLAN).

3.6.2 Multi-path IP (MPIP)

MPIP proposed in [74], which not only supports the legacy TCP and UDP but also works with MPTCP protocol. Researchers have claimed that MPIP can efficiently utilize all available interfaces and applications can benefit from multi-path transmissions in terms of higher aggregate throughput. It is resilience to failures and traffic variations on individual paths and can seamlessly work across different networks. There is no need to change the upper-layer protocols. In a coordinated way, it can adjust the transmission strategies to satisfy the diverse application needs. It can customize its multi-path routing depending on the application and user needs. Thus, it can significantly improve QoS and QoE using diverse strategies for multi-path routing, which are easily configurable. Like MPTCP, it only requires changes at end devices. It also works with nodes having a single network interface. If the two MPIP capable communicating nodes have, say, a and b interfaces respectively, the number of possible paths is a × b. The MPIP protocol stack is illustrated in Fig. 3.20.

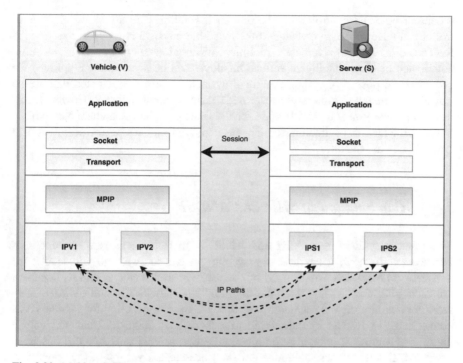

Fig. 3.20 Multi-path IP

MPTCP and MPIP can replace vertical handover technologies, allowing vehicles to use different RATs simultaneously. However, these technologies are not much explored in the V2I context. MPIP was introduced in 2017; thus, there is an excellent opportunity for its use and improvement in V2I communication.

3.6.3   Software-Defined Vehicular HetNet

The main goal of vehicular HetNet solution is to guarantee seamless and ubiquitous V2I connectivity to ensure QoS and QoE of diverse vehicular networking applications and services. The vehicle and occupants both must always be connected to optimal RAT that can ensure high throughput, ultra-low latency in a cost-effective way. However, the HetNet that consists of diverse RATs requires intelligent vertical handover decisions to ensure seamless mobility. For vertical handover to be fast and seamless, it should execute with a minimal possible delay. Thus, intelligent decisions need to be taken to determine the handover necessity, selection of target best target network out of the available networks, and conditions for handover-triggering. Such intelligence is possible only if you have a bird's eye view of the entire vehicular network architecture. Software-defined networking that de-couples the data plane and the control plane and takes the network intelligence to a centralized SDN controller can be one such solution [75]. Since the SDN controller can have global knowledge of the available physical radio resources of each RATs along with the vehicle's mobility information such as speed, location, anticipated heading, intelligent vertical handover can be accordingly employed for switching purposes [76]. The rules for flow installation of anticipated data traffic can also be installed proactively to provide faster content delivery. However, this domain is totally new and wide open for the research. The researchers can explore how such SDN-based intelligence can be deployed and tested for seamless V2I connectivity in HetNet.

3.6.4   Combination of MIH and ANDSF

The primary goal of both MIH and ANSF is to implement vertical handover. However, until now, it has been a race, and no one knows who will lead it for seamless connectivity. The demand of connectivity from vehicles and its occupant are increasing day-by-day; thus, the time has come for SDOs and researchers to work for a common goal and deploy the optimal and best solution for seamless V2I connectivity in HetNet by combining their beneficial features. There are only a handful of studies in which a combination between MIH and ANDSF has been proposed. However, these solutions have not been tested yet for V2I connectivity. There is an excellent research scope in this domain to decide its integration, features, components, and vertical handover algorithms.

3.6.5   Combination of ANDSF and Hotspot

Hotspot 2.0 has been introduced by the Wi-Fi Alliance and specified as the IEEE 802.11u standard, which is another emerging initiative for Wi-Fi roaming. The aim of introducing IEEE 802.11u is to improve the users' experience while roaming in Wi-Fi. It assists Wi-Fi-equipped mobile node in automating the process of network discovery, network selection, authentication, association, registration, and connection to the most suitable Wi-Fi hotspots in its service region. In the V2I context, this could make Wi-Fi a suitable underlying RAT, once it is deployed ubiquitously.

The combination of ANDSF and Hotspot is another popular initiative, which is gaining momentum these days. ANDSF and Hotspot 2.0 have complementary aspects, and it has been found that these two technologies can solve individual problems and provide a lot better solution for seamless connectivity in HetNet. Thus, their combination is expected to be one of the leading contenders who can enhance both QoE and QoS. Here, how to harmonize these two 3GPP-driven and Wi-Fi driven technologies becomes an interesting problem. Although 3GPP has specified three different policies for it, still there is a long way to go for an optimized, fast, secure, and seamless solution V2I connectivity in HetNet.

3.6.6   5G-Enabled Internet-of-Vehicles

The use cases envisioned by 5G for the automobile include eMBB, URLLC, and mMTC. We, therefore, need different advanced solutions in the vehicular network architecture, for example, for radio access networks and relevant core networks. We can refer to these as 5G-enabled internet of vehicles (IoV) that may consist of emerging technologies such as cloud-RAN (CRAN), mobile-edge computing (MEC), SDN, fog-computing, the co-existence of RATs, and vehicular cloudlets.

The IoV can be considered as an evolved form of the VANET [77]. There is a need for a universal model that combines these technologies to fulfill the diverse QoS and QoE requirements. This will open new doors for researchers to further contribute. For example, key areas could be: (1) efficient deployment of these technologies, (2) provisioning of resources for uninterrupted services to high-speed vehicles, (3) ensuring fairness in radio link aggregation, (4) ensuring reliability, availability, and mechanism for fault tolerance, (5) ensuring scalability in dense scenarios, (6) maintaining data consistency across edge, fog and cloud nodes, (7) ensuring security and privacy, (8) content prefetching, (9) inclusion of intelligence for vertical handover [78], and (10) streaming of content [79], fetching and multi-path technologies, and so on.

3.7   Conclusion

In this chapter, the authors have discussed the concept of V2I communication and its requirements. The details of the suitable radio access technologies and their enhancements for V2I connectivity are also provided. Since handover management plays a significant role in homogeneous and heterogeneous networks, their types and details for V2I in Wi-Fi, LTE, DSRC, 5G have been discussed. Then state-of-the-art standard solutions such as ANDSF and MIH and their amendments are described in detail. The vertical handover mechanisms implemented via MIH and ANDSF with proper use cases have been discussed. Various research proposals including SRC, SDR, SDN, transport layer protocols, and other solutions for HetNet connectivity are also presented.

The chapter further discusses not only the technologies and solutions but also the challenges associated with them. This chapter has also discussed the possible solutions to those challenges. Finally, the authors provided future research directions in emerging multi-path technologies such as MPTCP and MPIP, hybrid approach to combine standard solutions for the optimal solutions, SDN-based mechanism for a unified solution in HetNet, and 5G-enabled solutions for seamless V2I connectivity in HetNet.

References

1. Dey KC, Rayamajhi A, Chowdhury M, Bhavsar P, Martin J (2016) Vehicle-to-vehicle (V2V) and vehicle-to-infrastructure (V2I) communication in a heterogeneous wireless network–performance evaluation. Transp Res Part C: Emerg Technol 68:168–184
2. Zheng K, Zheng Q, Chatzimisios P, Xiang W, Zhou Y (2015) Heterogeneous vehicular networking: a survey on architecture, challenges, and solutions. IEEE Commun Surv Tutor 17 (4):2377–2396
3. Kenney JB (2011) Dedicated short-range communications (DSRC) standards in the United States. Proc IEEE 99(7):1162–1182
4. Uzcátegui RA, De Sucre AJ, Acosta-Marum G (2009) Wave: a tutorial. IEEE Commun Mag 47(5):126–133
5. Osseiran A, Monserrat JF, Marsch P (eds) (2016) 5G mobile and wireless communications technology. Cambridge University Press, Cambridge
6. Sjoberg K, Andres P, Buburuzan T, Brakemeier A (2017) Cooperative intelligent transport systems in Europe: current deployment status and outlook. IEEE Veh Technol Mag 12(2):89–97
7. Erl T, Mahmood Z, Puttini R (2013) Cloud computing: concepts, technology & architecture. Pearson Education, India
8. Mueck M, Ivanov V, Choi S, Kim J, Ahn C, Yang H, Piipponen A (2012) Future of wireless communication: radioapps and related security and radio computer framework. IEEE Wirel Commun 19(4):9–16
9. IEEE (2019) IEEE Std 802.1X, I: IEEE standard for port-based network access control
10. IEEE (2019) IEEE Std 802.11i, IEEE standard for wireless LAN medium access control (MAC) and physical layer specifications: amendment 6: medium access control security enhancements
11. IEEE (2019) IEEE Std 802.11r/D01.0, draft amendment to standard for information technology—telecommunications and information exchange between systems—LAN/MAN specific requirements part 11: wireless medium access control (MAC) and physical layer specifications: amendment 8: fast BSS transition
12. Singh PK, Vij P, Vyas A, Nandi SK, Nandi S (2019) Elliptic curve cryptography based mechanism for secure wi-fi connectivity. In: International conference on distributed computing and internet technology, January 2019. Springer, Cham, pp 422–439
13. IEEE (2019) IEEE 802.11u-2011—IEEE standard for information technology-telecommunications and information exchange between systems-local and metropolitan networks-specific requirements-part II: wireless lan medium access control (MAC) and physical layer (PHY) specifications: amendment 9: interworking with external networks
14. IEEE (2019) IEEE Std 802.11-2012 IEEE standard for information technology–telecommunications and information exchange between systems–local and metropolitan area networks–specific requirements–part 11: wireless lan medium access control (MAC) and physical layer (PHY) specifications amendment 6: wireless access in vehicular environments, IEEE Std 802 (11)
15. Singh PK, Nandi SK, Nandi S (2019) A tutorial survey on vehicular communication state of the art, and future research directions. Veh Commun 100164
16. Abboud K, Omar HA, Zhuang W (2016) Interworking of DSRC and cellular network technologies for V2X communications: a survey. IEEE Trans Veh Technol 65(12):9457–9470
17. IEEE (2019) IEEE P802.11-Task group BD (NGV) meeting update. http://www.ieee802.org/11/Reports/tgbd_update.htm. Accessed 1 May 2019
18. Naik G, Choudhury B, Park J (2019) IEEE 802.11bd & 5G NR V2X: evolution of radio access technologies for V2X communications. CoRR https://arxiv.org/abs/1903.08391. Accessed 1 May 2019

19. Evolved Universal Terrestrial Radio Access (E-UTRA) (2009) LTE physical layer; general description, 3GPP TR 36.201
20. Araniti G, Campolo C, Condoluci M, Iera A, Molinaro A (2013) LTE for vehicular networking: a survey. IEEE Commun Mag 51(5):148–157
21. GPP (2019) 3GPP TR 36.885, Study on LTE-based V2X services (Release 14), 3GPP technical specification group radio access network, v14.0.0, June 2016
22. Papathanassiou A, Khoryaev A (2017) Cellular V2X as the essential enabler of superior global connected transportation services. IEEE 5G Tech Focus 1(2)
23. GPP (2019) RP-181480: new SID: study on NR V2X. In: Proceedings of 3GPP planery meeting, vol 80, pp 1–10, June 2018
24. Naik G, Choudhury B, Park JM (2019) IEEE 802.11 bd & 5G NR V2X: evolution of radio access technologies for V2X communications. IEEE Access
25. Márquez-Barja J, Calafate CT, Cano JC, Manzoni P (2011) An overview of vertical handover techniques: algorithms, protocols and tools. Comput Commun 34(8):985–997
26. Chan PM, Sheriff RE, Hu YF, Conforto P, Tocci C (2001) Mobility management incorporating fuzzy logic for a heterogeneous IP environment
27. Kassar M, Kervella B, Pujolle G (2008) An overview of vertical handover decision strategies in heterogeneous wireless networks. Comput Commun 31(10):2607–2620
28. Zdarsky FA, Schmitt JB (2004) Handover in mobile communication networks: who is in control anyway? In: Proceedings. 30th Euromicro conference, August 2004. IEEE, pp 205–212
29. Mishra A, Shin M, Arbaugh W (2003) An empirical analysis of the IEEE 802.11 MAC layer handoff process. ACM SIGCOMM Comput Commun Rev 33(2):93–102
30. Shin S, Forte AG, Rawat AS, Schulzrinne H (2004) Reducing MAC layer handoff latency in IEEE 802.11 wireless LANs. In: Proceedings of the second international workshop on mobility management & wireless access protocols, October 2004. ACM, New York, pp 19–26
31. Singh PK, Chattopadhyay S, Bhale P, Nandi S (2018) Fast and secure handoffs for v2i communication in smart city wi-fi deployment. In: International conference on distributed computing and internet technology, January 2018. Springer, Cham, pp 189–204
32. Dutta A, Famolari D, Das S, Ohba Y, Fajardo V, Taniuchi K, Schulzrinne H (2008) Media-independent pre-authentication supporting secure interdomain handover optimization. IEEE Wirel Commun 15(2):55–64
33. Johnson D, Perkins C, Arkko J (2004) RFC 3775: mobility support in IPv6. IETF, June 2004, pp 1–165
34. Koodli R (2005) Fast handovers for mobile IPv6 (No. RFC 4068)
35. Soliman H, Castelluccia C, El Malki K, Bellier L (2005) Hierarchical mobile IPv6 mobility management (HMIPv6) (No. RFC 4140)
36. Al-Hashimi HN, Bakar KA, Ghafoor KZ (2010) Inter-domain proxy mobile IPv6 based vehicular network. Netw Protoc Algorithms 2(4):1–15
37. Benamar N (2017) Transmission of IPv6 packets over IEEE 802.11 networks in mode outside the context of a basic service set (IPv6-over-80211ocb)
38. Chekkouri AS, Ezzouhairi A, Pierre S (2015) Connected vehicles in an intelligent transport system. In: Vehicular communications and networks. Woodhead Publishing, Cambridge, pp 193–221
39. Alexandris K, Nikaein N, Knopp R, Bonnet C (2016) Analyzing x2 handover in lte/lte-a. In: 2016 14th international symposium on modeling and optimization in mobile, ad hoc, and wireless networks (WiOpt), May 2016. IEEE, pp 1–7
40. Gódor G, Jakó Z, Knapp Á, Imre S (2015) A survey of handover management in LTE-based multi-tier femtocell networks: requirements, challenges and solutions. Comput Netw 76:17–41
41. Ndashimye E, Ray SK, Sarkar NI, Gutiérrez JA (2017) Vehicle-to-infrastructure communication over multi-tier heterogeneous networks: a survey. Comput Netw 112:144–166

42. Lucent A (2009) The LTE network architecture—a comprehensive tutorial. Strategic Whitepaper
43. GPP TS 38.300 NR; NR and NG-RAN overall description, stage 2, (release 15), April 2019
44. ETSITS124312 (2015) Universal mobile telecommunications system (UMTS), LTE; access network discovery and selection functions (ANDSF) management object (MO), in 3GPP TS 24.312 version 12.10.0 Release 12, October 2015
45. De La Oliva A, Banchs A, Soto I, Melia T, Vidal A (2008) An overview of IEEE 802.21: media-independent handover services. IEEE Wirel Commun 15(4):96–103
46. ETSITS124312 (2015) Universal mobile telecommunications system (UMTS), LTE; access network discovery and selection functions (ANDSF) management object (MO), in 3GPP TS 24.312 version 12.12.0 Release 12.12, October 2015
47. Pasca TV, Tamma BR (2019) Traffic steering in radio level integration of LTE and Wi-Fi networks. Doctoral dissertation, Indian institute of technology Hyderabad
48. 3GPP (1999) Evolved universal terrestrial radio access (E-UTRA); LTE-WLAN radio level integration using Ipsec tunnel (LWIP) encapsulation; protocol specification. Technical report 29.060
49. SaMOG (2019) Study on S2a mobility based on GPRS tunnelling protocol (GTP) and wireless local area network (WLAN) access to the enhanced packet core (EPC) network (SaMOG). Technical report 23.852, 2013
50. Network-Based IP (2015) Flow mobility (NBIFOM)
51. De la Oliva A, Melia T, Banchs A, Soto I, Vidal A (2008) IEEE 802.21 (media independent handover services) overview. Interface 3:3GPP2
52. Xing P et al (2013) Multi-RAT network architecture, wireless world research forum, no. 9, November 2013
53. Haziza N, Kassab M, Knopp R, Härri J, Kaltenberger F, Agostini P, Aniss H (2013) Multi-technology vehicular cooperative system based on Software Defined Radio (SDR). In: International workshop on communication technologies for vehicles, May 2016. Springer, Berlin, pp 84–95
54. Ku I, Lu Y, Gerla M, Gomes RL, Ongaro F, Cerqueira E (2014) Towards software-defined VANET: architecture and services. In: Med-Hoc-Net, June 2014, pp 103–110
55. Zheng K, Hou L, Meng H, Zheng Q, Lu N, Lei L (2016) Soft-defined heterogeneous vehicular network: architecture and challenges. IEEE Netw 30(4):72–80
56. He Z, Cao J, Liu X (2016) SDVN: enabling rapid network innovation for heterogeneous vehicular communication. IEEE Netw 30(4):10–15
57. Stewart RR (2007) Stream control transmission protocol, September 2007. https://rfc-editor.org/rfc/rfc4960.txt. Accessed June 2019
58. Ford A, Raiciu C, Handley M, Barre S, Iyengar J (2011) Architectural guidelines for multipath TCP development. IETF, Inf RFC 6182:2070–1721
59. De Coninck Q, Bonaventure O (2017) Multipath quic: design and evaluation. In: Proceedings of the 13th international conference on emerging networking experiments and technologies, November 2017. ACM, New York, pp 160–166
60. Mena J, Bankole P, Gerla M (2017) Multipath tcp on a vanet: a performance study. ACM SIGMETRICS Perform Eval Rev 45(1):39–40. ACM
61. Singh PK, Sharma S, Nandi SK, Nandi S (2019) Multipath TCP for V2I communication in SDN controlled small cell deployment of smart city. Veh Commun 15:1–15
62. Williams N, Abeysekera P, Dyer N, Vu H, Armitage G (2014) Multipath TCP in vehicular to infrastructure communications. Technical report 140828A Swinburne University of Technology Melbourne
63. Huang CM, Lin MS (2010) RG-SCTP: using the relay gateway approach for applying SCTP in vehicular networks. In: The IEEE symposium on computers and communications, June 2010. IEEE, pp 139–144
64. Katsaros K, Dianati M (2017) A cost-effective SCTP extension for hybrid vehicular networks. J Commun Inf Netw 2(2):18–29
65. Mogensen RS, Reliability enhancement for LTE using MPQUIC in a mixed traffic scenario

66. Ferrus R, Sallent O, Agusti R (2010) Interworking in heterogeneous wireless networks: comprehensive framework and future trends. IEEE Wirel Commun 17(2):22–31
67. Barooah M, Chakraborty S, Nandi S, Kotwal D (2013) An architectural framework for seamless handoff between IEEE 802.11 and UMTS networks. Wirel Netw 19(4):411–429
68. Lampropoulos G, Passas N, Kaloxylos A, Merakos L (2007) A flexible UMTS/WLAN architecture for improved network performance. Wirel Pers Commun 43(3):889–906
69. Phiri FA, Murthy MBR (2007) WLAN-GPRS tight coupling based interworking architecture with vertical handoff support. Wirel Pers Commun 40(2):137–144
70. Li Y, Lee KW, Kang JE, Cho YZ (2007) A novel loose coupling interworking scheme between UMTS and WLAN systems for multihomed mobile stations. In: Proceedings of the 5th ACM international workshop on mobility management and wireless access, October 2007. ACM, pp 155–158
71. Kassar M, Achour A, Kervella B (2008) A mobile-controlled handover management scheme in a loosely-coupled 3G-WLAN interworking architecture. In: 2008 1st IFIP wireless days, November 2008. IEEE, pp 1–5
72. Wischik D, Raiciu C, Greenhalgh A, Handley M (2011) Design, implementation and evaluation of congestion control for multipath TCP. In: NSDI, March 2011, vol 11, pp 8–8
73. Barré S, Bonaventure O, Raiciu C, Handley M (2011) Experimenting with multipath TCP. ACM SIGCOMM Comput Commun Rev 41(4):443–444
74. Sun L, Tian G, Zhu G, Liu Y, Shi H, Dai D (2018) Multipath IP routing on end devices: motivation, design, and performance. In: 2018 IFIP networking conference (IFIP networking) and workshops, May 2018. IEEE, pp 1–9
75. Ghafoor KZ, Kong L, Rawat DB, Hosseini E, Sadiq AS (2018) Quality of service aware routing protocol in software-defined internet of vehicles. IEEE Internet Things J 6(2):2817–2828
76. Mahmood A, Zhang WE, Sheng QZ (2019) Software-defined heterogeneous vehicular networking: the architectural design and open challenges. Future Internet 11(3):70
77. Sadiq AS, Khan S, Ghafoor KZ, Guizani M, Mirjalili S (2018) Transmission power adaption scheme for improving IoV awareness exploiting: evaluation weighted matrix based on piggybacked information. Comput Netw 137:147–159
78. Sadiq AS, Bakar KA, Ghafoor KZ, Lloret J (2013) Intelligent vertical handover for heterogeneous wireless network. In: Proceedings of the world congress on engineering and computer science, vol 2, pp 774–779, October 2013
79. Sadiq AS, Bakar KA, Ghafoor KZ, Lloret J, Khokhar R (2013) An intelligent vertical handover scheme for audio and video streaming in heterogeneous vehicular networks. Mob Netw Appl 18(6):879–895

Chapter 4
Integrating Vehicular Technologies Within the IoT Environment: A Case of Egypt

Aya Sedky Adly

Abstract The evolution of internet of things over the past few years has impacted on a new era of connected vehicles. Telecommunications, smartphones, and advancement of new technologies have given the ability for vehicles to easily access the internet and have communication capabilities in an internet of vehicles (IoV) scenario. Many factors are involved in the development, realization, and application of the IoV while, at the same time, establishing smart industrial environments. At present, deploying various kinds of infrastructures in smart cities is considered highly promising in order to improve road safety, reduce pollution, regulate traffic, and manage collaboration with other technological platforms. However, these platforms are useful to further support IoV services, including vehicular cloud platforms, data handling architectures, and network security, which are essential ambitions for developing countries. There are also numerous inherent challenges. Smart cities, which are complex integrated network systems, can use data captured by IoV objects and process it using artificial intelligence to provide intelligent transport system with intelligent components (such as intelligent traffic control, intelligent traffic lights, and intelligent objects) that are all connected together and also with people who use these vehicles. For a developing country such as Egypt, an efficient architecture that is compatible with the feasible technologies, user requirements, and market demands is currently considered mandatory for integrating internet of things with vehicular standards and technologies. This chapter addresses these concerns along with the emerging applications and challenges for the future.

Keywords Vehicle · IoV · IoT · VANET · MANET · Evolution · Integration · Developing country · Egypt · Challenges

A. S. Adly (✉)
Faculty of Computers and Artificial Intelligence, Helwan University, Cairo, Egypt
e-mail: ayasedky@yahoo.com

© Springer Nature Switzerland AG 2020
Z. Mahmood (ed.), *Connected Vehicles in the Internet of Things*,
https://doi.org/10.1007/978-3-030-36167-9_4

4.1 Introduction

Smart cities design and development is the outcome of the growing technological advancements in the field of information technology, mainly as a result of the internet of things (IoT) vision. All objects in a smart city would have the ability to communicate with each other, through either the wired or wireless connections, deploying embedded processors [1, 2]. Increasing interconnection and interoperability between growing number of intelligent objects would introduce new smart environments, consisting of smart building, smart road, small provision of relevant facilities, and so on. One of the main goals of the IoT is to include vehicles connectivity with an aim to provide an essential role in proper and safe traveling, which will lead us to the concept of internet of vehicles (IoV) [3].

Recently, there has been a dramatic growth in the number of vehicles on the roads worldwide [4]. In Egypt, the majority of main cities during peak hours involve heavy traffic. Road repairs or even a small accident can lead to a massive traffic jam and more accidents. Every 30 s or so, someone somewhere dies because of a road accident, and nearly 10 persons are seriously wounded. According to the World Health Organization (WHO), currently, about 1.24 million persons die all over the world due to road accidents, and it is predicted that deaths may rise up to 40% by 2030 [5, 6].

Although developing countries including Egypt have nearly only 50% of the world's vehicles, it is reported that more than 90% of all the world deaths take place in developing countries. It is found that the majority of global road deaths (about 59%) are youth and middle aged between 15 and 44 years where external factors such as weather conditions are also probably the reason for some of these road accidents. As a result, studying the role of IoV on driving behavior, particularly in the developing countries, is considered a mandatory step toward reducing these casualties [5].

At a regional level, it is reported that the top death rate all over the world occurs within the eastern Mediterranean region including the African region. Egypt is situated in the north-east part of Africa, and has more than 32 deaths per a population of 100,000. Egypt is also one of the 10 countries where related research was conducted by WHO Road Safety in collaboration with a consortium of six international partners over a period of 5 years [7, 8]. Both governmental and non-governmental Egyptian national partners were involved in the project, and thus, they cooperated with WHO to make an agreement in March 2011, to develop a national 10 years plan for the Egyptian government, starting in May 2011. In fact, the Decade plan was initially envisaged in 2007 [8].

Road traffic injuries in Egypt are considered as the main reason of peoples' hospitalization and life losses. The Ministry of Health and WHO also carried out a community-wide survey to predict and assess Egypt's burden of injuries and deaths. This survey indicated that more than 62% of deaths on roads were caused by fatal injuries and approximately 70% of all road traffic injuries were in the economically productive age group.

The global status report, published in 2009 by WHO on road safety, has outlined the first global assessment of the situation of road safety for 178 countries. In this report, Egypt was one of the countries where death rate was more than 40 deaths per population of 100,000 [7, 8]. In this report, lack of early system warnings and driver's exhaustion was considered among the common causes for accidents and crashes [9]. In such situations, recommendations from the nearby vehicles observations could be considered essential to assist in improving safety for the vehicle users [3]. The report also suggested setting up of national priority actions for Egypt for addressing some core issues, including regulating road traffic, strengthening road safety, working out economic cost calculations, and developing national safety measurable targets [8]. In such situations, IoV can play a useful role to provide support in handling these issues.

Currently, in Egypt, vehicles are being equipped with novel technologies [10, 11] that give them the ability to connect with neighboring vehicles by forming vehicular ad hoc networks (VANET) [12]. In addition, the aim for building vehicular social network (VSN) is also to develop means for emergency-related data communications to be provided to passengers; as well as to provide enjoyable and entertaining driving-related services [13–15]. Thus, overcoming the said issues (as mentioned earlier) may become a reality in the near future. The category of mobile social network (MSN) may include this type of social network in which mobile devices are used for sharing fundamental data between mobile users [3, 16].

As the social internet of things (SIoT) can collaborate very well with smart cities vision, intelligent objects can become social entities for the benefit of people [17–19], in which smart things communicate with other smart things such as vehicle to vehicle (V2V) communication; as well as other smart devices to employ social network relationships and achieve demands of different similarly interested groups [3].

Vehicles are the future intelligent mobile objects that can offer leading services to customers. Industry and academia are recently investigating the vehicles as a resource model of computation [20]. The internet of things, at the same time, is acting as enabler for achieving additional technologically-based services, such as energy management, intelligent traffic management, and automated vehicle driving [21].

The chapter is organized as follows: Sect. 4.2 introduces the current state-of-the-art of the IoV. In Sect. 4.3, architecture, components, and limitations are illustrated. Section 4.4 introduces intelligent transportation system (ITS) as an emerging IoV-related application. Next, in Sect. 4.5, we briefly discuss the vehicular cloud mechanism. The possible solutions are then presented in Sect. 4.6. Finally, opportunities and challenges for the future are discussed in Sect. 4.7. Section 4.8 concludes the chapter.

4.2   Vehicular Technologies: State-of-the-Art

Over the recent couple of years, the wireless technologies and the internet of things (IoT) paradigm have rapidly advanced. In this scenario, cars are being vastly computerized and interconnected, as part of the internet of vehicles (IoV) vision using both the wired and the wireless technologies such as a remote system and personal digital assistants. Intelligent devices embedded within the systems and networks accomplish evident outcomes for both the drivers and the passengers [22]. Smartphones, mobile internet, cloud and edge computing are shifting gradually toward being active part of the IoV. The progression of IoV with the coexistence of cloud and edge computing has also raised so many challenges, including security of transmission of data, integration of data processing and services management, vehicular sensors interoperability, heterogeneity of smart devices, and handling of different operation, hardware and software platforms.

During the previous decade, customer anticipation has resulted in significant changes in the vehicular industry. Furthermore, powerful onboard units (OBUs) and vehicle to anything (V2X) connections are other influences that have further promoted the IoV evolution [23]. Besides, expectations of customers of vehicles are currently influenced by many current technologies such as electronic mobile devices, Bluetooth technologies, and increased internet bandwidth. In addition, new demands and requirements are evoked by urbanization and ecological sustainability. Communities that are based on internet are the drivers for the new digital sharing economy. It is well known that both conventional vehicles that are entirely mechanical (Auto 1.0) and electronic vehicles with assistance systems (Auto 2.0) are not currently capable of meeting customer hopes and demands [24].

In the current climate, smart city developments demands and requirements cannot be met with Auto 1.0 or Auto 2.0 due to the absence of appropriate standards, powerful OBUs, V2X hardware, and so on. The forthcoming Auto 3.0 evolution is currently having a major impact on the automotive industry that is now shifting toward intelligent transportation systems (ITSs) with V2X communications, information exchange between customer resources, vehicular gateways, cloud and edge servers, as well as data processing, collection, and storage for vehicular resources through web interfaces.

The upcoming Auto 3.0 is not yet much progressed in developing countries such as Egypt; however, it is hoped to become a vital element of a large connected intermodal chain that is linked to other modes of transport as well as the surrounding environment. The aim is to be able to fulfill the mobility requirements of consumers by delivering information in time without delays while supporting decision-making. Moreover, data should be filtered, referenced geographically, and be comprehensive. Auto 3.0 has widely enabled automatic vehicle data exchange and discovery. Relying on the network technologies, vehicular sensor data computations, and enhanced access, vehicles are assuming a vital role in the IoT vision and are therefore considered as a resource for the IoT-based systems [20]. This can enable the usage of many vehicular sensors to gather data in order to regulate traffic flow, monitor pollution and CO_2 levels, manage road intersections and more services that are important for initiatives of a smart city. The major objective of IoV is to enable interoperability among vehicles, smart devices, and different computational platforms such as cloud and edge servers, while reaching a collaborative awareness between customers and improving extensibility, sustainability, and computability of complex vehicular network systems.

Egypt is a country with plenty of land, sunny weather, and high wind speeds. This makes it an excellent place for renewable energy sources, especially electricity [25]. With the potential to reduce costs and provide safety, IoV can accelerate the transition of renewables in the industry by electrifying vehicles.

Electric vehicles (EV) can be connected, intelligent, practical, economical, which can reduce harmful emissions, as well as cost. EV charging stations can even provide the option to control charging from a smartphone application. The prospect of charging EVs with renewables is particularly exciting when the IoV model can guide the most efficient transition possible to electric vehicles.

In Egypt, IoV can provide many benefits such as regulating traffic and avoiding congestion and bottlenecks by providing real-time information that can help in taking the right decisions. Furthermore, pollution and CO_2 emissions can be reduced by anticipating engines adaption and different factors that affect CO_2 and pollutants emission such as traffic conditions, vehicle configuration, and driving behavior. Opening up the platforms and systems, while establishing market places, would be very beneficial, especially for the Auto 3.0 industry which allows multiplicity of services from many stakeholders. Changing the traditional IoT services that were previously just a part of a single contribution, which could not easily merge with other systems and platforms, is now an essential step and core requirement. This interconnection of different platforms, systems, and participants can provide unlimited potentials for future smart cities.

4.3 Vehicular Technologies: Architecture, Components, and Limitations

In 1996, vehicles communicating with external world started and continued to progress. The core goal of connecting vehicles was to avoid accidents, provide road safety, decrease traffic congestion, and reduce fuel consumption [26]. To sense the embedded physical assets in connected vehicles, many sensors were used.

Wireless connectivity enables vehicles to interact with internal (within vehicles) as well as external environments with many types of interactions, including vehicle to vehicle (V2V), vehicle to sensor (V2S), vehicle to road (V2R), vehicle to internet (V2I), and vehicle to anything (V2X) interactions [26]. The IoV is the junction between connected vehicles and the IoT which can also involve data communication, location tracking, shipping tracking, and energy conserving [27].

Mobile ad hoc networks (MANET) are networks that do not rely on a pre-existing infrastructure where each note can move freely in any route or direction. During nodes mobility, they can disconnect from one component or device and reconnect to another in their range. The major challenge in forming these networks is that they should be highly dynamic and both self-forming and self-healing, enabling peer-level communications between mobile nodes without reliance on centralized resources.

Vehicle ad hoc networks (VANET) are similar in some aspects to MANET and are used for smart transport systems. Some issues and solutions were previously investigated by Barba et al. [28] with respect to vehicle accidents, pollution, congestion, and safety in smart cities such as warning messages, traffic statistics, and intelligent traffic lights which can be very beneficial in Egypt. Using VANET, a communication infrastructure may be established that supports solving the identified problems, for example, by placing intelligent traffic lights (ITLs) at all intersections. Although this approach is promising and highly attractive, some inherent challenges still need to be addressed [29].

There are many characteristics common between generic vehicular networks and mobile ad Hoc network (MANET). These are all wireless networks that consist of nodes which do not have a predefined topology, bandwidth restriction, fast movements, and multihop communication requirements. On the other hand, there are also many obvious dissimilarities between them. These refer to node properties, mobility properties, mobility pattern, and other different aspects [30, 31]. These are briefly elaborated below:

- Node properties: Vehicles need to be able to connect with each other in VANET networks by fixed infrastructure or through sensors inside the vehicles. Thus, communication devices within the vehicles are required to have the capacity to achieve the desired tasks. Here the problem of power consumption does not exist as it is supplied by the vehicles' batteries [30, 32]. In general, the necessity for introducing new protocols and architectures for VANET environments is increasing due to the inherent properties of VANET.
- Mobility properties: Vehicles move quicker than the MANET network nodes. This can be considered a major challenge due to the possibility of the connection time between vehicles of being very short which may result in recurrent disconnection. This may cause other difficulties, for example, route planning and maintenance which will be very challenging [30, 33].
- Mobility pattern: Vehicles' movement is restricted by the physical map of roads, highways, traffic lights, and junctions. This can have effect on the nodes' mobility pattern. The density of nodes in the region can also affect the network performance because of collisions that may occur in the wireless network when the density is high; while in the low density regions, network fragmentation may occur [33].

On highways, short range signals are usually used for communication. Local area network connections can be used if the connectivity is between the vehicle and its internal components. Wi-Fi, Bluetooth, and Zigbee may be used if the communication is between one vehicle to another, vehicle to the internet, vehicle to the road, or vehicle to anything in the external environment. Yet, Wi-Fi range is 32 m, Bluetooth 100 m, Zigbee 50 m, and WiMAX range is 30 miles [26].

A few years ago, Egypt started to consider the WiMAX as a wireless connection to have nation-wide coverage since various potential users were unable to subscribe with internet service providers due to relatively limited capacity. WiMAX is an advanced technology that is still not widely used in developing countries for vehicle communication. It can simplify vehicle information exchange such as position, status, and speed which can be exchanged by an ad hoc network; nevertheless, many challenges and problems are still present.

Owing to many limitations that developing countries face, IoV may suffer from the following inherent issues:

- Frequency of change in network topologies of vehicles that can be very high and fast because of continuous mobility.
- Frequent interruption of data links of connected vehicles caused by the existence of different obstacles.
- Doppler shifts that may occur since the majority of vehicles are usually moving on the roads while being near to each other which can also cause multiple path fading and shadows.
- Data loss at building intersections and unconnected large vehicles.
- Occasional disconnection of data communication channels due to limited network range and dynamic network topology that change often fast and frequently.

Other challenges include the following:

- Synchronization of sensors data may not be uniform across all system components.
- Elements of IoT platforms may be heterogeneous, thus integrating them would not be a simple task.
- The same problem that may exist when integrating different smart devices into an intelligent transportation system.
- Back up of the intelligent transport system may not be appropriate—an alternative to cloud platform would be needed.

4.4 ITS—An Emerging IoV Application

Traditionally, vehicles have been just transportation objects that follow the drivers' commands. This vision has been changed with the advancements of systems, platforms, communications, and controls which opened the road toward the intelligent vehicle grids. Currently, a vehicle is a platform with a set of sensors that are collecting data from their surroundings or other vehicles, processing this data, and then offering it to the drivers or systems in order to support traffic management, safe navigation, and pollution control, to gain benefits that can provide solutions to countless problems that are especially faced by developing countries.

Like other applications of the IoT (e.g. smart buildings), the IoV also has storage, connections, intelligence, and learning abilities to fulfill the consumer demands [34]. Some forms of IoV technology have already being adopted by intelligent transportation systems (ITS) to further process the gathered information. This often needs a petabyte-scale information processing system. For such huge amounts of data, cloud computing would be most appropriate [26]. The vehicular cloud has greatly helped in the transition to IoV, providing all the services needed by the autonomous vehicles. It is considered as the equivalent of internet cloud for vehicles. Evolution of the urban fleet of vehicles to a network of independent vehicles that exchange their sensor data between them for optimizing a well-defined utility function and provide required information to drivers have progressed very fast from just uploading filtered sensor inputs to the cloud as GPS location [34].

With respect to the ITS, there are many categories of IoV-related cloud services which may be classified as follows:

- Infrastructure as a Service (IaaS): IoV and traffic-related computing services are based on the cloud framework, including vehicle monitoring, vehicle status, traffic status, real-time traffic analysis, vehicle safety status, data storage, and access billing and settlement. Open APIs are accessible by any third-party application developer to assist them in providing related services quickly.
- Platform as a Service (PaaS): Includes cloud storage, GPS data processing, ITS holographic data processing, data buses, information mining, information analysis, and information security.
- Software as a Service (SaaS): Here, developers can create applications that support IoV and ITS from various terminal PCs or mobile devices through third-party service resources and basic cloud services.

A successful intelligent transportation system is the result of integration of IoT, connected vehicles, and the cloud [26].

4.5 Vehicular Clouds

In the IoV scenario, independent vehicles should be able to cooperate to preserve traffic flow balance on the roads. It is expected that the vehicles will behave better than drivers by handling more traffic with minor delays, superior comfort for both driver and passengers, and with a reduced amount of pollution. Yet, the difficulty of controlling a huge number of vehicles cannot be easy. Thus, the vehicles must be able to harmonize the evacuation of critical regions in a quick and systematic manner in the presence of a natural catastrophe such as an earthquake. Accordingly, this will involve efficient communication and the ability to determine where the required resources (such as information about escape routes, police vehicles, ambulances, etc.) reside.

In addition, preventing malicious attacks is crucial as it could be literally fatal in the case of vehicles. Thus secured communication is obligatory as there is no standby control or second chance of intervention for the driver [34].

As the cloud computing systems are becoming increasingly more sophisticated, service applications are getting improved and better. However, services are now more distributed over various domains, using virtualization and distributed computing technology [35, 36]. Cloud services are ordered mostly from particular cloud providers, and provisioned cloud systems are in charge of ensuring the service quality [36–38].

Vehicular cloud is a new network and computing paradigm, specifically intended for vehicles which can provide efficient communication and distributed processing environment. Numerous vital services for vehicle applications such as spectrum sharing, attack protection, routing, content search, and dissemination can be offered by this mobile cloud approach. This could be realized via open standard and open interfaces that are shared by all manufacturers.

In case of cloud computing, developers are able to deploy applications automatically through utilization of distributed computing technologies on diverse servers during task allocation and storage distribution [39, 40]. The cloud computing architecture is usually composed of three layers: The bottom layer is service oriented, thus it would be considered IaaS. The middle layer is a service platform that permits developers to place their own applications, thus it would be considered PaaS. The top layer permits users to access services with a way that meets their needs, thus it would be considered SaaS.

Server power cost and hardware cost can be minimized by cloud computing through provision and use of virtual devices. Several data storage databases can be accessed from countless cloud serving systems. For the fulfillment of the SLA (service-level agreement), heterogeneous cloud computing supports the on-demand assignment of cloud resources [41, 42]. Providers of heterogeneous cloud computing services can respond to highly efficient service requests by joining an integrated cloud computing framework device [36].

A vehicular cloud can provide efficient vehicular services which cannot be served with collaborative efforts alone between participants. The vehicular cloud is temporarily created by interconnecting resources available in the vehicles and road side units (RSUs). The idea is different from the internet cloud that is created and managed by a cloud provider. These networked resources work as a common virtual platform that accentuates efficiency of cooperation [34].

Vehicles that contribute toward cloud services may register in the RSUs in a way resembling directory-based discovery which is maintained by constrained application protocol (CoAP). The vehicular cloud integrates traditional RSUs with microscale data centers to address the progressing vehicular scenarios in a generic approach and to employ software-defined networking, dynamic instantiation, replication, and migration of IoV services [23].

In general, RSU can be a good candidate for the negotiator role by joining the cloud as a stationary member. Resource-constrained RSUs as cameras may have consistent connections to vehicles even though having limited computing power and storage. Thus, they can still manage and store data and information indexes for effective and efficient content discovery [34].

4.6 Possible Solutions

The major objective of IoV-based services is to provide easier and safer driving experience. However, studies relating to integrating IoV with other infrastructures are still in their initial stages [43]. In order for the IoV to be beneficial, many services such as traffic management, pollution control, remote monitoring, road navigation, urban surveillance, business intelligence, in-car entertainment, and so on need to be available on intelligent transportation systems.

Many challenges such as reliability, security, scalability, quality of service, and the absence of universal standards are of concern as well [44, 45]. A systematic approach and collaboration among researchers, vehicles companies, government authorities, law enforcement, standardization institutions, and cloud service providers is needed to get over the complexity involved in implementing vehicular clouds and integrating various devices and systems with vehicular clouds. Thus, IoT and cloud computing may provide opportunities for technology innovation in the vehicular industry and can assist as enabling infrastructures for implementing vehicular data clouds [45].

In Egypt, as the number of vehicles is growing, obtaining data about vehicles is becoming harder to achieve. Although the intelligent transportation systems (ITSs) are advancing as a solution with distinctive capabilities and features which use IoT and wireless sensor networks (WSNs), they are not fully deployed or used in Egypt yet. ITSs can handle many situations such as tracking vehicle location and ambulance intimation when an accident happens. WSN that consists of very small devices that cooperate to sense the vehicle parameters can provide the required data for such situations.

In [46], a network model has been proposed that can validate the integrity of shared data and consider signature refreshment by the support of a third-party auditor (TPA) in the cloud which can be suitable for Egypt resources and environment. This model has the following entities: vehicle (for use or to offer IoV services), gateway (as access point of/for the vehicle), TPA, and the vehicular data cloud. The cloud could be used by people (drivers, drivers' family, drivers' friends, passengers, cyclists, foot-travelers, etc.) or by any intelligent component (intelligent sensing devices). The access point acts as a gateway between the network of the vehicle and the road. Different groups can be formed between vehicles; for instance, according to car type or common routes which can also request to verify integrity to share data from the TPA.

IoT, cloud computing, and emergence of big data have driven many user demands in the last several years. IT enterprises and developers have also provided several services and applications. Although VANET and vehicle telematics lack the processing capacity for handling information globally, yet they can be used in short-term applications or small-scale services. However, this limits the applications development on consumer vehicles. Hence, the conventional VANET, vehicle telematics, and other connected vehicle networks have to evolve into the IoV as there is a crucial requirement for an integrated and open network system [47].

Nowadays, as the time spent by people in their vehicles is increasing, demands for providing more entertainment, comfort, and safety are increasing. Accessing internet from vehicles to use applications such as Google Maps, YouTube, WhatsApp, Facebook, and other popular applications is now considered essential for both drivers and passengers. In addition, mobile technologies and services are advancing and their use is growing which make them a major choice as a device type for IoV and vehicular applications. In vehicles, people's expectations are to have the same connectivity they can attain within buildings. Thus, vehicles infrastructure should understand and satisfy user demands in order to be able to manage these intelligent contexts [45].

Although IoV may have common features that are similar to IoT, there are still many features that are different. For example, most people, when using wireless mobile networks, follow a random walk model, whereas in IoV they are subjected to many factors such as the surrounding environment, road structure, road distribution, as well as neighboring vehicles. In general, internet pays attention to people and how to provide their demanding services. Whereas IoT attention is directed to smart devices and data awareness of connecting them together, the IoV is directing its attention to intelligently integrating people and vehicles together. Generally, the vehicular network, people behavior, and service models are different for IoV in comparison to IoT and the internet [47].

Although smart cities use vehicular clouds to download and store data using mobile devices as gateways, these are focused on data sharing and exchanging data between network nodes. Hence, totally distributed networks can face difficulties due to selfish nodes. However, vehicles that are willing to cooperate by becoming a gateway may be used as a solution, and their anonymity may still be preserved [48].

Owing to vehicular networks' dynamic topology and vehicles' highly mobility, an adjacent group management for the moving vehicles is needed to support data transmission by efficient and stable routes. Thus, vehicular grouping can be a promising solution that can assist in vehicular network connectivity between neighboring vehicles that are single or multiple hops apart. However, this method requires frequent updates of nodes which are joining or leaving the groups [49].

Dividing a vehicular network into less number of nodes or into subgroups may simplify and stabilize routing, simplify vehicular network management, and preserve load balance. Grouping vehicles in developing countries may open new insights for drivers and passengers while enhancing driving conditions and environments.

4.7 Opportunities and Challenges for the Future

New technologies and applications are needed for the future mobility of vehicle industry, including different aspects of automation, electric powering, and new models for provisioning of IoT services [50]. It is estimated that the services of internet-integrated vehicles will increase dramatically from about 30 to 90% by the year 2022. Encouraging opportunities can be availed by better deployment of internet of things capabilities as well as cloud computing provision to the vehicular platforms and services while new efficient connectivity solutions are provided, over time, by the innovations of the automotive industry [51].

Recently, the high-speed internet services demand in Egypt has radically stepped up and, in accordance with this, the demand for IoV services will continue to grow. Thus, vehicular data clouds should be reliable, efficient, and secured while mechanisms and algorithms should be at least satisfactory to meet the majority of these requirements.

IoV is aiming to enable exchange of data between the vehicles as well as between the vehicles and smart objects inside and outside the vehicles. Efficiently exchanging information between the vehicles (and with other smart devices) is a difficult challenge due to topology being extremely dynamic, vehicles being randomly distributed, and the existence of huge masses of information in IoV. For the safety of driving, it is essential to sense and transmit many types of data in different formats. Additionally, for vehicles in similar situations or related conditions, exchanging the safety information or road conditions and incidents would be necessary. In general, vehicle numbers are considerably large which may dramatically accentuate the mass of data, thus it should be cached and transmitted between vehicles in an economical, effective, and efficient way after being abstracted, classified, processed, and organized within an accepted period of time. However, not only the requirement of quickness of time is needed to be fulfilled but also the spatial constraints as well as the mechanism of information transmission which ought to be employed according to the information format, quantity, and characteristics.

For the IoV, awareness of real-time requirements is vital and crucial. Entertainment (e.g. availability of music, social media offerings) also becomes much more needed for people especially for those having long vehicle journeys. Consequently, huge quantity of data including audio and video should be transferred in time. Finally, due to IoV topology being extremely dynamic, and vehicles being randomly distributed, a scalable data distribution procedure is needed to adjust with rapid changes in topology and irregular fluctuations in node density in order for real-time requirements to be fulfilled [52].

In general, many issues still need to be solved in developing countries, such as the following:

- Intelligent route planning: Some of the important challenges that face the current state of IoV and need to be resolved are route planning and next position prediction which are highly essential in decision making in real time. High mobility in VANET and topological frequent changes can cause predictions to be harder to make. Cars should be able to find the most economical, safe, fast, efficient, and beneficial routes to arrive to their destinations. Improving effective path planning algorithms would be a useful significant step for IoV further development.
- Interoperability: Enhancements in network architectures interoperability is essential as well as algorithms and communication protocols in order to simplify mobility management in the IoV domain. Currently, satisfying all the IoV requirements and constraints, while maintaining interoperability, is still a complicated task.
- Real-time data processing: Processing large amounts of data, acquired at the same time, is one of the important requirements of IoV. Balancing between parallel and sequential processing to obtain the needed results in time is highly challenging that requires effective solutions.
- Intelligent sensing devices: Interactions of vehicles with their environment is normally realized through a set of interconnected sensors. These sensors gather data that needs to be combined and analyzed through an intelligent system before being used for decision-making purposes. Intelligent systems, based on artificial intelligent and machine learning, are the pillars of the IoV vision that still needs to be fully tested in real situations prior to being confidently used by citizens.

4.8 Conclusion

Technologies are now merging from IT to electronics to networking to ease many aspects of human life. One such getting together is in the connectivity of vehicles on the roads in what is known as internet of vehicles (IoV). Powerful sensors and actuators are embedded in modern vehicles along with communication devices that can acquire and exchange data between themselves and other smart objects in vehicular networks such as VANET and MANET. Vehicular components are also evolving to be capable of connecting to other components either inside the vehicles or in surrounding environments including neighboring vehicles, buildings, smart roads, and traffic lights. With vehicular cloud technology and continuous improvement in vehicular equipment and mobile devices, peoples' ways of interaction with vehicles is constantly shifting. Several interfaces can be integrated to help the drivers accessing the internet or using mobile devices especially during driving. Speech recognition systems can remove the distraction of looking down at the mobile device while driving, as drivers can keep their eyes on the road if a heads-up display were used.

In general, vehicular interfaces should also provide integration methods for different types of devices to be used which can provide services to the drivers or passengers. Consequently, future vehicles will result in raising communication technology requirements and, thus, become part of the wider internet of things.

Considering the case of Egypt, this chapter has discussed the relevant vehicular technologies in the context of IoV and looked at the architectures and related components with some suggestions for further future development in this area.

References

1. Al-Hader M, Rodzi A, Sharif AR, Ahmad N (2009) Smart city components architecture. In: 2009 international conference on computational intelligence, modelling and simulation. IEEE, pp 93–97
2. Su K, Li J, Fu H (2011) Smart city and the applications. In: 2011 international conference on electronics, communications and control (ICECC). IEEE, pp 1028–1031
3. Alam KM, Saini M, El Saddik A (2015) Toward social internet of vehicles: concept, architecture, and applications. IEEE Access 3:343–357
4. Warner M (2013) General motors sales up in December. Technical Report. Fox Business, New York, NY
5. Organization WH (2015) Global status report on road safety 2015. World Health Organization
6. Kumar A, Prakash J, Dutt V (2014) Understanding human driving behavior through computational cognitive modeling. In: International conference on internet of vehicles. Springer, pp 56–65
7. Violence WHODo, Prevention I, Violence WHO, Prevention I, Organization WH (2009) Global status report on road safety: time for action. World Health Organization
8. Organization WH (2011) Egypt: a national decade of action for road safety 2011–2020
9. Administration NHTS (2009) Traffic safety facts: 2007 data: pedestrians. Ann Emerg Med 53 (6):824
10. Plotkin S, Stephens T, McManus W (2013) Vehicle technology deployment pathways: an examination of timing and investment constraints
11. Schütz M, Dietmayer K (2013) A flexible environment perception framework for advanced driver assistance systems. In: Advanced microsystems for automotive applications 2013. Springer, pp 21–29
12. Morgan YL (2010) Notes on DSRC & WAVE standards suite: its architecture, design, and characteristics. IEEE Commun Surv Tutor 12(4):504–518
13. Abbani N, Jomaa M, Tarhini T, Artail H, El-Hajj W (2011) Managing social networks in vehicular networks using trust rules. In: 2011 IEEE symposium on wireless technology and applications (ISWTA). IEEE, pp 168–173
14. Fei R, Yang K, Cheng X (2011) A cooperative social and vehicular network and its dynamic bandwidth allocation algorithms. In: 2011 IEEE conference on computer communications workshops (INFOCOM WKSHPS). IEEE, pp 63–67
15. Smaldone S, Han L, Shankar P, Iftode L (2008) Roadspeak: enabling voice chat on roadways using vehicular social networks. In: Proceedings of the 1st workshop on social network systems. ACM, pp 43–48
16. Falk H (2011) Applications, architectures, and protocol design issues for mobile social networks: a survey. Proc IEEE 99(12):2125–2129
17. Atzori L, Iera A, Morabito G (2011) Siot: giving a social structure to the internet of things. IEEE Commun Lett 15(11):1193–1195
18. Atzori L, Iera A, Morabito G, Nitti M (2012) The social internet of things (SIOT)–when social networks meet the internet of things: concept, architecture and network characterization. Comput Netw 56(16):3594–3608

19. Ortiz AM, Hussein D, Park S, Han SN, Crespi N (2014) The cluster between internet of things and social networks: review and research challenges. IEEE Internet Things J 1(3):206–215
20. Abdelhamid S, Hassanein HS, Takahara G (2015) Vehicle as a resource (VaaR). IEEE Netw 29(1):12–17
21. Datta SK, Da Costa RPF, Bonnet C, Härri J (2016) Web of things for connected vehicles. In: International world wide web conference, pp 1–3
22. Stepień K, Poniszewska-Marańda A (2018) Towards the security measures of the vehicular ad-hoc networks. In: International conference on internet of vehicles. Springer, pp 233–248
23. Datta SK, Haerri J, Bonnet C, Da Costa RF (2017) Vehicles as connected resources: opportunities and challenges for the future. IEEE Veh Technol Mag 12(2):26–35
24. Grote C (2014) Keynote: IoT on the move: the ultimate driving machine as the ultimate mobile thing. In: 2014 IEEE international conference on pervasive computing and communications (PerCom). IEEE, pp 50–50
25. Adly AS (2019) Climate change and energy decision aid systems for the case of Egypt. In: Climate change and energy dynamics in the Middle East: modeling and simulation-based solutions. Understanding complex systems. Springer International Publishing, Cham, pp 79–107. https://doi.org/10.1007/978-3-030-11202-8_4
26. Devi YU, Rukmini M (2016) IoT in connected vehicles: challenges and issues—a review. In: 2016 international conference on signal processing, communication, power and embedded system (SCOPES). IEEE, pp 1864–1867
27. Adly AS (2019) Technology trade-offs for IIoT systems and applications from a developing country perspective: case of Egypt. In: The internet of things in the industrial sector: security and device connectivity, smart environments, and industry 4.0. Computer communications, networks. Springer International Publishing, pp 299–319. https://doi.org/10.1007/978-3-030-24892-5_13
28. Barba CT, Mateos MA, Soto PR, Mezher AM, Igartua MA (2012) Smart city for VANETs using warning messages, traffic statistics and intelligent traffic lights. In: 2012 IEEE intelligent vehicles symposium. IEEE, pp 902–907
29. Keertikumar M, Shubham M, Banakar R (2015) Evolution of IoT in smart vehicles: an overview. In: 2015 international conference on green computing and internet of things (ICGCIoT). IEEE, pp 804–809
30. Fonseca A, Vazão T (2013) Applicability of position-based routing for VANET in highways and urban environment. J Netw Comput Appl 36(3):961–973
31. Guoqing Z, Dejun M, Zhong X, Weili Y, Xiaoyan C (2008) A survey on the routing schemes of urban vehicular ad hoc networks. In: 2008 27th Chinese control conference. IEEE, pp 338–343
32. Mohandas BK, Nayak A, Naik K, Goel N (2008) Absrp-a service discovery approach for vehicular ad hoc networks. In: 2008 IEEE Asia-Pacific services computing conference. IEEE, pp 1590–1594
33. Albraheem L, AlRodhan M, Aldhlaan A (2014) Toward designing efficient service discovery protocol in vehicular networks. In: International conference on internet of vehicles. Springer, pp 87–98
34. Gerla M, Lee E-K, Pau G, Lee U (2014) Internet of vehicles: from intelligent grid to autonomous cars and vehicular clouds. In: 2014 IEEE world forum on internet of things (WF-IoT). IEEE, pp 241–246
35. Papagianni C, Leivadeas A, Papavassiliou S, Maglaris V, Cervello-Pastor C, Monje A (2013) On the optimal allocation of virtual resources in cloud computing networks. IEEE Trans Comput 62(6):1060–1071
36. Chang Y-C, Chen J-L, Ma Y-W, Chiu P-S (2015) Vehicular cloud serving systems with software-defined networking. In: International conference on internet of vehicles. Springer, pp 58–67
37. Li A, Yang X, Kandula S, Zhang M (2010) CloudCmp: comparing public cloud providers. In: Proceedings of the 10th ACM SIGCOMM conference on Internet measurement. ACM, pp 1–14

38. Ngan LD, Kanagasabai R (2012) Owl-s based semantic cloud service broker. In: 2012 IEEE 19th international conference on web services. IEEE, pp 560–567
39. Dikaiakos MD, Katsaros D, Mehra P, Pallis G, Vakali A (2009) Cloud computing: distributed internet computing for IT and scientific research. IEEE Internet Comput 13(5):10–13
40. Ahlgren B, Aranda PA, Chemouil P, Oueslati S, Correia LM, Karl H, Söllner M, Welin A (2011) Content, connectivity, and cloud: ingredients for the network of the future. IEEE Commun Mag 49(7):62–70
41. Bakshi K (2011) Considerations for cloud data centers: framework, architecture and adoption. In: 2011 aerospace conference. IEEE, pp 1–7
42. Shan C, Heng C, Xianjun Z (2012) Inter-cloud operations via NGSON. IEEE Commun Mag 50(1):82–89
43. Pacheco J, Satam S, Hariri S, Grijalva C, Berkenbrock H (2016) IoT security development framework for building trustworthy smart car services. In: 2016 IEEE conference on intelligence and security informatics (ISI). IEEE, pp 237–242
44. Yan G, Wen D, Olariu S, Weigle MC (2012) Security challenges in vehicular cloud computing. IEEE Trans Intell Transp Syst 14(1):284–294
45. Kim Y, Oh H, Kang S (2017) Proof of concept of home IoT connected vehicles. Sensors 17 (6):1289
46. Ruan Z, Liang W, Luo H, Yan H (2015) A novel data sharing mechanism via cloud-based dynamic audit for social internet of vehicles. In: International conference on internet of vehicles. Springer, pp 78–88
47. Yang F, Wang S, Li J, Liu Z, Sun Q (2014) An overview of internet of vehicles. China Comm 11(10):1–15
48. Alouache L, Nguyen N, Aliouat M, Chelouah R (2018) Credit based incentive approach for V2 V cooperation in vehicular cloud computing. In: International conference on internet of vehicles. Springer, pp 92–105
49. Al-Hamid DZ, Al-Anbuky A (2018) Vehicular grouping and network formation: virtualization of network self-healing. In: International conference on internet of vehicles. Springer, pp 106–121
50. Underwood S (2014) Automated, connected, and electric vehicle systems. Expert forecast and roadmap for sustainable development. Graham Institute for Sustainability, University of Michigan, Ann Arbor
51. Coppola R, Morisio M (2016) Connected car: technologies, issues, future trends. ACM Comput Surv (CSUR) 49(3):46
52. Kanter T, Rahmani R, Li Y, Xiao B (2014) Vehicular network enabling large-scale and real-time immersive participation. In: International conference on internet of vehicles. Springer, pp 66–75

Chapter 5
Protocols and Design Structures for Vehicular Networks

Kelvin Joseph Bwalya

Abstract With the ever-increasing demand for pervasive processing of big data from heterogeneous devices and networks, there is an increased demand for optimal networks in Vehicle-to-Vehicle (V2V) communication models. Effective communication of heterogeneous devices with non-identical communication consoles demands that the devices are connected with a well-marshalled software and hardware abstraction to facilitate seamless interaction and information interchange, between the technologies of different nature supplied by diverse vendors. In today's information-intensive environments, processing of multi-dimensional data from heterogeneous smart devices in a networked environment can be a huge challenge. The type of network supporting V2V communication is critical to overall functionality of connected vehicle communication. Although there is a significant body of knowledge on different V2V networks, there is limited research on components for the topology and architectural arrangement for optimal vehicular networks. There is also a knowledge gap on what should influence topologies and routing mechanisms in vehicular networks. This chapter is as a result of research focusing on optimal designs for vehicular networks which culminate from a synthesis of knowledge in different contexts. This research brings to the fore the principles, protocols, topologies and storage options that underpin vehicular networks. The chapter proposes a conceptual framework articulating a basic architecture configuration that can be used in Mobile Ad hoc NETworks (MANET) with an emphasis on resource-constrained contexts.

Keywords Protocols · Data · Network · MANET · VANET · V2V · V2P · V2X · P2P · WAN · IaaS · NaaS · STaaS

K. J. Bwalya (✉)
School of Consumer Intelligence and Information Systems,
University of Johannesburg, Johannesburg, South Africa
e-mail: kbwalya@uj.ac.za

© Springer Nature Switzerland AG 2020
Z. Mahmood (ed.), *Connected Vehicles in the Internet of Things*,
https://doi.org/10.1007/978-3-030-36167-9_5

5.1 Introduction

Because of constant mobility with varying speeds, density and ever-changing positions of vehicles on the roads, it has always proved difficult to come up with global network structures and topologies for dynamic information management. Therefore, it is important to constantly consider what network attributes need to be established for any given scenario. Understanding the network attributes that can be used as input to design an optimal network may well be cardinal to designing efficient information and vehicular networks; as, due to highly dynamic information infrastructures in vehicular networks, it is very difficult to ensure that nodes have the same correct and current information at any given time. Since vehicles are highly mobile and do not maintain one location throughout their information sharing life cycles, they rely on Vehicular Ad hoc Networks (VANET). These networks are considered as clusters of networks that vehicles form at any given moment in time.

An essential component of VANET is Vehicle-to-Vehicle (V2V) communication that involves the exchange of information between two or more vehicles in a vehicular network. There are two types of vehicles in such communication: traditional vehicles which are controlled by human drivers and autonomous vehicles which are (possibly driverless) and controlled by technology. In traditional V2V communication, vehicles have embedded sensors sharing information with sensors within the vehicles as well as with sensors (smart devices) in the environment called Roadside Units (RSUs). The in-vehicle sensors share information among one another to know the vehicle dynamics such as level of fuel, operational status of brake, status of headlights and vehicle speed. With regard to communication among in-vehicle sensors, one challenge is to have all the numerous embedded sensors to synchronically communicate with each other to maintain the current information. In case of V2V communication for autonomous vehicles, the vehicle is equipped with many more sensors, smart devices, cameras and actuators to instantly capture and share information to aid its mobility through the environment.

In contemporary vehicles, the number of smart devices is increasing exponentially, with time. Allen and Overy [1] have posited that there will be more than 21 million autonomous vehicles on the roads, connected to the Internet by the year 2025. To achieve efficient navigation of the different information resources in vehicular networks, it is important to consider the concept of spatial–temporal distributed data and information management. This concept entails that there is cognition that information in vehicular networks is highly dynamic and changes with space, i.e. changing location of vehicles.

Because of the increase in the number of sensors in vehicles, distributed traffic management systems are also been used in real environments to manage traffic routing scenarios [17]. These systems reduce the occurrence of accidents on the roads and make general mobility less stressful. In many countries around the world, it is now mandatory for vehicles to be installed with special location-based sensors in order to make the road transport systems intelligent and more efficient.

For example, in Russia, all vehicles are required to be installed with ERA-GLONASS as an accident emergency response system.

With the many developments in V2V communication, there is an increased push to achieve Vehicle-to-Everything (V2X) communication scenarios. V2X enables a vehicle to communicate with any digital agent in the environment. A digital agent is a device or gadget in the environment which is able to exchange information with vehicles as they traverse the environment in which they exist. V2X presents scenarios where a vehicle is able to communicate with any information host.

Using the V2X communication model, vehicles are able to capture current dynamic information and make intelligent instantaneous decisions. In order to achieve appropriate V2X communications, there is need to ensure that technical and non-technical issues are addressed. Some of these issues and related topics include the following: (1) need for heightened privacy and security of information; (2) need for better network topologies and routing mechanisms that are intelligent enough to read the situations on the ground and help vehicles have the relevant information to make intelligent decisions; and (3) network configurations that regard Information as a Service (IaaS). Because of higher dynamic nature of Mobile Ad hoc Networks (MANET), a relevant question is: how intermittent network clusters (formed by highly mobile networks) can be used as Network as a Service (NaaS) and how we can develop adaptive network storage repositories in the realm of Storage as a Service (STaaS).

The above issues and questions prompt researchers and practitioners to keep investigating so that long-lasting solutions or conceptual underpinnings can be found. In this chapter, we intend to present the relevant research currently in progress. The conceptualisations presented in this chapter are mainly espoused upon applied information management thinking where technology is used as an enabler to achieve effervescent information access, storage, sharing and utilisation in diverse contextual settings. We hope to explore: (1) the different networks that can be deployed in VANET in the V2V context to achieve true ubiquitous sharing of and access to a variety of digital resources; (2) discuss different network options to achieve different levels of functionality in connected vehicles configurations; and (3) critical issues in relation to design and deployment of existing networks.

This chapter is arranged as follows. The next section presents the background to discuss the key motivation for this research. Subsequent sections discuss the concepts of vehicle communications, network principles, protocols and topologies in ubiquitous network. The last parts of the chapter explore data storage and privacy issues in MANET. Specifically, the chapter proposes a simplified architectural configuration for MANETs especially in a developing world context. Before the conclusion, we present ideas for future research.

5.2 Background

The emergence of the Internet of Things (IoT) has opened opportunities where any object, connected to another object, has a chance of being allocated a unique Internet identification. Coupled with advances in object-oriented programming, people are not very far from having machines autonomously communicating with one another and making intelligent information-based decisions. In this regard, vehicles have a realistic chance of communicating with other vehicles and making decisions autonomously.

Autonomous vehicles are now being produced in large quantities, owing to the rapid advances in V2V communication and IoT. Because of IoT, many nodes, e.g. Roadside Units (RSUs) can be installed in the environment so that spatial information is shared as various events occur. The different developments in the V2V landscape make further demands that vehicles not only communicate with fellow vehicles but also with the objects in their surroundings. For example, vehicles have to communicate with pedestrians (Vehicle-to-Pedestrians—V2P), with infrastructure (Vehicle-to-Infrastructure—V2I) and with every other related object (Vehicle-to-Everything—V2X).

The realisation of V2X will culminate in less environmental impacts and collisions, thus improving overall safety and reducing gridlock (traffic congestion). It will enable vehicles to communicate with other vehicles, the environment (roadside equipment), pedestrians, and the Internet. With the rapid technological advances anchored on the IoT, it cannot be overstated that the realisation of V2X is no longer a far-fetched dream. It is already enabling access to huge sets of information available in the cloud that may make it possible to have information such as sensor and high-definition environmental mapping data, and real-time traffic data.

Vehicular networks are highly dynamic given the ever-changing positions of vehicles, from one information and network cluster to the other. Although no universally agreed definitions exist, these Vehicular Ad hoc Networks (VANET) have specific characteristics. With the extension of the original concept in [10], a basic definition of a vehicular network can be stated as follows:

Lemma *Any ad hoc vehicular network (VANET) is presented by a graph $G = f(\alpha, \beta, \Omega)$, where α is a set of vehicles (nodes), β is the set of the possible information links between the nodes, and Ω is the set of protocols or information routing mechanisms used by the vehicles in exchanging information.*

To clearly understand the different entities of the MANET, it is important to explore the composition of the vehicular networks. Some of the key entities that make up a vehicular network include: (1) nomadic wireless personal devices (such as smartphones and personal communication devices), (2) roadside infrastructure devices (RSUs) capturing information such as electronic signs and traffic signals, (3) control centres such as traffic management centres, (4) the cloud environment and (5) intra-vehicle sensors and communication devices. The different sensors and devices both inside and outside the vehicle enable spatial intelligence to be attained and the vehicles to generally make decisions beforehand to avoid catastrophes. Spatial intelligence requires that sensors are adequately distributed in the environment to capture events as they occur, and mobile devices are able to capture current information. When spatial intelligence is achieved, current information is readily accessible via mobile devices and intelligent decisions made instantly.

The use of different types of technologies in vehicular information networks allows various types of information to be shared without considerable problems. As a result, Gehrsitz et al. [9] posit that technology has been used in vehicles in the following development trajectories: (1) diagnostics, (2) tracking, (3) infotainment, (4) cooperative systems (V2V, V2I), (5) predictive analytics, (6) autonomy and (7) car sharing. Starting with mere diagnostics, technology in vehicles is now used for communicating with vehicles' surroundings, predictive analysis where vehicles are able to analyse current information to come up with future scenarios and enabling autonomous vehicles.

Because of the many sensors included in contemporary vehicles, it is now possible to monitor individuals' driving behaviour or general vehicle interaction with the environment. Monitoring vehicle data will change many different business models, starting with insurance. For example, many insurance companies are now tracking miles covered by the vehicle and offering 'Pay As You Drive' (PAYD) insurance options. Future insurance models may be based on 'Pay How You Drive' (PHYD) model. With the different opportunities unlocked by V2X communication, there are many aspects of vehicular communication that need to be examined. This is further articulated in the following section.

5.3   Communication in Vehicular Networks

The general topological arrangement in vehicular networks is usually in a client-server (with a central authority, the server) or Peer-to-Peer (P2P) configuration.

- In a client-server arrangement, the client moderates the communication as it is the requesting agent. The server is the agent providing the answer to the requests sent to it. Servers usually have a huge reservoir of information which clients seek from time to time. In vehicular networks, the server will have all the necessary information stored in the cloud.
- The P2P communication model allows each agent in the network to send out requests or provide information responses to other peers.

In order to have a more reliable communication arrangement, a typical configuration is that of Server Farms where vehicles act as servers. In this configuration, vehicles can share resources such as computing power and network bundles in order to effectively process the vast information in the MANET. When vehicles cooperate and coordinate in this way, the emerging network configuration forms swarm intelligence. The vehicles need to announce their presence and readiness to initiate communication and share information they have captured using their sensors when they reach a vicinity. That means, if a vehicle acts as a server, it is the routing channel for communication. Alternatively, the P2P network does not require a central authority to coordinate the communication in the vicinity. All the participants in such a network are considered equals and have flexible interconnectivity and have equal communication opportunities. Given the dynamic nature of vehicular networks, ad hoc connectivity in the P2P networks allows a dynamic establishment of communication links between hosts when desired. Such a network topology allows high flexibility in link management and allows spontaneous establishment of communication at any given time [4].

The key tenet of MANET is that vehicles have to exchange information on the status of the environment which they are in. To do that, different communication modules exist in vehicular networks. There are different forms of topologies, routing protocols, network configurations and architecture that define the design of a communication network. When vehicles are moving, they constantly change positions and connect or reconnect to different networks. By so doing, the network topologies are constantly changing. Because of the dynamic nature of contemporary information systems, there is an increasing demand that information networks are agile, intelligent and include early warning systems for accident situations so that the probability of vehicles involved in an impending accident is reduced [19]. Further, it is a fact that vehicular networks are esposued on strict Quality of Service (QoS) requirements such as the need for prioritised routes and high data latency rates [9]. With the strict QoS requirements in vehicular networks, a lot of new and foreseen challenges need to be considered when designing the communication networks.

As a result of the aforementioned challenges, a lot of technological conceptualisations and innovations have been proposed. The first issue that attracted much attention is the use of different modes of wireless communications and attending to different challenges with regard to information packets exchange. In typical wireless communications, issues such as packet collisions, channel fading and communication obstacles in the environment can impede the effectiveness of communication in vehicular networks [19]. In many contexts, communication in VANET is achieved by Dedicated Short-Range Communication (DSRC) which is proposed as part of the IEEE 802.11p standard. The DSRC technology is ready to be implemented in real environments achieving a significant ration of communication. Here, vehicles can exchange up to 10 messages within a second. The basic architecture of DSRC, conceptualised on the idea presented in [3], is shown in Fig. 5.1.

As can be seen in Fig. 5.1, Robust Header Compression (ROHC) is used in the compression of packets so they conform to the throughput of the communications channel. Specifically, ROHC is a standard that compresses IP, UDP, UDP-Lite, RTP and TCP headers of Internet packets. Once the headers have been compressed, packets are able to move up the communication channel through the Media Access Control (MAC) and Physical (PHY) layers.

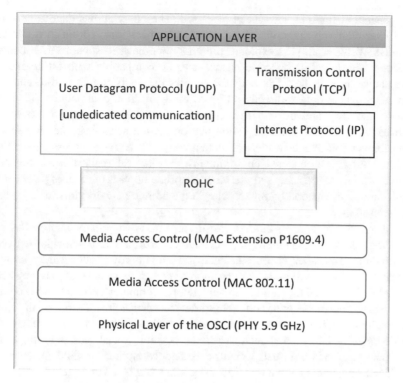

APPLICATION LAYER

User Datagram Protocol (UDP)

[undedicated communication]

Transmission Control Protocol (TCP)

Internet Protocol (IP)

ROHC

Media Access Control (MAC Extension P1609.4)

Media Access Control (MAC 802.11)

Physical Layer of the OSCI (PHY 5.9 GHz)

Fig. 5.1 DSRC protocol architecture

Wireless technologies and DSRC enable vehicles close to one another to exchange information which may be critical in avoiding deadly accidents [3]. The IEEE 802.11a MAC scheme is a de facto standard for DSRC targeting Vehicle-to-Road (V2R) and is currently being extended to include V2V communication especially at the MAC, link and radio layers.

Given today's advanced information-based environments, it cannot be overemphasised that the complexity in design of both in-vehicle and Vehicle-to-Everything (V2X) continues to increase. There has also been a rise in electronic control units in vehicles pushing the agenda of automation in vehicles. The drive towards efficient and dynamic information management directly culminates into an increase in the complication of desired in-vehicle, intra-vehicle and information or communication infrastructure [9]. As a result, modern cars come with built-in equipment which has made it easier for wireless communication to be used at a large scale in vehicular networks so as to manage highly dynamic information. Some of these devices may be a Personal Travel Assistant (PTA) or Personal Digital Assistant (PDA). The PTA and the PDA may be integrated into the vehicle navigation systems so that autonomous decisions can be made [3].

The question of dealing with data is an important one in vehicular networks, and as such, concepts such as Named Data Networking (NDN), which is espoused on new generation Internet, have been proposed. The NDN allows data to be propagated over relatively huge distances in vehicular networks (10 km in a matter of seconds). Wang et al. [17] explored the NDN approach to support V2V communications. The NDN architecture is meant to be compatible with future Internet. A key advantage of the NDN is that it enables the realisation of end-to-end communication where data is directly retrieved by application data names. This means that the mobility and session management as well as the service recovery as required in the TCP/IP communication platform is not needed. In the long run, the direct implication of such a scenario is that there will be faster access and retrieval of information in connected environments. A practical realisation of the NDN communication model may require an appropriate application naming convention which can be understood by all vehicles in the ad hoc network: this requires some standardisation.

With rapid advances in the IoT paradigm, multi-disciplinary approaches have been proposed to enable intelligent exchange of information in vehicular networks. For example, computer vision has been used to improve traffic safety allowing vehicles to use vision to obtain information to detect obstacles that may negatively interfere in the travelling path of the vehicle [4]. Computer vision can also come in handy when local sensors are not able to detect vehicles that are out of the line of sight such as vehicles entering an intersection. Active sensors may be able to collect a wide range of information such as vehicle position, speed, heading, slipperiness coefficient, etc. This information can be exchanged among the different entities in the V2X arrangement so that intelligent decisions can be taken in real time [4]. Wu et al. [18] proposed the multi-tier, multi-access edge clustering methodology which, at different times and depending on the density of vehicles in the network, generates different levels of clusters (made possible by wireless LANs, WLAN) with different levels of throughputs. The proposed methodology uses hierarchical clustering and combines different sets of communication to increase the efficiency of the network. Specifically, Mobile Edge Programming (MEC) is applied for in vehicular networks to achieve short delays and allow higher throughputs.

5.4   Vehicular Network Principles

Cloud Network topologies and architecture define the level of efficiency with regard to the exchange of information in vehicular networks. VANETs have highly dynamic topologies and are characteristic with, for example, frequent disconnected networks, congested networks, delay in the relay of packets, security and privacy issues. Information networks that go out of the way to avoid collisions and fatal roadway accidents are desired in vehicular networks [19]. Further, as vehicles traverse through clusters in MANETs, there is a challenge to keep same quality of Internet connectivity. Kristiana et al. [12] propose Vehicle-to-Vehicle Urban Network (V2VUNet) as a filtering methodology which filters the nearby vehicles to identify possible loss of connection. This enables the different vehicles to share the Internet with the vehicles so that they maintain connectivity.

The various types of communication networks used in vehicular networks are generally wireless. For the purposes of this chapter, key communications networks usually considered in vehicular networks include Local Area Network (LAN), Metropolitan Area Network (MAN), Wide Area Network (WAN), Wireless, and Inter Network (Internet). Topologies may take each of the following forms: vertical network, circular network (Ring), chain network, wheel network, star network, bus network or a mesh network. Vehicular networks are more concerned with the network topologies as they define the information architecture to be deployed in dynamic environments. Desired network configuration in vehicular networks is influenced by the mesh network which is a synthesis of two or more networks. In a mesh network, heterogeneous gadgets are able to share information using multiple connection channels.

There are different types and designs of wireless network configurations. Some of the types have been discussed above. The type of network model used in VANET is determined by the functional requirements of the network to achieve network intelligence [14]. The following determines the decisions with regard to network types:

- Degree of tolerance to path loss, signal interference, signal noise levels as measured by, for example, the Signal-to-Noise plus Interference (SNIR). Such attributes may culminate in the need for a network topology with multi-path configuration such as a mesh network.
- Low latency requirement of a network that may use a topology model which accounts for the MAC and PHY properties of the OSI protocol stack so as to increase latency. The increment of latency will translate into increased throughput of packets.

There is need to investigate more efficient MAC schemes that can be used for efficient data exchange in vehicular networks. With regard to information management, there is need for more improved versions of techniques to transmit and forward data to different nodes in vehicular networks. Another important topic, where a lot of effort needs to be devoted, is information filtering. It is important to ensure that only authentic information is forwarded for auctioning at different nodes. Knowing which information is credible and trustworthy is still an open issue that needs to be clearly investigated. So far, there is no optimal network organisation, topology or architecture that has been proposed for this. Investigation in this domain needs to continue towards the design of a global network architecture that can be used in vehicular networks.

Most vehicles' internal systems are wired using a BUS network topology configuration with common bus-systems such as LIN, CAN and FlexRay being widely deployed. These bus structures increase the wiring challenges in modern vehicles. Therefore, modern vehicles would consider using Power Line Communication (PLC) instead of using traditional bus-systems. There are many commercial network configurations that are already in practice in vehicular networks. Some of the more common ones are the following:

(i) LIN: a cost-effective vehicular networking solution based on Universal Asynchronous Receiver–Transceiver (UART)-based communication configuration. It uses a single-master and multiple-slave networking architecture which can be ideal in in-vehicle communication.

(ii) CAN: a serial bus network based on the asynchronous communication mode, connects sensors and actuators in a vehicle. CAN is used as a protocol for generic embedded communications.

(iii) Intelligent Distributed Control Solutions (IDCS): an in-vehicle embedded distributed system that aims to facilitate information sharing with nodes in the same network. It comes with a SMARTMOS™ analogue control integrated circuit that makes it possible to autonomously distribute data in concerned networks.

(iv) FlexRa: an in-vehicle communications system that can support a data rate of 10 Mbps for vehicle applications. The advantage of using FlexRay is that it culminates in increased throughput and facilities-distributed computing of information using a global clock.

5.5 Protocols in Ubiquitous Networks

There are a variety of network topologies, protocols and configurations forming vehicular networks [13]. Protocols are set of rules that guide the passage of information in networks. There are hundreds of potential protocols that can be used in vehicular networks. However, the most common protocols used in vehicular networks are based on IEEE 802. Vehicular networks require short-range communications so that vehicles entering a given environment are able to quickly connect, exchange information and make decisions. Short-range communication technologies and protocols such as 3rd Generation Partnership Project (3GPP) LTE-V2X PC5 (also known as LTE side-link) and IEEE 802.11p are important in reducing accidents in VANETs. Another important standard guiding information exchange in vehicular networks is the IEEE 802.11p which is being used to achieve multiple access (Carrier Sense Multiple Access protocol with Collision Avoidance, CSMA-CA) [5].

Well developed vehicular protocols need to consider the key characteristics of the envisaged network. It is a requirement that vehicular communication protocols take into consideration the nature of ad hoc networks for transmission and forwarding data among vehicles. Therefore, a majority of routing mechanisms and protocols in VANET have been designed upon table-driven and demand-driven protocols. Examples of table-driven protocols include the Wireless Routing Protocol (WRP) and the Destination-Sequenced Distance-Vector protocol (DSDV). The table-driven protocols require hosts to have tables that contain routing information, i.e. routes the data packets have to take to reach other hosts in the network. Each host should have its own view of the network; and every time it informs the other hosts, there is a change in the network view. To be used in this communication protocol, each host comes with a route discovery mechanism which has to identify the route structure immediately it is connected to a network.

Because of ever-increasing quantities of data and information exchanged in vehicular networks, there is usually some congestion en-route. Contemporary vehicular networks need congestion detection and prevention mechanisms. This is important to ensure that vehicles do not have to wait for information indefinitely when they enter a vehicular network. With advances in wireless communication designs, network protocols now include congestion control policies and dissemination procedures for emergency warnings [11].

Since information in vehicular networks is provided on-demand and is needed without delay to avoid catastrophes, on-demand protocols have also been proposed. The Dynamic Source Routing (DSR) protocol, the Ad hoc On-demand Distance-Vector protocol, and the Temporally Ordered Routing Algorithm (TORA) are some of the common protocols in this regard. These protocols are able to quickly create routes instantaneously when one of the hosts requests information from the network. The created route is discarded immediately as it is not needed going forward after information transmission is concluded.

Route optimisation is very important in vehicular networks which are mostly designed with nested topologies. For optimisation purposes, there have been research efforts to create multiple routes; and at the same time, to determine the shortest route chosen using the Open Shortest Path First (OSPF) protocol [4]. Traditionally, Dynamic Source Routing (DSR) and Ad hoc on-Demand Vector (AODV) have failed as protocols in inter-vehicle networks because of higher mobility.

It is well established that it is better for a vehicle to receive information from a vehicle that is not too far from the requesting vehicle. Protocols have been proposed that estimate the location of the requesting vehicle using the vector coordinates and/ or measuring the Euclidean distance. An example of such a protocol is the Location-Aided Routing (LAR). It uses the combination of the identifier and the position of the destination and host (using the previous location and the movement characteristics).

Because of the highly dynamic nature of vehicular networks, routes and distances of vehicles are constantly changing. Therefore, optimal routing of information is not easy to achieve. The ad hoc distributed communication protocols aim to contribute to minimising the effect of congestion in distributed vehicular networks. These protocols are used by vehicles to notify other nodes of impending congestions and therefore traffic jams. When vehicles heading in the same direction experience sudden reduction in speed, they form a virtual group which then automatically transmits messages to other nodes that a traffic situation or incident is imminent. The other vehicles can now automatically consider rerouting upon receiving those messages from other vehicles. The key disadvantage of the ad hoc distributed communication protocol is that it uses a flooding-like mechanism (best effort as used in datagram routing protocol) where many messages are sent out with the hope that one of them will reach its intended destination. To avoid such a problem, systems are being proposed that use multi-hop transmissions.

Information architecture in vehicular networks has been influenced by the Software-Defined Networks (SDN); that have been implemented to solve problems that slow the system by separating the control plane from the data plane. This introduces flexibility which is a key requirement for dynamic networks and provides excellent programmability which is cardinal to vehicular networks [6]. Most routing protocols for SDNs have used an OpenFlow protocol which utilises a packet-relaying machine with multiple flow entries. The machine is made to check whether the in-coming information packets match some flow entry on the machine. If not, the packet is sent to a controller for further processing. Another commonly used standard is the ForCES protocol which has both the Control Elements and the Forwarding Element on the same device but with a further Logical Function Block [6].

5.6   Issues and Emerging Themes in VANET

A vehicle's Internet connection, provided by wireless technologies, can be obscured due to elevated land such as mountains or hills surrounding an area, moving through tall buildings or through a tunnel. With the V2VUNet, filtering can be done so that Internet is shared by nearby vehicles that have an Internet connection [12]. As a result, the design of the topologies and the routing mechanism need to take note of possible environmental obstacles and design technological solutions that do not fall short on this regard. For environments where the network can easily be obscured, Power Line Communication (PLC) has been proposed. Distinct from the requirements in Home Area Networks, PLC in vehicular networks needs to service a lot of nodes with high latency but with comparatively low data rates.

The following are some of the key challenges that need to be considered when designing information networks in vehicles networks:

Network Data Storage

One of the key questions in the design of contemporary VANET networks refers to data storage. Vehicles cannot be used as storage repositories due to their higher mobility. However, stationary vehicles can be used as temporary storage and network infrastructure so that vehicles in the vicinity are able to experience the true meaning of spatial intelligence. Using this thinking, vehicular clouds can be formed comprising different clusters of parked vehicles. When vehicles are used in this way, a vehicular cloud computing network is formed, and data is stored in this network based on Storage as a Service (STaaS).

The idea of creating a vehicular cloud network is to put in place an information management infrastructure where ubiquitous access to information is enabled for the heterogeneous agents. Thus, true spatial–temporal infrastructure is required that can be at the centre of ensuring that the benefits of intelligent road infrastructure are realised [2]. For this to be achieved, the following considerations are needed:

- Design of the information system in such an environment needs to be based upon a dynamic routing protocol such as the Virtual Cord Protocol (VCP). The VCP comes with service capacities such as the Distributed Harsh Tables (DHT) which may be cardinal to delivering information traffic in a clustered distributed network.
- Only two functional operations, f(LOOKUP) and f(PUBLISH) are used to make information available in the system and to access the information put in the public domains to the authenticated nodes with respect to a given system.
- Encouraging optimal network resource utilisation by facilitating the decomposition of big data into smaller interconnected blocks is required, so that they can be pervasively delivered on-demand.

The design of a distributed systems for data storage which allow ubiquitous access to data is cardinal in the design of contemporary VANET systems. Vehicular networks need to have access to a central data store as they traverse through the environment to make intelligent decisions as and when required. Although there are many requirements for the design of such a system, a fault-tolerant distributed system is cardinal for information-intensive environments. Thus, it is important to design the distributed system based on Disruption-Tolerant Networks (DTNs) embedded onto a carefully designed mesh network where disruption in the network does not translate into overall disruption of the network [2].

Big Data and Predictive Analytics

With the advances in information capturing tools and information-intensive environments generating all sorts of information such as videos, photos, numbers and text, the question of data management (access, storage, transmission, processing and sharing) comes onto the scene. This data and information are needed to be used as reference or historical data in making on-the-go decisions. Therefore, contemporary network designs need to have provisions for innovative data storage and processing of big data timeously when called upon [2]. The question of data storage is big in VANETs. Large amounts of data cannot easily be transferred through mobile wireless network infrastructure. This is because it is very expensive to construct network infrastructure that would handle higher data rates [2].

The emerging trend is that big data and predictive analysis are driving the new innovations in in-car technology. In the near future, applications needing human interventions will be minimal. The improvements in the capabilities of applications to aggregate, combine and simultaneously analyse structured and non-structured data to identify patterns will culminate into increased completion of automated tasks in vehicles [1]. The identified patterns will enable the prediction of behaviour and make intelligent decisions before the event occurs. The result of advancement in big data and predictive analysis is that information shared within and across the vehicles (intra-vehicle application communication), as well as between devices and sensors in the environment, will be already processed and ready to be applied in specific contexts. This will therefore increase the likelihood for the achievement of spatial intelligence everywhere [1].

In addition, there are so many advantages for sharing processed data with predicted insights for events likely to happen in the immediate or near future. Context-based analysis of data creates the following value:

- By monitoring the exact locations of vehicles and driving habits by analysing sensor data, businesses can design ways to increase efficiency in their business process value chains.
- Over time, the collection and analysis of more detailed and accurate performance data for vehicles will translate into the design of more reliable, advanced and safer vehicles going forward. These vehicles will eventually be intelligent with in-vehicle technologies enabling autonomous parking and traffic routing.
- Enable market segmentation to meet customers' needs.

Privacy in VANET

The bid to understand privacy issues in vehicular networks is one of the most current topics that has not been given much attention in the VANET research domain. Because of its importance, privacy and security requirements have been standardised in the IEEE 1609 and ETSI C-ITS for the goal of achieving message authentication by using asymmetrical key authentication digital signatures in vehicular networks [7]. When vehicles share information, some of it may be susceptible to unwarranted attaches increasing the probability of it landing in wrong hands. Because vehicles share networks and it is anticipated that information needs to be shared automatically, some personal information may end up being shared unnecessarily or distorted inadvertently to cause confusion on intelligent networks [8]. Research has proved that some acceptable levels of privacy can be achieved in practice. However, there is limited research with regard to design of secure information networks.

Modern automobiles can collect vast amounts of information such as vehicle locations (using GPSs), drivers' health information over time, driving habits, environmental information and owners' residence locations. Such information can be used to target the vehicle owners by criminals or people who aim to put them at a disadvantage. For example, by observing their locations over a period of time, vehicle owners may be stalked or discriminated against owing to their perceived political affiliations due to, for example, areas visited. Therefore, the ability to design privacy-friendly systems is needed to protect vehicle owners from sharing sensitive information that can put their lives at stake [8].

VANET allows vehicles to share information so that drivers can make decisions on how to react to the impending danger. Questions of the degree of trustworthiness come in. Knowing the trustworthiness of the source is important for vehicles so that information from unknown sources can be ignored. In autonomous vehicular networks, advanced cryptographic mechanisms are used in end-to-end communication to assure the receiver of the information of its worthiness [16]. Schmidt et al. [16] have proposed a Vehicle Behaviour Analysis and Evaluation Scheme (VEBAS) which can be used to authenticate the sender of the information. VEBAS may pave way for design of systems which can be used to automate decision-making in vehicles by observing vehicles with negative behaviour.

5.7   Proposed MANET Structure

To limit the effect of the challenges in vehicular networks, different researchers and practitioners have suggested innovative designs to be used in MANET. For the purposes of this chapter, we focus on the Vehicle Edge Computing (VEC) which was recently introduced for use in vehicular networks to argument the computing capacity in highly dynamic networks [15]. VEC is specifically used to reduce latency and achieve overall improvement in QoS.

Because of the high densities of vehicles in the vehicular networks, the topology or design of the information architecture of vehicular networks may be influenced by cloud and/or edge computing. This allows data to be stored in a centralised location in remote or virtual servers sharing the same computational power [15]. If cloud computing architecture is not designed properly, it can culminate in unavoidable delays in signal propagation and transfer of information packets from the server to the vehicles as they pop up in the network clusters. Contemporary designs in cloud computing have seen the data analysis and processing take place close to the vehicles.

Based on the VEC model [15], Fig. 5.2 shows the proposed structure for MANET application which includes only the absolute necessary entities ensuring that the desired QoS is retained. Only the more active layers of the OSI protocol stack need to be considered for edge computing applications. Figure 5.2 is conceptualised upon the architectural components of Fig. 5.1. Both figures are further conceptualised upon the OSI protocol stack. Based on Fig. 5.1, four layers of the OSI are involved:

1. physical layer (RSUs, in-vehicle nodes),
2. transport layer (logic link control, the MAC layer enabling processing and analysis),
3. network layer (protocols, network configuration and network channels), and
4. application layer.

One of the key requirements in the design of highly dynamic information networks is an effervescent storage configuration which allows intermittent access by heterogeneous devices. In order to achieve that, a low-cost MANET structure is designed in such a way that the informed is stored as Storage as a Service (STaaS) where information is stored in dedicated storage. The information is sensed from the environment and stored in the cloud. As vehicle approaches a given environment, it is able to access information from the cloud through a dedicated network layer, preferably a mesh network which allows multiple connections. The multiple

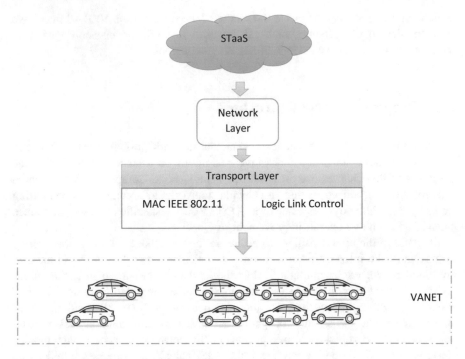

Fig. 5.2 Proposed low-cost MANET structure

connections are important to enable heterogeneous vehicles and devices access the cloud. The information is finally dispatched to the vehicles or other agents in the vehicular networks using appropriate transports protocols as defined by the Media Access Control (MAC) and the Physical layer (PHY) Wi-Fi communication. The captured information is then presented on vehicle information display units to guide vehicle navigation, and to share with other vehicles, etc. This is a basic information infrastructure that uses the fundamental and core entities needed in vehicular communication networks.

The configuration above ensures that there is better QoS (less packets are discarded from the system) and low latency culminating into overall improved communication. Edge computing uses servers that can act both as storage repositories and have massive computational power. For the PHY and MAC layers of vehicular networks, standards such as the DSRP Wireless Access for Vehicular Environment (DSRP WAVE), IEEE 802.11p will play a major role in the realisation of efficient network connections for MANETs [14].

Since there are still known limitations in the edge computing layer with regard to processing huge sets of data, the cloud layer allows data aggregation and enables batch processing, advanced data mining and processing of big data (both structured and unstructured). The cloud layer has the capacity to enable the processing of complex data within the shortest possible time. The processing of the data in

Fig. 5.2 works in such a way that data that does not immediately require processing has to be sent to the cloud to enable a more detailed analysis. Data that needs immediate processing can be processed at the edge layer using edge computing empowered by SDN. The incorporation of a decoupled network and data layers of the edge layer using SDN significantly reduces the delay in processing and analysis of immediately needed data. The edge layer facilitates intermittent communication with the in-vehicle nodes and devices using common communication protocols such as IEEE 802.11p, 4G (and now 5G), Long-Term Evolution (LTE) and the proposed 3GPP [15]. The processing of information is so fast that on-demand intelligent decisions can be made immediately thus increasing the safety of vehicles on the roads. With such a detailed and complex requirement set in the analysis and processing of data to achieve spatial intelligence in vehicular networks, the need for an intelligent, agile and highly dynamic network cannot be overemphasised.

With a majority of the protocols based on the cloud, the proposed MANET structure is less costly and can easily be embedded into a real environment. As developing countries are busy designing different forms of intelligent information networks, the proposed vehicular information architecture can be used as a blueprint.

5.8 Future Research

One of the key open issues that could influence future research include the drive to solve challenges in extreme dynamic information-intensive vehicular networks towards network resource optimisation. Network resource optimisation conceptualises a highly integrated network (e.g. mesh) where nodes are able to share applications and software through the network middleware so that tasks are executed cooperatively. In vehicular networks (with more than 200 sensors per vehicle), huge amount of data is generated, considering the huge vehicular interactions. This data comes from applications based on augmented reality, AI, high-definition maps and location-based services. The key question, here, refers to balancing the stringent QoS requirements in vehicular networks with limited network resources and capacity.

As a result, vehicular systems need to be designed in such a way that they are scalable and employ reusable off-the-shelf network components. There are many open-source reference points and blueprints that can be referenced in the design of VANETs. Regional organisations within the EU have understood the need for next generation communication such as 5G which is being aggressively pursued in the region. To this end, an open development ecosystem has been established where different researchers meet to discuss innovations [5]. For example, the 3GPP V2X Cellular Solutions Long-Term Evolution (LTE) Release 14 has been standardised as a blueprint for technology design targeting V2X communications based on 5G.

Although advances have been made in the applicability of the data-driven approach induced by machine learning in vehicular networks, there is still a lot of work to be done in order to seamlessly integrate it into the design of vehicular network modules [20].

5.9 Conclusion

This chapter explores the key issues at the forefront of research and practice in vehicular networks and suggests different principles and network options, for effective MANET in the context of the design of networks, as well as the different principles and technologies that guide better V2V communication. In this respect, consideration needs to be made with regard to how heterogeneous devices with non-identical communication are designed to seamlessly exchange information.

The chapter posits that the different principles explored need to be considered in the design of agile, adaptive and intelligent vehicular networks in real environments.

Future research directions are presented to guide the practitioners, designers and researchers on what aspects of V2V communication design still need to be considered when designs are conceptualised. It is worth noting that with the rapidly changing technologies upon which V2V communications are designed, it is important to keep looking into the different and alternative options for different contextual settings.

References

1. Allen C, Overy LLP (2017) Autonomous and connected vehicles: navigating the legal issues. http://www.allenovery.com/SiteCollectionDocuments/Autonomous-and-connected-vehicles.pdf. Accessed 4 July 2019
2. Balen J (2015) Spatio-temporal distributed background data storage and management system in VANETs. In: Proceeding of the 3rd GI/ITG KuVS Fachgespräch inter-vehicle communication (FG-IVC 2015), Ulm, Germany, vol 13(3), pp 22–25. https://bib.irb.hr/datoteka/787684.Balen-spatio-temp.pdf. Accessed 4 July 2019
3. Bechler M, Schiller J, Wolf L (2014) In-car communication using wireless technology. https://www.researchgate.net/publication/242429992_IN-CAR_COMMUNICATION_USING_WIRELESS_TECHNOLOGY/download. Accessed 4 July 2019
4. Chisalita L (2006) Communication and networking techniques for traffic safety systems. Unpublished dissertation, Linköpings universitet, Sweden
5. EU Report (2017) An assessment of LTE-V2X (PC5) and 802.11p direct communications technologies for improved road safety in the EU, 5 December 2017. http://5gaa.org/wp-content/uploads/2017/12/5GAA-Road-safety-FINAL2017-12-05.pdf. Accessed 4 July 2019
6. Fan Y, Zhang N (2017) A survey on software-defined vehicular networks. J Comput 28(4): 236–244. https://doi.org/10.3966/199115592017062803025
7. Feiri M, Petit J, Kargl F (2015) The case for announcing pseudonym changes. http://fg-ivc.car2x.org/proceedings-fg-ivc-2015.pdf. Accessed 4 July 2019

8. Förster D, Bosch R (2015) Discussing different levels of privacy protection in vehicular ad-hoc networks. In: Proceeding of the 3rd GI/ITG KuVS Fachgespräch inter-vehicle communication (FG-IVC 2015), Ulm, Germany, vol 13(3), pp 29–31
9. Gehrsitz T, Kellerer W, Kellermann H (2014) In-car communication based on power line networks. http://mediatum.ub.tum.de/doc/1200475/702617.pdf. Accessed 4 July 2019
10. Hajj HM, El-Hajj W, Kabalan KY, El Dana MM, Dakroub M, Fawaz F (2009) An efficient vehicle communication network topology with an extensible framework. https://pdfs.semanticscholar.org/8985/a4ee452f25bcd2d22451fbff3417f0870c53.pdf?_ga=2.92291477.942827893.1558614756-2097276758.1558614756. Accessed 4 July 2019
11. Kim T (2015) Assessment of vehicle-to-vehicle communication based applications in an urban network. Unpublished PhD thesis, Virginia Polytechnic Institute and State University, USA
12. Kristiana A, Schmitt C, Stiller B (2015) Investigating a reliable inter-vehicle network in a three-dimensional environment. In: Proceeding of the 3rd GI/ITG KuVS Fachgespräch inter-vehicle communication (FG-IVC 2015), at Ulm, Germany, vol 13, issue 3, pp 16–20
13. Mai A, Schlesingerv D (2011) A business case for connecting vehicles. https://www.cisco.com/c/dam/en_us/about/ac79/docs/mfg/Connected-Vehicles_Exec_Summary.pdf. Accessed 4 July 2019
14. Mishra KS, Chand N (2012) Applications of VANETs: present & future. Commun Netw 5:12–15. https://doi.org/10.4236/cn.2013.51B004)
15. Raza S, Wang S, Ahmed M, Anwar MR (2019) A survey on vehicular edge computing: architecture, applications, technical issues, and future directions. Wirel Commun Mob Comput 1–19. https://doi.org/10.1155/2019/3159762, https://www.hindawi.com/journals/wcmc/2019/3159762/. Accessed 4 July 2019
16. Schmidt RK, Leinmüller T, Schoch E, Held A, Schäfer G (2008) Vehicle behavior analysis to enhance security in VANETs. In: Fourth workshop on vehicle to vehicle communications (V2VCOM 2008), Eindhoven, The Netherlands
17. Wang L, Afanasyev A, Kuntz R, Vuyyuru R, Wakikawa R, Zhang L (2012) Rapid traffic information dissemination using named data. In: Proceedings of the 1st ACM workshop on emerging name-oriented mobile networking design—architecture, algorithms, and applications, pp 7–12
18. Wu C, Liu Z, Zhang D, Yoshinaga T, Ji Y (2018) Spatial intelligence towards trustworthy vehicular IoT. IEEE Commun Mag 56(10):22–27. https://doi.org/10.1109/mcom.2018.1800089
19. Yang X, Liu J, Zhao F, Vaidya NH (2004) Vehicle-to-vehicle communication protocol for cooperative collision warning. In: Proceedings of the first annual international conference on mobile and ubiquitous systems: networking and service (MobiQuitous'04), 22nd–26th August 2004, Boston, USA. ISBN 0-7695-2208-4
20. Ye H, Liang L, Li GY, Kim JB, Lu L, Wu M (2018) Machine learning for vehicular networks. IEEE Veh Technol Mag. https://arxiv.org/pdf/1712.07143.pdf. Accessed 4 July 2019

Part II
Frameworks and Methodologies

Chapter 6
Intelligent Traffic Management Systems for Next Generation IoV in Smart City Scenario

V. Vijayaraghavan and J. Rian Leevinson

Abstract With ever-growing number of vehicles on roads, traffic congestion is becoming a major problem in big cities around the world. Traffic congestion leads to pollution, time delays, excessive fuel consumption, financial losses and can severely disrupt normal human life. Conventional traffic management systems that rely on predecided traffic signal timings and pneumatic actuators are woefully inadequate in handling current traffic scenarios. Therefore, there is a pertinent need for a modern and intelligent overhaul of conventional traffic management systems and the introduction of such systems in modern smart cities. Intelligent traffic management systems are an ensemble of networks and systems integrated with each other to ensure optimum user commuting experience. They use a variety of advanced techniques such as reinforcement learning, Q theory, RFID tagging, IoT, IoV and local context awareness. Intelligent traffic management systems also help with safety issues, route optimization, delay reduction, and pollution control. This chapter explores conventional traffic management systems and their drawbacks, intelligent traffic management systems and recent advancements in them such as the introduction of reinforcement learning and local context-aware systems. This chapter aims to provide a glimpse of how intelligent traffic management systems can help drive smart cities of the future.

Keywords IoV · Traffic management · IoT · Q theory · RFID · Reinforcement learning · Context awareness · Intelligent traffic management systems

V. Vijayaraghavan (✉)
Infosys Limited, Bengaluru, India
e-mail: Vijayaraghavan_V01@infosys.com

J. Rian Leevinson
Infosys Limited, Chennai, India
e-mail: rian.leevinson@infosys.com

6.1 Introduction

The ever-increasing number of motor vehicles on the road has led to massive traffic problems in major cities around the world. Traffic congestion in the USA makes drivers lose 97 hours per year stuck in traffic, costing the country over $87 billion in time, with an average of $1348 per person. Drivers in London spend over 227 hours stuck in traffic each year [1]. In New Delhi, India, over $1.6 million worth of fuel is wasted every day, by vehicles idling in traffic [2]. Traffic congestion can lead to fuel wastage, excessive time delays, economic losses, and environmental pollution and thereby has serious consequences on travel, national and local economy, businesses, and the citizens.

Before the use of automated traffic management systems, traffic flow was regulated by policemen situated at intersections and key junctions. With the growth in urban population and expanding cities, implementing and coordinating such a system became a massive logistical and workforce management challenge. Eventually, signaling systems such as semaphore traffic signals and traffic towers saw widespread implementation, especially in cities of the USA [3].

However, traffic towers were big, expensive and obstructed traffic while semaphores were not easily visible to drivers, especially in the night. Therefore, they were subsequently replaced by interconnected traffic signals that were controlled by single switches. The emergence of computers in the 1950s led to the development of automated traffic signals [4]. These systems were based on phase-controlled traffic signal control, pneumatically-actuated traffic signal control, and manually-operated traffic signals. They were not designed to handle high volumes of dynamically varying traffic that is often encountered in major cities today.

The advent of the Internet of Things (IoT) is now paving the way for a connected future where vehicles, devices, and other individual entities communicate with each other in integrated networks. Moreover, massive strides made in the world of big data analytics, cloud computing, and computer vision are driving transformation across industries. These modern technologies can be leveraged to develop advanced traffic management systems. Such systems will be capable of tracking and monitoring the flow of vehicles through the road network using a variety of sensors, RFID tags, Computer Vision (surveillance cameras, vehicle cameras) and GPS tagging. Vehicles and the distributed traffic management system communicate with each other using the Internet of Vehicles (IoV). Cameras and sensors placed in important intersections and junctions monitor the flow of traffic and predict imminent congestions and traffic jams by analyzing traffic movement.

Intelligent traffic management systems are especially important in smart cities and in existing cities that have other means of transport integrated with road systems, such as trams, railroads, and cycling lanes. Moreover, the planning of infrastructure in smart cities is made easier by these integrated systems that enable the construction of key bridges, flyovers, tunnels, underpasses, and junctions are based on the traffic demand. These intelligent traffic management systems are used to efficiently plan and allocate parking spaces based on the demand in different areas of the city. The system automatically allocates parking slots based on a variety of factors such as the type of vehicle, duration, and the location.

Furthermore, intelligent traffic management systems help emergency vehicles navigate through congested cities and urban areas to reach their destinations on time. When the system detects unforeseen road events such as vehicle breakdowns, accidents, or road blocks, it automatically alerts the authorities. The system subsequently provides the authorities with information about the best route to ground zero, the type of situation, and severity of the event. The system may also additionally restrict traffic flow in and around the critical corridor through which the emergency vehicles pass through. This considerably reduces the time taken by emergency personnel to respond to critical situations, thereby improving the response time.

Modern intelligent traffic management systems can be upgraded by incorporating reinforcement learning into the system. Reinforcement learning uses a reward and penalty system to optimize a cost function that is often used to improve a metric such as queue length in signal or time delay. Such systems can automatically learn new changes in the cost function and can efficiently account for dynamic variations in the traffic flow. Moreover, local context awareness can further enhance the performance of intelligent traffic management systems. By studying the local environment and situation around each individual vehicle, the traffic management system can cater to the need of each individual vehicle by redirecting them through a new route or guiding them to the nearest hospital or vehicle garage or dispatching emergency services to vehicles in imminent distress.

Smart cities serve as excellent platforms to harbor intelligent traffic management systems due to the advanced infrastructure, high-speed communication networks and well-planned roadways. Smart cities need smooth flowing traffic to prevent disruptions in operations especially since they are designed to be socioeconomic hubs.

Overall, the implementations of intelligent traffic management systems substantially decrease the travel duration, improves safety, reduces traffic jams, management costs, and financial losses. They also aid in reducing air pollution and in managing noise pollution, thereby paving way for sustainable and cleaner smart cities.

The flow of this chapter is as follows: Sect. 6.2 introduces conventional traffic management systems, their types, and drawbacks. Section 6.3 briefly explains the concept of intelligent traffic management system and explores some of its functionalities. Section 6.4 explains recent advancements in intelligent traffic management systems such as reinforcement learning and local context-aware services in detail. Section 6.5 explores the challenges and future perspective of intelligent traffic managements and Sect. 6.6 concludes the chapter.

6.2 Conventional Traffic Management Systems and Practices

Ever since the advent of vehicles, there has been a need to regulate their movement to ensure smooth traffic flows. Traffic signals are the simplest and the most effective way to control and facilitate the flow of vehicles, especially in junctions and busy intersections.

Conventional traffic management systems mostly use fixed-timing traffic light control systems [5]. In such systems, the signal lights are designed to work in specific patterns based on historical data. The traffic signal's timings are predetermined and follow strict cycles, regardless of the flow of vehicles. These systems are intended for ideal scenarios where the traffic flow is smooth and regular. However, such situations rarely occur in real time as traffic flow is often dynamic and highly irregular. This causes excessive delays and traffic congestions especially in intersections with high volume of vehicle flow. In the context of traffic management, time delay is defined as the time taken by a vehicle to cross an intersection with a traffic signal to that of a vehicle that crosses the same intersection without a traffic signal. Such delays increase travel times and disrupt normal human life in cities. Traffic congestion is generally measured by the amount of delay experienced by each individual vehicle transiting through a junction. Time delay and queue length are also considered important metrics to measure the operating efficiency of a traffic management system [6].

The following sections explore the two most popular types of conventional traffic management systems that were used or also currently being used in some parts of the world. Furthermore, their drawbacks and the challenges faced by these systems in handling growing traffic needs are explored.

Pre-Timed Signal Control

Historically, conventional traffic management systems have utilized delay models in intersections that consist of a deterministic component and a stochastic component to represent both the steady flow component of traffic and to incorporate random variations and fluctuations. The deterministic component of traffic treats the flow of vehicles and signaling operations as continuous variables that are governed by flow rates that vary over space and time [7]. The stochastic component of time delays is based on the steady-state queueing theory. This component incorporates random fluctuations and variations in the arrival of vehicles into intersections and vehicle movements over a time period. Conventional systems use models that use a combination of both deterministic models and stochastic models to address a wide range of traffic problems. Such models became popular and were widely adopted due to the pioneering work, as reported in [8], on traffic signal settings.

Vehicle-Actuated Signal Control

Vehicle-actuated signal control systems work by using pneumatic pads near intersections to detect vehicles moving over them. Unlike pre-timed signals,

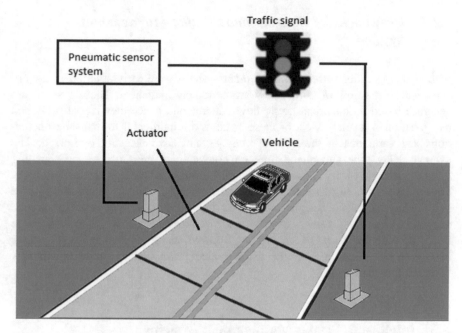

Fig. 6.1 Pneumatically actuated traffic management system

actuator-based systems can account for pedestrians and bicycles in intersections as well. Actuator-based traffic signaling systems use intervals of green and red signals that are controlled in response to the number of vehicles detected by the pneumatic pads. The system can alter the order and phase of the traffic signal along with the duration of the state of the signal [9].

Figure 6.1 depicts a pneumatic actuator system that tracks vehicles that pass over it and uses that information to control the traffic signal in the junction. These systems are designed for small intersections without large fluctuations in traffic. They are capable of operating uninterrupted at low-volume conditions. Another added benefit of pneumatic systems is pedestrian crossing assistance. Actuators placed in footpaths monitor the number of pedestrians arriving at a given intersection and by changing the state of the traffic signal, the system allows them to cross the road.

Vehicle-actuated systems are expensive to install, operate, and maintain. They require regular inspection and maintenance. The actuator pads placed on the roads for vehicles to pass over are difficult to install and repair and their breakdown may cause traffic disruptions.

6.2.1 Drawbacks of Conventional Traffic Management Systems

Although conventional traffic management systems were early attempts to govern and regulate the flow of traffic, they are extremely inefficient. These systems are designed based on historical traffic flow patterns and hence they repeat pre-timed red-green phase cycles. Most of these systems are indifferent to real-time fluctuations and variations in the flow of vehicles and are hence not optimized. The average delay time and queue time on conventional pre-timed traffic signaling systems are high. Moreover, excessive engine idling leads to increase in emissions. This contributes to further deterioration of air quality in major cities. Besides, these systems have simplistic designs and hence do not account for special scenarios such as road blockades, movement of emergency vehicles, accidents, and sudden surge in traffic. With the advent of smart cities, conventional traffic management systems have to be revamped and upgraded using modern technologies to cope up with new demands and challenges.

6.3 Intelligent Traffic Management Systems

Intelligent traffic management systems are an integrated ensemble of technologically advanced entities that enable efficient and optimized traffic flow. Technologies such as IoT, 4G/5G, IoV, RFID, GPS, Video analytics together encompass an intelligent traffic management system. Such systems can automatically and dynamically adapt to unexpected real-time scenarios in which conventional traffic management systems generally fail. To deal with ever-growing traffic demands in modern cities, intelligent traffic management systems tend to have self-adapting systems with decentralized architecture.

Stable and high-speed wireless Internet connectivity throughout the city is one of the most essential components of an intelligent traffic management system. Such a wireless network helps integrating the city's existing infrastructure with the various components of the intelligent traffic management system such as CCTV cameras, traffic signals, automatic road barriers, toll booths, and IoT sensors. The wireless connectivity is achieved through high-speed 4G/5G and Low Power Wide Area Networks (LPWAN) connectivity systems [10]. These low-powered networks are especially useful with the advent of small IoT sensors that do not need the traditional high-bandwidth connectivity.

Figure 6.2 shows the components of an intelligent traffic management system. The data about the state of the vehicle is gathered using onboard sensors, traffic cameras, RFIDs, or infrared sensors and the data is transmitted through WiFi and LPWAN to the traffic management system to control the traffic signals and regulate traffic.

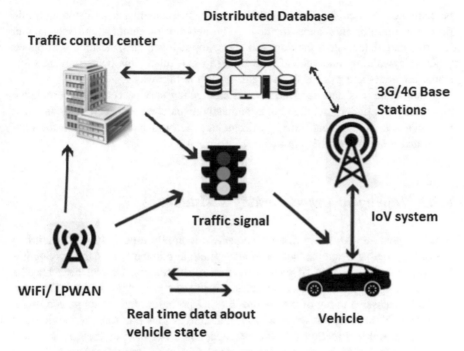

Fig. 6.2 Intelligent traffic management system

Traffic flow through busy urban intersections can be monitored using advanced video and image analytics. Along with the flow of vehicles, the movement of pedestrians and cyclists can be monitored and analyzed to improve traffic in dense cities. The large number of vehicles and difficult weather conditions such as snowfall or fog found in certain cities can be a challenge as this will push the resolving power of the cameras and object identification algorithm in the system. Thermal imaging cameras are used to distinguish cyclists and pedestrian from vehicles and improve their road safety [11]. Such powerful cameras and sensors can swiftly pinpoint anomalous events on public transportation systems such as fires in tunnels, passenger intrusion into railway tracks, vehicle flow at railway crossings, malicious activities onboard buses, and trains.

The functionalities of an intelligent traffic management system may include the following key aspects.

6.3.1 Incident Detection and Emergency Response

Incident detection and management are extremely important in highways and expressways. The high-speed movement of vehicles through these roadways has to be continuously monitored and regulated to prevent accidents, ensure smooth traffic flow, and monitor anomalous incidents. The network of cameras and sensors can identify and detect slow-moving vehicles, stopped vehicles, accidents, dropped objects, jaywalking, and vehicle jaywalking in real time. This helps prevent additional accidents and pileups from occurring in the highway.

If an incident is detected, the nearest emergency service center is alerted; the emergency responders can then be immediately dispatched. This automatic incident detection can save critical time by eliminating the human component in the flow chain and thereby send help quickly.

6.3.2   Intelligent Urban Parking Assistance

Parking spots are extremely difficult to search manually especially in dense urban areas. High demand and the lack of robust system to efficiently manage parking has led to a massive scarcity of parking spots in city centers. Crucial time may be wasted, navigating around city blocks to identify open parking spaces. This also results in excessive idling of the engine, causing pollution, fuel wastage, and driver frustration. Moreover, this can lead to traffic congestions due to slow-moving and stagnant vehicles. Intelligent traffic management systems can incorporate local context awareness to identify and reroute the drivers to nearby parking spaces in real time.

A database of all the available parking spots in each city block is maintained with their availability status being updated to the server using IoT, RFID, and using image analytics to dynamically identify parking areas in real time [12]. The live status of parking lots is classified as either available, reserved, out-of-service, and occupied. This system allows the users to reserve parking spots near them using mobile and web-based applications.

Figure 6.3 shows a parking lot that is intelligently managed. The system tracks available, reserved, and occupied parking spaces and accordingly guides incoming vehicles to the appropriate parking spot. The parking lots are monitored using cameras and RFID sensors that relay information through WiFi/LPWAN to the traffic management system.

The city planners and administrators can be informed about areas with chronic parking problems. These hotspots can be identified using geo-tagging and adequate measures can be taken to provide more parking spaces. Furthermore, IoV is used to identify the type of the vehicle to efficiently manage spaces in highly constricted areas. Historical data can be analyzed to identify specific periods of the day with high parking demands and vehicles can be rerouted to appropriate parking zones accordingly.

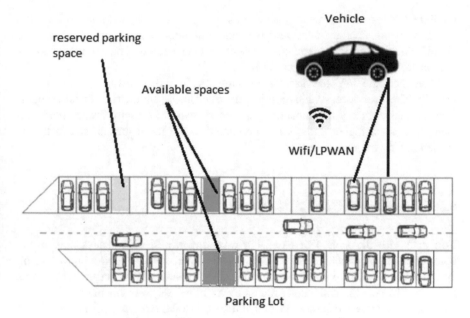

Fig. 6.3 Urban parking management

6.3.3   Route Optimization

Time delay is considered an important factor when evaluating the efficiency of traffic management. It is imperative that vehicles reach their destination safely in the least time as possible. Intelligent traffic management systems actively monitor and regulate the flow of traffic by dynamically controlling traffic signals at intersections. Such a system can identify traffic congestions in the area and inform drivers about alternate routes. This helps in decongesting the junctions and help vehicles reach their intended destinations on time.

During emergency situations such as road accidents and mishaps, time is a critical factor. Traffic congestions between the emergency center and the location of the accident can severely increase response time. Intelligent traffic management systems can analyze the traffic situation in the route ahead and can automatically redirect the emergency vehicle through the optimum route. The system can also warn about local situations and road blocks that may cause additional time delay.

6.3.4   Vehicle Theft Identification and Detection

Intelligent traffic management systems can help authorities in identifying and tracing stolen vehicles. When the owner alerts the authorities about the theft, the intelligent traffic management system can search its vast integrated network for the stolen vehicle. This includes searches via toll gate transactions, number plate identification using traffic cameras present in intersections, and additional information from onboard IoT sensors [13].

Moreover, the IoV has user-based authentication that restricts access to the vehicle. The user access validation is performed using biometrics, facial recognition, or password/key combinations. Such a system is normally faster, more efficient, and has a higher success rate compared to the usual manual search that is carried out by the authorities.

6.3.5 Automated Toll Management

Toll gates are the source of extensive queueing times, time delays, and traffic congestions in highways and expressways. It is a place where vehicles have to stop and proceed, thereby wasting crucial time. Intelligent traffic management systems help automate the toll collection process by enabling easy e-payment of the toll fee. Advanced cameras and sensors placed at the start of the toll road, RFID tags placed on the vehicle, automatically identify the type of vehicle, the registered vehicle number, and generate the appropriate fee to be paid by the driver [14].

With the advent of smart cities, conventional toll booths may be completely overhauled by automated electronic toll systems that are much more efficient. They do not cause traffic congestions as the vehicles do not have to stop, they do not require man power to issue the toll tickets.

6.4   Advancements in Intelligent Traffic Management Systems

Recently, Intelligent traffic management systems have seen an overhaul by emerging technologies such as AI and context-aware systems. These advanced techniques help improve the scope further, increase efficiency, and enhance the operational capabilities of traffic management systems. Although most of them are still in the research and development phase, they are extremely promising approaches [15].

Reinforcement learning has emerged as an excellent technique for a variety of applications including traffic management. Such systems work on a reward and penalty basis where the model learns from real-time data by rewarding correct decisions while penalizing incorrect ones. This reinforces and optimizes the model to accurately represent the real-time traffic conditions. Such a system can dynamically adapt itself to varying traffic conditions without human intervention.

Local context awareness is another approach that can be integrated with traffic management systems to personally cater the needs of individual vehicles in the overall system. Local context awareness is used to study the situation in and around the vehicle to take decisions. This data can be transmitted back to the traffic management system to aid in local micromanagement of vehicles according to their needs by rerouting them, suggesting alternate routes, or dispatching emergency services.

In the following sections, two of the most promising advancements in intelligent traffic management systems are explored in detail: (1) reinforcement learning approach to optimize traffic flow; and (2) local context awareness in IoV to aid intelligent traffic management systems.

6.4.1   Reinforcement Learning Approach

Reinforcement learning techniques are a viable option to enable smart management of traffic systems. Traffic patterns at particular junctions are analyzed, assessed, and the traffic signals are operated accordingly. Reinforcement learning is a type of machine learning where the system learns from historical data and the model is constantly reinforced by real-time data. The system is offered a reward for right decisions and is penalized for wrong decisions. The context of the decision is decided based on reducing the time delay. A decision that leads to a reduction in time delay is rewarded while a decision that increases time delay is penalized by the model. Here, the optimization may be governed by factors such as number of vehicles passing through a junction, position of vehicles and waiting time. The reinforcement learning model strives to optimize a cost function that is based on the mentioned factors.

Reinforcement techniques help in addressing dynamic variations in the traffic flow in real time. Such algorithms have consistently outperformed conventional systems with fixed time and manually-operated traffic lights. These algorithms generally consider the road as the state and the operation on light as the action. Based on the traffic situation on the road, corresponding action is taken by changing the state of the traffic light to red or green (To stop or enable the flow of traffic). However, the discrete state-action pairs require extensive storage capacities that render them impractical to be applied in most scenarios. Therefore, advanced techniques such as deep Q learning methods using continuous state representations are explored. These algorithms work by learning a Q-function to try to map the state and action to rewards. These functions are built by considering the average travel time, queue length, rate of flow of vehicles, direction of vehicle movement, and average delay.

The framework suggested by Wei H [16] consists of an online component and an offline modeling part. The offline component consists of fixed light timings and the traffic flow through such a junction is monitored and the data is collected. Once the model is trained on this data, it is moved into the online section. In the online part, the model will observe events and take appropriate action to whether turn the signal to green or red. The reward is calculated based on the feedback received from the reinforcement system and after many such iterations, the model is updated.

The state of the model is considered as a 4-way intersection with multiple lanes. The operation of the traffic signal with state change from switch on to switch off or vice versa is considered an action. The reward for each action is calculated using the weighted sum of a variety of factors such as follows:

- Sum of queue length over all approaching lanes in the intersection
- Sum of delay over all approaching lanes
- Sum of updated weighting time over all approaching lanes
- Light switch indicator
- Total number of vehicles passing through the intersection
- Total travel time of vehicles through the intersection

Given the current state of the traffic condition, the goal of the agent is to find the optimum action that may lead to the maximum reward r in the long run. To determine the optimum reward based on the state and action, the model has to use a Deep Q-Network. This is to deal with the complex nature of the real-world traffic scenarios. The design of conventional reinforcement learning techniques to estimate the Q-function is rather simple and such models experience difficulties and ambiguities when applied on complex real-life traffic scenarios. To overcome these difficulties, the proposed network contains a sub-structure called "Phase Gate".

The images obtained from the traffic cameras are analyzed using two convolutional layers. The output of these convolutional layers is concatenated with the four extracted features, namely total number of vehicles, queue length, waiting time, and phase. The concatenated features are subsequently fed into fully-connected layers to map the traffic conditions to possible rewards by optimizing the traffic flow. A separate learning process is designed for each phase and these phases are selected using phase-controlled gates. This helps in the decision process by distinguishing the decisions for different phases. It ensures that the network is not biased towards any specific action and the overall performance of the model is enhanced.

The system uses the experience relay technique where samples from memory are used to periodically update the network. New data samples are added to the network and old data samples are removed. Popular reinforcement learning models tend to store all the state-action-reward samples in the same location in memory. Therefore, this memory location will prominently consist of the state-action-reward pairs of the most frequent phases only. This creates an imbalanced scenario where the system is able to identify the most frequent phase scenarios well but underperforms in the case of rarer scenarios. To overcome this balancing issue, the state-action-reward samples are stored in different memory locations. This is broadly based on the Memory Palace Theory [17] from cognitive psychology. It ensures that different training samples are stored in different memory and hence, the training process is balanced and unbiased.

6.4.2 Local Context Awareness

Local context awareness is the ability of a network or a service to be aware of its surrounding environment and take decisions dynamically and automatically. A context-aware system is a service that uses local context awareness to deliver important information regarding the local state of the vehicle to the traffic management system. Context-aware services are designed to observe the vehicle, the driver, and the surrounding environment using onboard sensors and cameras. The traffic management system uses this contextual information to take decisions based on the situation in the vehicle's vicinity.

Data from a variety of such context-aware entities is combined together, analyzed, and transmitted over IoV gateways to the cloud. This compiled information is used to derive insights about the traffic situation and incidents in different areas. Subsequently, drivers in the vicinity are alerted and updated about developing situations and unfavorable conditions. This helps in refining the traffic management system by accounting for local factors that could potentially disrupt traffic.

Context-aware systems can be broadly classified into location-aware services and object-aware services based on the type of stimulant used to trigger the system. The location-aware services work by observing changes in the vehicle's environment. This includes situations where there is poor weather in the vicinity of the vehicle or when the vehicle enters an accident prone zone or when there is a road blockade ahead. Object-aware services work by observing changes in the service object such as low fuel in the vehicle, oil change, and tyre change. When context-aware services are activated, the context information from the various sensors and detectors will be recorded and analyzed to diagnose the situation. Based on the information gained, necessary action will be taken by the system to address the given situation. By understanding various contexts, a context-aware service dynamically adapts itself to changes in its local environment [18].

The architecture of context-aware IoV systems in intelligent traffic management systems consists of five layers based on hierarchical spatial regions: Physical layer, IoT layer, communication layer, context-aware service layer, and application layer as shown in Fig. 6.4.

The lower physical layer primarily consists of key entities such as city infrastructure, vehicles, driver, and pedestrians. Data is generated at this layer with the interaction of these entities at junctions with the intelligent traffic management system (ITMS). The IoT layer with its assortment of sensors, cameras, and detectors gathers this data. This data is transmitted over a variety of communication routes present in the communication layer of the system such as WiFi, LPWAN, 4G/5G networks, and Transmission Control Protocol/Internet Protocol (TCP/IP) networks. The final application layer consists of the various components of the ITMS system such as toll management, smart parking, integrated public transport system and fleet management.

Moreover, vehicular ad hoc networks (VANETs) are also a viable way of transmitting contextual data. VANETs are an application of mobile Ad Hoc

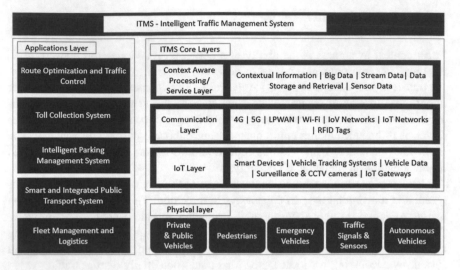

Fig. 6.4 Architecture of a local context aware system (adapted from [19])

networks (MANETs) that use dedicated short-range communication (DSRC) to allow vehicles to communicate with each other [20]. By integrating VANET service with the Internet, interconnected IoV networks can be formed. The vehicles in these networks use local context awareness to help the intelligent traffic management system make decisions.

The context awareness service layer consists of sensors and data processing units that derive contextual information from raw data. This contextual information is used by the vehicle and the traffic management system to make informed decisions. Furthermore, this data is passed onto the application layer. The application layer applies the derived contextual information and observations to manage applications such as smart parking, intelligent traffic control, incident identification, and controlling road congestion. These applications are an integral part of smart cities and modern IoV-based traffic management systems.

When the intelligent traffic management system has acquired all the necessary data, the overall contexts of the drivers and vehicles are studied using contextual reasoning in the context-aware service layer. This helps identifying anomalies and abnormal contextual situations such as vehicle repair, distracted driver, and varying vital signs of the driver [21]. The context-aware IoV system reacts to the identified situation by suggesting the driver with appropriate services. This helps in proactively identifying and preventing imminent mishaps by alerting the driver and taking corrective action on time. Furthermore, Pre-built services can be invoked in different situations, such as fuel management, selecting nearby hospitals and fuel stations to assist the driver, give suggestions about the optimum route based on the traffic, weather conditions, and other factors. As the system is exposed to different scenarios, and attached with more sensors, the volume and quality of data acquired by the systems increases. This helps the IoV context-aware system learn new scenarios and help anticipate hazardous situations especially when combined with reinforcement learning and machine learning models.

Contextual information is generally classified into two types—low-level (direct raw context) information and high-level (indirectly deduced) information. Low-level contextual information is directly obtained from sensors and cloud web services. This information includes the vehicle's location, speed, and direction of the vehicle, vehicle's oil and fuel levels, and local weather conditions. This low-level contextual information is directly derived from the local environment as raw data and is subsequently processed.

Indirect context information is inferred from context reasoning based on the data from the sensors and detectors. This information is not directly available and is rather derived from low-level data. This includes information such as the overall health status of the driver, the current route condition, or the weather status of the upcoming route. This enables the system to proactively identify and anticipate situations before they are actually encountered.

One of the most recent and advanced method of context modeling is Ontology modeling. It is used to describe the context of the information using reason, logic, and semantics. It has many advantages including strong expressive capability, support for logical reasoning, ease of interpretability, and convenience for knowledge reuse and sharing. More abundant context information can be deduced via user-defined reasoning. By acquiring raw context information of the driver, vehicle, and environment, the statuses of the system can be deduced through reasoning based on pre-defined rules such as health status rules, in-car environment rules, vehicle speed status rules, and road condition rules. This information is used to reason and understand the context logically.

Moreover, a survey by Strang et al. [22] suggests that ontology modeling is one the most promising assets for context modeling when compared with other techniques such as Key Value Models, Graphic Models, Object Oriented Models, and Logic Based Models. Therefore, ontology modeling is suitable for the description and definition of context and its relations in vehicles. This modeling technique helps intelligent traffic management system make better decisions by giving additional inputs on contextual information, thereby enabling the traffic management system to be aware of the local context of individual vehicles and thereby, guiding and managing the vehicles based on their specific needs.

6.5   Challenges and Future Perspective

Intelligent traffic management systems may be considered an inevitable necessity to meet the demands of the ever-growing number of vehicles on road. Conventional systems have to be intelligently revamped to handle modern urban traffic conditions and also to account for incidents and special conditions. This is especially true in the case of smart cities developments where traffic systems have to be highly refined and optimized to operate at maximum efficiency. However, there are a variety of challenges faced by organizations and government bodies in widespread implementation of intelligent traffic management systems, as discussed below.

6.5.1 Data Integration

Due to extensive use of sensors and IoT devices in intelligent traffic management systems, they suffer from data integration problems. The lack of interoperability and standardization between IoT devices has often been highlighted. On large distributed networks such as intelligent traffic management systems that could span over the area of an entire city. Hence, integrating all the IoT sensors, IoV networks, traffic signals, and CCTV cameras is a major challenge. The heterogeneous data that is generated from all the devices should be efficiently collected, processed, and analyzed in real time. Moreover, the data from different sources may be in different formats and employing encryptions that need to be standardized throughout the system to ensure seamless data processing and analysis.

6.5.2 Security and Privacy

Intelligent traffic management systems are a large-scale ensemble of different networks, services, and devices. Such systems may consist of millions of individual distributed IoT devices. Due to lack of standardized security guidelines and safety protocols, such devices are vulnerable to external attacks and security breaches. Furthermore, the use of cloud to store and process data factors in as another vulnerability. With such vast number of interconnected devices, all the nodes have to be monitored and secured as a breach in one device can compromise the integrity of the entire network. Intelligent traffic management systems tend to collect and process sensitive data related to the personal details of people. This may include images, personal details, live location of people and vehicles, personal identification numbers, vehicle registration numbers, etc.

Intelligent traffic management systems control almost all the aspects of road transportation including traffic signals, toll collection, emergency dispatch, and parking management. Malicious entities may shut down the system, bringing the entire road traffic to a halt, or they may create fake warnings or alerts, thereby disrupting road traffic almost through the entire city. If such a system is compromised, it can severely disrupt normal human life.

6.5.3   Investment and Operating Costs

Intelligent traffic management systems have high initial investment costs. A system of such complexity and scale is extremely difficult and expensive to set up. Such systems generally consist of millions of sensors, devices, servers, and networks that require installation, configuration, and testing. Moreover, the installation of sensors and cameras on private properties could require legal permissions and undertakings that could further hamper the implementation of such systems.

The sensors, cameras, and network devices used in intelligent traffic management systems are expensive to operate and maintain. These networks require continuous electricity and stable internet connections over wide geographical areas. Malfunctioning sensors and devices could send out wrong readings and false alarms that could lead to serious consequences. Hence, the health of the overall system needs to be continuously monitored.

6.5.4   Sabotage and System Evasion

Physical components of Traffic management systems such as sensors, cameras, and other devices are prone to physical damages and tampering. Drivers have been observed to intentionally sabotage and damage equipment to evade law. Such damages are difficult to repair and the offender often manages to get away with the offense [15].

Drivers tend to disable the onboard system such as GPS or IoT sensors to disconnect themselves from the traffic management system. This causes anomalous behavior in the traffic management system because of the unidentified vehicle. This could lead to potentially catastrophic events as the vehicle is not sufficiently registered or tracked by the system and hence its decision-making ability is compromised. Furthermore, Drivers have also been observed to mask their vehicle registration number plates or use magnetic numbers to change their number plates dynamically. These techniques have been used to evade the authorities and cause public discourse.

An ideal intelligent traffic management system should be able to efficiently identify and handle such offenses and acts of sabotage. This can be done using anomaly detection and entity-based behavior analysis. This may be implemented in the form of complimentary cameras with cross-vision that cover each other in their field of vision and more stringent vehicle identification systems with heftier fines for offenders. It has also been noted, e.g., by [15], that the education and awareness among the public plays an important role in the realization of intelligent traffic management systems.

6.6 Conclusion

The ever growing number of vehicles on roads has created major traffic, safety and pollution issues around the world. Conventional traffic management systems are not capable of addressing these problems adequately. Moreover, lack of optimization in traffic signals has led and is still contributing to excessive traffic congestions, high fuel consumption due to idling, elevated pollution levels, time delays, and economic losses. Intelligent traffic management systems use a variety of modern techniques such as IoV, IoT, reinforcement learning, and context-aware reasoning techniques to optimize traffic flow, improve road safety and make transits more efficient. They serve as the backbone of commuting and transportation in smart cities. These systems utilize the data from the wide networks of traffic cameras, IoT sensors, and detectors available in smart cities. Furthermore, reinforcement learning and local context-aware services are emerging as viable solutions to enhance and improve intelligent traffic management systems. This will further help in realizing a future where transits and commuting using road vehicles is much more efficient, optimized, safe, and sustainable.

References

1. INRIX (2019) INRIX: congestion costs each American 97 hours, $1,348 a year (2018). Accessed Apr 2019
2. Davis N, Joseph HR, Raina G, Jagannathan K (2017) Congestion costs incurred on Indian roads: a case study for New Delhi
3. Sessions GM (1971) Traffic devices: historical aspects thereof. Institute of Traffic Engineers, Washington, p 3
4. McShane C (1999) The origins and globalization of traffic control signal. J Urban Hist 25 (3):79–404 at p 382
5. Özkul M, Capuni I, Domnori E (2018) Context-aware intelligent traffic light control through secure messaging. J Adv Transp **2018**(Article ID 4251701)
6. Dion F, Rakha H, Youn-Soo K (2004) Comparison of delay estimates at under-saturated and over-saturated pre-timed signalized intersections. Transp Res Part B: Methodol 38:99–122. https://doi.org/10.1016/S0191-2615(03)00003-1
7. Rouphail N, Tarko A, Jing L (2019) Traffic flow at signalized intersections
8. Webster FV (1958) Traffic signal settings. Road Research Technical Paper 39
9. Tom MV (2014) Vehicle actuated signals, IIT Bombay, Transportation systems engineering 39
10. Zhang L (2018) Intelligent traffic systems: implementation and what's down the road? https://www.cleantech.com/intelligent-traffic-systems-implementation-and-whats-down-the-road/. Accessed Apr 2019
11. Flir (2019) Optimized traffic flows for motorists, pedestrians and bicyclists. https://www.flir.com/traffic/optimize-traffic-flows-for-motorists-pedestrians-and-bicyclists. Accessed Apr 2019
12. Barone RE, Giuffrè T, Siniscalchi SM, Morgano MA, Tesoriere G (2014) Architecture for parking management in smart cities. IET Intell Transp Syst 8(5):445–452. https://doi.org/10.1049/iet-its.2013.0045
13. Bali P, Parjapati G, Aswal R, Raghav A, Gupta N (2015) Intelligent traffic monitoring system (ITMS). Progr Sci Eng Res J

14. Washington State Comprehensive Tolling Study (2016) Final report—volume 2, background paper #8: toll technology considerations, opportunities, and risks. Accessed Apr 2019
15. Behruz H, Chavoshy P, Lavasani A, Mozaffari G (2013) Challenges of implementation of intelligent transportation systems in developing countries: case study—Tehran. WIT Trans Ecol Environ 179:977–987. https://doi.org/10.2495/SC130832
16. Wei H, Yao H, Zheng G, Li Z (2018) IntelliLight: a reinforcement learning approach for intelligent traffic light control. In: Proceedings of the ACM SIGKDD international conference on knowledge discovery and data mining, pp 2496–2505
17. Eric LG, Christopher MR, Enoch Ng T, Jeremy CT (2012) Building a memory palace in minutes: equivalent memory performance using virtual versus conventional environments with the method of Loci. Acta Physiol 141(3):380–390
18. Ozkul M, Capuni I, Domnori E (2018) Context-aware intelligent traffic light control through secure messaging. J Adv Transp 2018:1–10. https://doi.org/10.1155/2018/4251701
19. Rehena Z, Janssen M (2018) Towards a framework for context-aware intelligent traffic management system in smart cities, pp 893–898. https://doi.org/10.1145/3184558.3191514
20. Souza DAM, Brennand CA, Yokoyama RS, Donato EA, Madeira ER, Villas LA (2017) Traffic management systems: a classification, review, challenges, and future perspectives. Int J Distrib Sens Netw
21. Chang J, Yao W, Li X (2017) The design of a context-aware service system in intelligent transportation system. Int J Distrib Sens Netw 13
22. Strang T, Linnhoff-Popien C (2004) A context modeling survey. In: First international workshop on advanced context modeling, reasoning and management, UbiComp

Chapter 7
Smart Transportation Tracking Systems Based on the Internet of Things Vision

**W. K. A. Upeksha K. Fernando, Ruwani M. Samarakkody
and Malka N. Halgamuge**

Abstract In recent years, people have pursued smarter and faster options for fast-paced modern lifestyles. This is in response to many technologies that have recently emerged. Vehicle tracking systems based on the Internet of Things (IoT) technology is one of these. This chapter aims to examine the IoT-based cutting-edge vehicular tracking systems. For this purpose, 30 peer-reviewed research publications, from 2014 to 2018, were selected and raw data was extracted. Selection was based on certain critical parameters such as: measuring attributes of the moving vehicle, sensors and actuators used for data obtaining in tracking devices, data transferring methods for transmission, networks and protocols utilized for communication, utilized stock for data storage, programming languages or systems, and algorithms utilized for raw data analysis. The investigation demonstrated that (i) a large portion of the IoT sensors and actuators were centered on the primary location tracking system in cloud data centers that can be handled remotely by retrieving real-time data, (ii) The GPS sensors widely use in vehicle tracking systems were based on the RFID technology, (iii) Wi-Fi networks were the most popular networks while GSM/GPRS and TCP/UDP protocols were the best transport layer protocols, (iv) most used storage method was observed as the cloud for smart vehicle tracking systems, and (v) Kalman filter was the most popular algorithm in vehicular tracking systems. Moreover, the most critical advantage of using IoT for tracking systems was the effectiveness, security, and intrusion of protection for the passengers. The security administration could monitor students by remotely tracking their RFID sensor tags or any IoT sensor embedded in the tracking unit. This chapter also reviews and provides relevant information for road traffic officers and related experts, correspondence technologists, and technological innovation researchers on the IoT-based smart vehicle tracking frameworks.

W. K. A. U. K. Fernando · R. M. Samarakkody
School of Computing and Mathematics, Charles Sturt University, Melbourne, VIC 3000, Australia

M. N. Halgamuge (✉)
Department of Electrical and Electronic Engineering, The University of Melbourne, Parkville, VIC 3010, Australia
e-mail: malka.nisha@unimelb.edu.au

© Springer Nature Switzerland AG 2020
Z. Mahmood (ed.), *Connected Vehicles in the Internet of Things*,
https://doi.org/10.1007/978-3-030-36167-9_7

143

Keywords Internet of Things · IoT · Smart cities · Transport tracking system · Intelligent transportation system · Cloud

7.1   Introduction

The Internet of Things (IoT) acts as the main role in object connecting through interconnected networked connections. It was first introduced by Kevin Ashton in 1998 with the use of RFID sensor tags [1]. According to a report [2], the volume of IoT associated devices in 2020 will surpass more than seven times the current situation worldwide and Cisco anticipates that the number of IoT associated devices will reach about the 50 billion mark in 2020. IoT device connectivity provides an alliance to generate data and deal with other communication devices in the network.

In this context, as in any other technology, vehicle tracking systems have various forms of challenges. Many studies have highlighted the issues of handling a massive amount of data in IoT that can be managed effectively and efficiently in the cloud networking environment. Tracking is a process of taking note of the geographical locations and coordinates of the vehicles using the GPS technology [3]. Vehicle tracking is a rapidly growing area with the elevation of technology innovation.

A number of experiments have been conducted on vehicular tracking systems; however, some issues on the power system are yet to be resolved such as GPS signal loss, notification message delays, time delays at bus stops, safety and security issues, etc. This current study reviews the possible and feasible solutions to identify the locations of vehicle and related issues.

These days, vehicle tracking is very important because it helps to develop noteworthy security and other options in transportation systems. Many scholars have suggested that new technology-based tracking systems and such innovative systems are generally constructed using real-time data for tracking the location of the vehicle [4]. Almishari et al. [5] have proposed a portable tracking system that uses batteries for the operation by highly focusing on power reduction algorithms. The authors of the [3, 5–10] have collected, for their studies, sensor data from locations, speed, temperature, time, power, and velocity, by using radio frequency identification (RFID) tags, and GPS satellites. According to these studies, many tracking units consist of GPS sensors and temperature sensors. These tracking systems physically relate to ATmega328 and Arduino UNO R3 microcontroller through SIM 908 GPS/GPRS modules [5–7, 9, 11].

A rapid positioning and tracking system were built by Chang et al. [12] under the Ad Hoc network, to improve the precision and accuracy with GPS. They have implemented a procedure to cumulate data sensing time at different levels. Their proposed system transferred the data between the cloud and the sensors using Wi-Fi signals. Then, the tracking and estimation continue through α–β–γ filters on an Ad Hoc network. A few authors, e.g., [8, 12] have used RSSI to acquire signals from IEEE 802.15.1 Bluetooth, IEEE 802.15.4 ZigBee, IEEE 802.11b/g and Wi-Fi to resolve the relative distance issues.

Jisha et al. [11] considered developing a platform to ensure the safety of school children and reduce the missing rate of school children. The planned system can estimate the average waiting time for the school buses and, simultaneously, parents can monitor the location of their children continuously. In this system, the EM 18 RFID readers, Arduino Mega 2560 microcontroller and SIM28ML GPS sensors in the hardware layer communicate with GPRS technologies. The application layer consists of UBIDOTS cloud platform as cloud server while using Google map, and Apache Spark open source as clustering framework. Here, email notification and SMS are used for message transfer between the users. Signal processing follows radio waves and GPRMC signals along with the speed. This study has used Kalman filtering-based prediction algorithm because of weak GPS signals to enhance the effectiveness, while K-nearest neighbor (k-NN) algorithm was integrated to analyze clusters. Also, the distance between two points of latitudes and longitudes have been calculated using Haversine formulae. Similarly, research in [4, 12] has proved the effectiveness of the Kalman algorithm.

Raj et al. [3] also conducted a similar invention to develop a secure and trustworthy school traveling system for students. They implemented a system comparable to the one in [11], whereas parents can track the school bus to ensure their children's safety until reaching the home. The application enables to pursue the speed of the bus, delay of moving, schedule updates, contact the driver, or authorize people at all levels via the tracking system that was also installed on the running bus. MFRC522 RFID reader and Ublox 6 M GPS module were connected to the ESP8266 microcontroller to detect the signals. The data from the bus was transmitted using RF signals to the cloud server database. Then, Google Map APIs were used to plot the locations of the bus route while Firebase Cloud Messaging service was utilized to push the message notification.

A scalable public transport system is designed by Lohokare et al. [13] which is reliable and efficient with the use of MQTT instead of HTTP. The sensing attributes of the bus are obtained using GPS/GSM module on the bus and NFC tags RF transceivers. Load balancer cloud architecture, used in the design system, can automatically load the instance of a collection of requests. The system has developed an option for the clients called REST APIs to access the Android applications. The study has used Haversine formula to calculate the distance between locations of the bus stops while artificial neural network was used to predict the vehicle arrival time. The prediction was based on the historical data rather than real-time data. However, the authors have proposed to develop trained artificial intelligence by feeding historical data and utilize Google maps to provide real-time locations. Similarly, a smart vehicle tracking system was implemented by Uddin et al. [4] that can block the fuel lines of the engine and then, send notifications to the owner during the robbery. The main aim of the experiment was to protect the vehicle from robbery using IBM Bluemix and Google IoT cloud platforms.

Recent evidence suggests that social networks with GSM/GPS technology have a potential to design an efficient vehicle tracking system [6, 7]. An experiment conducted by [6] has used GPS module to get geographical coordinates of the vehicle at regular time durations. The data was collected by measuring the vehicle

ID while Google Earth utilized the monitoring of location and tracking the current status of the vehicle on the map. Moreover, Desai et al. [7] introduced an effective location tracking mechanism to test the monitor vehicle parameter. They have used open-source platform because it facilitates a dynamic, efficient, and cost-effective project. The fuel level, traveled distance, temperature of the environment, latitude, and longitude were selected as the main parameters for the study. It has conclusively been shown that a Bluetooth low energy technology has a potential to develop a model for the bus tracking system [8]. In this system, Bluetooth Low Energy (BLE) module on the bus and the bus stop communicate and transfer data to the Yellow API through the yellow application. They have used BLE capable device which has fully configured with AT command and Node MCU to send and receive data like Arduino with ESP8266.

A real-time cloud-based drone-based tracking system was proposed by Koubaa et al. [9] that can track the GPS location of moving objects. The system uses GPS sensors, 3DR Solo, DJI Phantom, Erle copter, and unmanned aerial vehicles (or drones) to collect data. The GPS signals transfer the data between a cloud server database and the sensors. Correspondingly, Pendor et al. [10] proposed a system to reduce the heavy traffic-related fatalities over the country. This system can track and monitor vehicles and passes messages to the drivers when they exceed the speed limit. Experimental data is collected using Battery Assisted Passive Tag. The laser, radar, radio frequency, and signals were used as the transferring methods while the information stored in the cloud. The authors have utilized the image processing, Euler's algorithms and the WAN network system.

A GPS-GSM predicted vehicle tracking system is designed by Mangla et al. [14] that can monitor vehicles using a mobile app based on Google Map. The system has potential to send SMS notification of vehicle coordinates and other sensing attributes to the users. Authors have used GPS antenna, GSM modem, GPS module, 9 V battery, USB Connector, DC power jack as sensors and actuators while TCP protocols are used in the project [14].

These kinds of tracking systems focus on the real-time location of objects, speed, and temperature of the vehicle as well as on changing attributes of vehicles. Google map APIs or short message service (SMS) can be used to identify the location. The microcontroller system unit mounted inside the vehicle continuously sends messages to Google map APIs through the cloud with details of vehicle location and other information at specific time intervals [5]. Also, this kind of tracking systems can be used for anti-theft systems which are designed to prevent the vehicles from being stolen.

Recently, researchers have shown an increased interest in vehicle tracking systems and related applications. However, the issues and challenges that arise in vehicular tracking systems have been neglected. Therefore, this chapter discusses the possible implications for the faults in vehicle tracking systems.

The main contributions of this chapter include the following:

- Gather raw data from 30 peer-reviewed journals and organize the data according to certain experimental parameters.

- Analyze the parameters using certain measuring attributes of the moving vehicle, viz, sensors and actuators used for data obtaining in tracking devices, data transferring methods for transmission, networks and protocols utilized for communication, data storage stock, programming languages or systems, and raw data analysis algorithms
- Design the experimental structure and summarize the raw data in tabulation format
- Adopt comprehensive data analysis technique to reveal the results and explain them in graphical format.
- Conduct extensive experimental evaluations to demonstrate the efficiency and effectiveness of the vehicle tracking system.

Furthermore, the present work elaborates recent IoT-based tracking systems in transportation by grouping the pilot data from various tracking systems based on different purposes. The scientific articles for data collection were published between 2014 and 2018 and sourced from IEEE journal databases. Most of the reviewed articles examine unique innovative components of IoT tracking devices. However, the past work focuses more on the mechanical parts of the developed tracking devices which are more useful for the experts involved in road development fields or construction. The multifaceted nature of the past investigations is hard to comprehend by customary individuals. We have attempted to resolve this issue in this chapter, where the multifaceted nature is much reduced when contrasted with other comparable works, and takes the idea of IoT in vehicle tracking to customary individuals. In the meantime, the work investigated here is the latest development to clear the path for new researchers to build up the procedures to design further developed tracking devices or new tracking devices that would help control more efficient tracking systems.

Vehicle tracking systems reflect various types of uses for the daily vehicle users. These systems help to find the lost and stolen vehicles. Some tracking systems have potential to memorize the parking position that helps users to identify the location in complex or large parking malls [14]. Also, the efficiency and the quality of the fleet management systems can be improved using tracking systems [6]. Sometimes the automobile industry utilizes tracking systems for validation of the testing vehicles [8]. Considering all limitations and facts, this chapter aims to suggest possible solutions for the identified issues in vehicle tracking systems and applications.

The remainder of this chapter is organized as follows. Section 7.2 develops a completed methodology for the experiment. Section 7.3 explains the experimental results. Section 7.4 summarizes the related work and compares the experimental results with the values available in the literatures. Finally, Sect. 7.5 outlines the conclusions and future work.

7.2   Data Collection and Analysis

The data for analysis was collected from 30 IEEE peer-reviewed research articles, publishes during 2014–2018. Table 7.1 presents the relevant raw data in a structural tabulation form. Data is summarised under the following headings:

- Measuring attributes (location, speed, etc.).
- Varieties of sensors and actuators.
- Data transfer methods.
- Networks and protocol standards.
- Data storage methods.
- Programming languages and software used.
- Algorithms used.

To assess data incorporation criteria, an information comparison table was drawn with the qualities recorded previously. All publications were related to IoT driven systems or tracking devices that supported vehicle tracking condition. The publications which gave lack of information to the comparison table qualities were avoided and the examinations that have not been distributed in reviewed scientific publications.

7.3   Data Analysis and Result

The data was gathered from scientific publications as shown in Table 7.1. Raw data was analyzed against the attributes of the tracking system obtained via the sensors. After the examination of raw data, information was condensed into rates of percentages and then seven distinctive investigation results were obtained. After the analysis, the results were illustrated in graphical format, as shown in various figures provided in this section.

7.3.1   Measuring Attributes of Using IoT Smart Sensors in Tracking Systems

Referring to Fig. 7.1, many authors (14%) have used location as the most important tracking measurement because location may be the easiest and most practical attributes to track vehicles. Time and speed were considered the second most important attributes in the vehicle tracking systems and it was around 11% for each parameter. Distance, longitude and latitudes were also selected as useful attributes to the tracking systems.

Table 7.1 Summary of data collected from 30 publications

#	Author	Measuring attributes	Sensors/Actuators	Data transfer methods	Networks/Protocols	Data storage	Programming languages/ system software	Algorithms
1	Almishari et al. [5]	– Location – Speed – Temperature – Time – Power	– RFID – Smartphone – GPS module – Satellite – Temperature sensor – ATmega328 – GPS/GSM antenna – Arduino UNO R3 – Battery	– SMS	– GSM/GPRS Network – National Marine Electronics Association (NMEA) – HTTP	– Cloud	– MySQL – HTML/XML – CSS/PHP – JS/Ajax – Linux cent OS – Android	– Power reduction Algorithms
2	Chang et al. [12]	– Time Differentials on Signal Arrivals – Time of Signal Arrivals – The angle of azimuth – Receiving strength from a signal indicator	– Satellite – NFC	– Wi-Fi	– Ad Hoc Networks – ZigBee – Wi-Fi – Bluetooth – 3G/4C	– Cloud – Fog	– Web Browser	– Kalman Filter – Central Limit Theorem – a-β-g filter
3	Jisha et al. [11]	– Location – Speed – Time – Velocity – Distance – Timestamp – Latitude – Longitude	– RFID – Smartphone – GSM/GPRS modem – GPS module – Arduino Mega 2560 – Speed sensor – Battery	– SMS – NMEA messages – Email – Radio frequency signal	– GSM/GPRS Network – National Marine Electronics Association (NMEA)	– Cloud	– Android – C Programming – Google map/APIs – UBIDOTS Cloud Platform – Python – Apache Spark	– Kalman Filter – k-nearest neighbor (k-NN) algorithm – Haversine formulae – Artificial Neural Network (ANN), – Support Vector Machine (SVM) – Root mean square error
4	Lohokare et al. [13]	– Location – Latitude – Longitude – Route Number – Bus Number – The direction of the bus – Timestamp – Station Number	– Satellite – GPS module – NFC – Smartphone – GSM/GPRS modem – RFID	– SMS	– Bluetooth – MQTT	– Cloud	– MySQL – HTML/XML – JS/Ajax – MQTT broker – CSS/PHP – Android – Google map/APIs – Load Balancer	– Haversine formulae – Artificial Neural Network (ANN),
5	Uddin et al. [4]		– RFID – NFC – GPS module – Satellite – Proximity card – GSM/GPRS modem – LinkIt One Development Board – MediaTekMT2502A Soc Microcontroller	– SMS	– Bluetooth – Wi-Fi – Amazon web-service – GSM/GPRS Network – MQTT – Constrained Application Protocol (CoAP) – GNSS	– Cloud	– Google Earth – Representational State Transfer (REST) APIs	– Kalman Filter

(continued)

Table 7.1 (continued)

#	Author	Measuring attributes	Sensors/Actuators	Data transfer methods	Networks/Protocols	Data storage	Programming languages/ system software	Algorithms
6	Raj et al. [3]	– Location – Speed – Time – Contacts	– RFID – Smartphone – GPS module – Satellite – ESP8266-06 module – Raspberry Pi	– Radio frequency signal	– HTTP – Wi-Fi	– Cloud	– MySQL – Java – HTML/XML – C programming – Arduino – Android – Google map/APIs	
7	Lee et al. [6]	– Distance – Time – Bus Number – Latitude – Longitude	– Satellite – Smartphone – GPS Module – GSM/GPRS modem – GPS/GSM antenna Arduino UNO R3 ATmega328	– SMS	– GSM/GPRS Network – HTTP – TCP/UDP	– Flash memory – MySQL database	– C programming – MySQL – CSS/PHP	– Kalman Filter
8	Charoenpom et al. [7]	– Route Type – Data – Location – Bus Number – Station Number	– RFID – Smartphone – GPS module – GSM/GPRS modem – Arduino UNO R3 – NodeMCU – ESP8266-06 module – Wi-Fi Module – Arduino Mega 2560	– SMS	– Wi-Fi – LAN – Bluetooth – HTTP – iBeacon – Bluetooth Low Energy	– Web server	– Android – Google map/APIs – iOS – open-source firmware	– AT Command
9	Dessai et al. [8]	– Speed – Location – Fuel level – Distance – Temperature – Latitude – Longitude	– RFID – NFC Speedometer – GPS module – SIM808 Arduino – GSM/GPRS modem – Arduino Mega 2560	– SMS	– Ethernet – Bluetooth – ZigBee – Li-Fi – HTTP – TCP/UDP – Wi-Fi – LTE	– Web server	– Arduino – Google map/APIs	– AT Command
10	Koubaa et al. [9]	– Acceleration – Speed – Time – Distance – Location	– GPS module – Drones	– JSON serialized messages	– TCP/UDP – USB protocol – ROSLink Protocol – MAVLink communications protocol – Cloud platform – LTE – 3G/4G – Optical fiber network – Wi-Fi	– Cloud – MySQL database	– Hadoop Map/Reduce model – MySQL – Robot Operating System – Drone map Planner (DP) – Google map/APIs – Android – GoNative – Cloudroid – iOS	– Network Calculus – Cumulative function – Pipeline model – Rate-latency service curve

(continued)

Table 7.1 (continued)

#	Author	Measuring attributes	Sensors/Actuators	Data transfer methods	Networks/Protocols	Data storage	Programming languages/system software	Algorithms
11	Rahul et al. [10]	– Speed – Time – Distance	– RFID – GPS module – Battery assisted passive tag (BAP) – Ethernet module	– SMS – Radio frequency signal – Laser – Radar	– Ethernet – WAM – Wi-Fi – TCP/UDP	– Cloud	– Arduino	– Euler's algorithms – Image processing Algorithms
12	Mangla et al. [14]	– Location – Longitude – Latitude	– GPS/GSM antenna – Arduino Uno R3 – GSM/GPRS modem – Atmega328 – GPS module – Ethernet module – Battery		– TCP/UDP – WAN – Wi-F	– Flash Memory – Cloud – RAM – EEPROM	– Arduino – Google map/APIs – C programming	– AT Command
13	Weimin et al. [15]	– Location – Temperature – Speed	– GPS module – GSM/GPRS modem – Temperature sensor – RFID – Smartphone	– Wi-Fi – Radio frequency signal – SMS	– 3G/4G – Wi-Fi – Ad hoc Networks – Intranet	– MySQL database – Fog	– MySQL – Android	
14	Anusha et al. [16]	– Location – Temperature – Latitude – Longitude – Distance	– GPS module – GSM/GPRS modem – Buzzer – Alcohol Sensors – Temperature sensor – Satellite – Battery – Eye Blink sensor – Motor – ARM-7 Microcontroller LPC2148	– Radar	– 3G/4G – GSM/GPRS Network		– C programming – Google map/APIs	
15	Evizal et al. [17]	– Location – Bus Number – Frequency	– RFID – FR4 material – Times Microwave	– Radio frequency signal	– National Marine Electronics Association (NMEA)	– Cloud	– 3D CST simulation software	– Friis Transmission formula
16	Tang et al. [18]	– Location – Cell ID – Receiving strength from a signal indicator – Longitude – Latitude	– GPS module – GSM/GPRS modem – Smartphone – Battery – Satellite	– SMS – Query messages	– HTTP – TCP/UDP – IP – iBeacon – Wi-Fi	– Cloud – Web server	– Representational State Transfer (REST) APIs – MySQL – Android	– Triangulation technique

(continued)

Table 7.1 (continued)

#	Author	Measuring attributes	Sensors/Actuators	Data transfer methods	Networks/Protocols	Data storage	Programming languages/ system software	Algorithms
17	Pratama et al. [19]	- Location - Receiving strength from a signal indicator - Longitude - Latitude - Distance - Frequency - Time	- Smartphone - RFID - Battery - GPS module - UHF transceiver - Raspberry Pi	- Radar	- Bluetooth Low Energy - iBeacon - Bluetooth - Cloud platform - UHF	- Cloud	- MySQL - Google map/APIs - Android - Web Browser	- Kalman Filter - Mean-Shift
18	Balbin et al. [20]	- Location - Latitude - Longitude - Time	- GPS module - Smartphone - Accelerometer - Battery - Ultrasonic Sensor - Relay - Alarm - UHF transceiver	- Wi-Fi - NMEA messages	- Wi-Fi - National Marine Electronics Association (NMEA) - USB protocol - Bluetooth - Wi-Fi - UHF - Pubnub - 3G/4G	- Cloud - Web server	- Python - HTML/XML - JS/Ajax - Google map/APIs - Raspberry Pi	
19	Yu et al. [21]	- Location - Speed - Bus Number - Vehicle type - Frequency - The direction of the bus	- RFID - Battery - Ethernet module - UHF transceiver - Alarm - Directional Antenna - Ground sense coil	- Radio frequency signal - Query messages	- LAN - TCP/UDP - IP - Ethernet - UHF	- EEPROM - Flash memory - Web server - Cloud	- Local Server - Firewall - Microsoft SQL Server - Microsoft Visual C++ - Windows OS	- Data clearing algorithm
20	Bojan et al. [22]	- Location - Time - Latitude - Longitude - Altitude	- GSM/GPRS modem - GPS module - Smartphone - Personal Computers - RFID - Mobile handsets - Satellite - ATmega328 - Arduino Mega 2560 - Arduino Uno R3	- SMS - Wi-Fi - NMEA messages	- GSM/GPRS Network - Wi-Fi - HTTP - National Marine Electronics Association (NMEA) - 2G - TCP/UDP	- Web server - Cloud - Flash memory	- SPSS - Keyhole Markup Language (KML) - Arduino - MySQL - CSS/PHP - Google map/APIs - C programming	

(continued)

Table 7.1 (continued)

#	Author	Measuring attributes	Sensors/Actuators	Data transfer methods	Networks/Protocols	Data storage	Programming languages/system software	Algorithms
21	Bojan et al. [23]	– Location – Temperature – Humidity – Longitude – Latitude – Altitude – Speed – Time stamp	– GPS Module – NFC – Temperature sensor – Humidity sensor – GSM/GPRS modem – RFID – Satellite – ATmega328 – Arduino Mega 2560 – Arduino Uno R3	– SMS – NMEA messages	– GSM/GPRS Network – HTTP – National Marine Electronics Association (NMEA) – TCP/UDP – IP – 2G – 3G/4G	– Web server – Flash memory – EEPROM	– Keyhole Markup Language (KML) – Google Earth – Web Browser – iOS – Arduino – C programming – MySQL – CSS/PHP	
22	Zambada et al. [24]	– Speed – Location – Time – Latitude – Longitude	– RFID – Speed sensors – GSM/GPRS modem – Location sensor – PIC18F45K20 microcontroller – Smartphone – Battery – Satellite – GPS/GSM antenna	– Radio frequency signal	– MQTT – GSM/GPRS Network – 2G – Cloud platform	– Cloud	– Microchips PICkit 3 – MPLAB – C programming – Microsoft Visual C++ – MPLAB XC compiler – Google map/APIs – JS/Ajax	– AT Command
23	Emrah et al. [25]	– Location – Longitude – Latitude	– Raspberry Pi – GPS module – GSM/GPRS modem	– SMS	– GSM/GPRS Network	– Web server – Cloud	– OpenCV – Web Browser – Android – Python – MySQL – JS/Ajax – HTML/XML – Microsoft Visual C++ – C#	
24	Khan et al. [26]	– Speed – Location – Time – Longitude – Latitude	– GPS module – Battery – Eye Blink sensor – RFID – GSM/GPRS modem – PIC18F45K20 microcontroller – Vibration sensor – IR sensor	– Radar – Radio frequency signal – SMS	– GSM/GPRS Network	– Web server – Cloud	– Google map/APIs – Windows OS	– Software algorithms

(continued)

Table 7.1 (continued)

#	Author	Measuring attributes	Sensors/Actuators	Data transfer methods	Networks/Protocols	Data storage	Programming languages/ system software	Algorithms
			- Buzzer - Gas sensor - Satellite - Smartphone - interfacing sensors					
25	Chellaswamy et al. [27]	- Location - Speed - Frequency	- Raspberry Pi - Accelerometer - GPS module - GSM/GPRS modem - Satellite - Frequency detectors	- SMS	- Ad Hoc Networks - GSM/GPRS Network - Cloud platform - IP	- Cloud	- Python - Matlab	- Fuzzy Logic - Honey bee optimization Algorithm - RLS filter
26	Sutar et al. [28]	- Location - Time - Frequency - Station Number - Bus Number - Distance - Longitude - Latitude	- RFID - GPS module - Smartphone - GPS/GSM antenna - GSM/GPRS modem - Arduino Mega 2560 - Arduino Uno R3 - Raspberry Pi - Battery	- SMS - JSON serialized messages	- 2G - 3G/4G - HTTP	- Web server - Cloud - LAMP server	- Android - MySQL - CSS/PHP - Google map/APIs	- Markov chain-based Bayesian decision tree algorithm
27	Chellaswamy et al. [29]	- Location - Frequency - Power - Speed	- RFID - GPS module - Alarm - Camera - Vibration sensor - Arduino Uno R3 - LPC2114 microcontroller - GSM/GPRS modem	- Radio frequency signal - SMS	- ZigBee - TCP/UDP - IP - ModBus Application Protocol - GSM/GPRS Network	- Cloud - RAM - ROM	- Matlab	- PSO algorithm
28	Seiber et al. [30]	- Speed - The direction of the bus - Temperature - Humidity - Time - Location	- Drones - GPS module - Linkit 7688 Duo Microprocessor board - Air quality sensor - Battery - Satellite	- Wi-Fi - Radar	- Wi-Fi - Bluetooth - National Marine Electronics Association (NMEA) - Bluetooth Low Energy - IP	- Cloud	- Arduino - Linux - 3D modeling software - Eagle design tool	- 3D modeling algorithm

(continued)

Table 7.1 (continued)

#	Author	Measuring attributes	Sensors/Actuators	Data transfer methods	Networks/Protocols	Data storage	Programming languages/ system software	Algorithms
29	Srinivasan et al. [31]	– Speed – Location	– GPS module – Speed sensor – Accelerometer – Camera – Crash sensor – SIM – Diagnosis module			– Cloud		
30	Costa et al. [32]	– Location – Frequency – Electromagnetic field – Longitude – Latitude – Receiving strength from a signal indicator	– RFID – GPS module – Battery – GSM/GPRS modem – Satellite – GPS/GSM antenna	– Radio frequency signal – SMS	– GSM/GPRS Network – IP – Navy Navigation Satellite System – Ubiquitous Sensor Network – MQTT	– Cloud		– Checksum

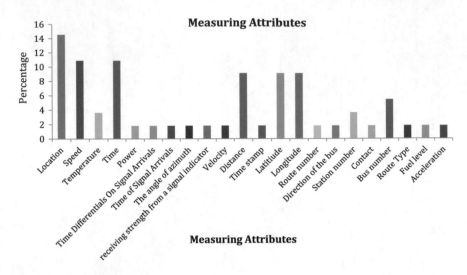

Fig. 7.1 Measuring attributes

7.3.2 Commonly Used IoT Sensors in Vehicular Tracking Systems

IoT sensors are the main part of IoT technology. Microcontrollers control the data transmission between the IoT sensors and other actuators. According to the results in Fig. 7.2, 15% of the experiments were interested in using GPS sensors for the tracking systems and the use of these sensor types is significant. RFID technology and satellites were popular with 8% of the users. However, a negligible proportion of studies were used as a temperature sensor, *Wi-Fi* modules, Node MCU, Drones, and proximity cards.

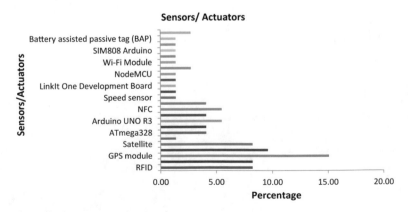

Fig. 7.2 Types of sensors and actuators

The GPS module has many plus points such as saving the end cost, improving the tracking skill, and mapping. These can act as security systems against the vehicle theft by sending information about the location of vehicles. These reasons might be influenced to enhance the significant use of GPS module than the other sensors for the tracking systems.

7.3.3   Data Transfer Methods Between Sensors and Actuators

The analyzed results of data transferring methods are shown in Fig. 7.3. The Short Message Service (SMS) was the most significant, effective, and efficient data transmission method for vehicle tracking systems and it was popular with 47% of the participants. Also, 17% of the studies were concerned with radio frequency signals to transfer data. However, all other data transferring methodologies that we have studied in this review were popular with 6% of the authors.

When considering the data transfer between a sensor and actuator, SMS data transferring was the most significant out of all the other data transferring methods. The reason may be that SMS data transfer can eliminate the need for Internet connectivity and this type of system only requires location services and telephony connectivity which is cost-effective and user-friendly. In particular, the SMS transferring system can eliminate the requirement of the Internet connection during the operation which keeps functioning the tracking system even in the rural areas.

7.3.4   Network and Protocols Utilized in Communication Methods

The analyzed data showed that significant percentage (12%) of the studies used Wi-Fi as a network for this kind of tracking systems. Also, 8% of the studies were focused on GSM/GPRS. However, TCP/UDP protocols were identified as the most appropriate transport layer protocol as shown in Fig. 7.4, and it has reported usage of 8% out of all other contributors.

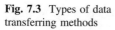

Fig. 7.3 Types of data transferring methods

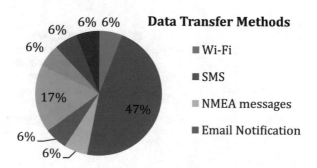

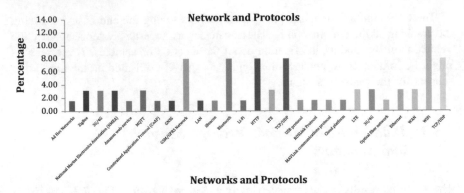

Fig. 7.4 Networks and protocols

The results revealed that Wi-Fi is the widespread network with a higher proportion of the authors. Wi-Fi allows the users to connect multiple-devices over one network. The wireless nature of this network facilitates the user to access their resources from any convenient location within the primary network environment. The mobility of the network and the productivity may be the other useful advantages of a Wi-Fi network. Wi-Fi users can maintain constant affiliation with the desired network even in different locations. The network deployment and expandability and the low cost of the setup are also identified as main advantages of a Wi-Fi network. All the mentioned facts may influence the popularity of the Wi-Fi in tracking system as compared to other networks.

7.3.5 Data Storage Approaches

Referring to Fig. 7.5, the most striking result that has emerged from the data is that cloud storage for the tracking systems is reported by 50% studies. Furthermore, Web server (11%), MySQL database (11%), and Flash memory (11%) were identified as other important data storage methods.

Fig. 7.5 Data storage methods

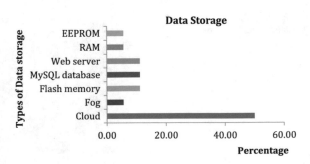

In the cloud data storage service, data can be maintained, managed, and backed up remotely. The online data storage facility allows the user to access the data from any location via the Internet. Also, the cloud storage allows the user to interchange the storing location cloud to local or vice versa. The bandwidth, cost-effectiveness, ability of disaster recovery and easy accessibility are the most common rewards of the cloud storage. Therefore, numerous vehicle tracking system providers use the cloud storage and this may be the main reason that 50% of the users for using cloud storage as opposed to other storage methods.

7.3.6   Use of Languages and Software Systems for Tracking

A significant proportion of the sample (13%) used Google Map/APIs as their programming language for tracking systems (Fig. 7.6). Android is the second popular language with a reported 12% of all the other programming languages used. However, use of Java, OS, Python, and Linux OS is insignificant.

The Google map/API is a reliable source that has easy documentation process and can handle numerous markers. Therefore, those facts may be the reasons for the considerable proportion of the scholars to use Google map/API for their experiment in tracking systems.

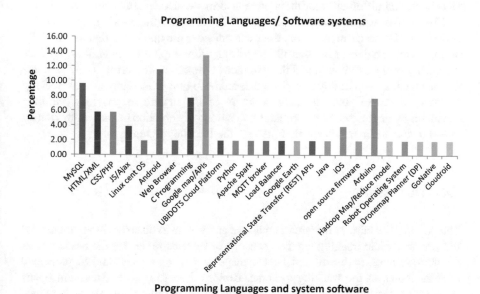

Fig. 7.6 Programming languages, systems, and software

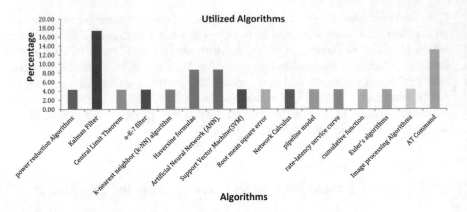

Fig. 7.7 Utilized algorithms

7.3.7   Algorithms Used in Vehicular Tracking Systems

The analyzed results in Fig. 7.7 show that Kalman filter was the most popular algorithm for tracking system with a reported 17% use out of all the other algorithms. A significant percentage (13%) of the authors used AT command algorithm for their studies. However, Network calculus, Cumulative functions, Euler's algorithm, and all other algorithms were also mentioned in some selected studies.

The statistical technique of Kalman filter explains the random structure of experimental measurements. This filter can consider the quantities that are partially or completely neglected in other filtering or algorithms. The Kalman filter provides the quality of the estimation and the variance of the estimation error. Therefore, this filtering technique is highly appropriate in online digital processing such as tracking systems. Moreover, the recursive structure of the technique provides real-time execution by neglecting past estimates and storing observations. Thus, authors may intend to use Kalman filter techniques for the tracking systems.

7.4   Discussion

The point of the current research and above analysis is to discover more about new IoT driven vehicle tracking devices, sensors, or systems to deliver the best solution for transportation authority and fleet management of a country; and to recognize relevant issues in the tracking systems. Besides, hopefully, the discussion opens avenues for information security specialists to further explore implications to reduce vulnerabilities of IoT tracking sensors and devices. With the development of the IoT vision, the transportation system analysts and field experts have expanded the tracking attributes of the vehicles using numerous IoT-driven sensors and actuators that have been executed. Some of these tracking systems are now being utilized successfully in many developed countries.

One of the main objectives of this study is to design energy-efficient vehicle tracking system; as there are different ways to provide power to the GPRS/GPS module. Some tracking systems utilize unnecessary large amount of power for storage and processing of unnecessary data sets [5]. There is also a limitation of GPS in some locations so it needs precise and much better tracking technique of locations when designing a tracking system. Therefore, this study also proposes the use of RSSI signal from the devices to compute the related distance between two objects while developing the assumptions of the position. Furthermore, precision of the RSSI signal functionality can be improved using the α–β–γ filter.

Kalman algorithm is an appropriate option to filter dynamic data by using more than one filtering algorithms for tracking systems [4, 6, 11, 12]. Therefore, the predicted results based on the Kalman algorithm have returned the most accurate values. Nevertheless, to support vector machines, artificial neural network algorithms can be suggested for improvement of the advance filtering methodologies [11]. The MQTT method is a scalable and demanding method for future solutions with more studies. The Lightweight transferring methods contribute to an efficient tracking system, as well. However, the MQTT has some drawbacks that can affect the centralized broker and this can, in turn, affect scalability. The MQTT operates over TCP and has "handshaking set up communication link" so that it increases a communication and wake-up time [13].

Controlling the vehicles' fuel system has lately become easy because of the IoT cloud platform. This is advantageous for vehicle owners to protect their vehicles from anti-theft situations. Mobile tracking application should be fast, secure, and accurate to control the movements of the vehicle remotely. However, there is a risk of the mobile application such that if the mobile phone is lost or becomes unavailable, the vehicle information can be reviewed by authorized personnel [4]. The system that develops tracking mechanism within itself may better ensure the safety of the user; as the individual tracking system monitors the geographical coordinates of the running vehicle and the immediate environment. As an example, Raspberry Pi based biometric authentication systems provide highly accurate and real-time identification functionality. The functionality of this type of system mainly depends on the installation point of the RFID reader. Precise installation of the RFID readers affords better communication and it should be the flexibility of selecting RFID readers for commercial purposes [3].

In order to analyze data, the cloud-based vehicle tracking system can monitor the moving vehicle in real time while updating the database with a very short time interval frequency [3, 6]. A few studies have used Arduino UNO R3; however, the major drawback of the Arduino UNO R3 is the limited number of digital inputs/output ports. This limitation has been addressed by introducing Arduino Mega 2560 because it has many inputs/outputs. So, this option can be utilized for tracking systems according to such purposes. The ESP8266-01 is a common and cost-effective Wi-Fi enabled chip; however, in real time, there may occur traffic overloading and the system is thus unable to send and receive the requests.

Tracking systems can also be organized by using NodeMCU with ESP8266. This option overwhelms the issues in ESP8266-01 and facilitates faster and accurate satisfaction percentage by using lightweight protocols just like MQTT [7].

Most developing countries require infrastructure facilities to reduce the road traffic. The IoT-based tracking systems will open new opportunities to facilitate more accurate, qualitative, and efficient performances to reduce the amount of traffic. These systems are also responsible for monitoring the real-time velocity of the vehicles and pass the messages to the relevant drivers by asking to reduce vehicle speeds, as situations dictate. RFID passive tags are acceptable technology for these kinds of systems. These findings of our review of literature mainly provides insights into RFID passive tags, NodeMCU with ESP8266 and associated technologies in vehicle tracking systems to be further investigated in future studies.

Considering our research findings and facts, it can be concluded that there is seamless connectivity between users, cloud, and IoT-driven tracking systems and there is reliability with respect to tracking of moving objects (i.e., vehicles). However, this study identifies numerous limitations regarding communication between moving objects. When the communication distances between two objects are less than 1-m, then there is influence to achieve the optimum speed, acceleration, network delay, processing time in the cloud, and responsiveness to the tracking system. Implication and governance of proper rules and regulations for the use of any tracking devices may help to improve the relevant technology in future. Some studies have introduced user friendly mobile application. However, those applications have limited functionality for the users. Developing a user-friendly environment can improve the commercial use of this technology. However, overall conclusion of the study emphasizes that more experiments should be conducted to enhance the quality of the IoT-driven tracking systems.

Some authors have explained that smart transportation systems must be incorporated with the Mobile Edge Computing and they demonstrated the drawbacks of the vehicle tracking system based on the traditional cloud platform. Furthermore, Mobile Edge Computing has more computing resources with additional storage capacity that is closer to end users and vehicles [33, 34]. Sophisticated vehicle tracking systems are highly susceptible to the processing in real time, qualitative computing experiences and the latencies [35]. Alternatively, they explained delays and unstable connection of the remote clouds present certain drawbacks over the Mobile Edge Computing for the IoT-based vehicle tracking systems.

Sasaki et al. [36] have proposed a system to control vehicles by using an infrastructure that allocates resources dynamically. Similarly, authors in [37] developed mobility tracking architecture for vehicles with complex computations that located the devices at the edge of the computation space. Typically, computation offloads the Mobile Edge Computing initially intended to offload tasks on demand and therefore, it enables to offload the tasks to Edge Computing Devices while lowering latencies and consumption of the energy [33, 37].

In the last few years, IoT-based vehicle tracking systems have achieved certain milestones and vehicles are beginning to incorporate automated driving techniques to ensure good communication between vehicles. IoT-based vehicle tracking system is not only a measure of the location, it can also be used to secure the passengers and drivers while using the vehicles. Virtual vehicle hijacking and vehicle data invading have also become common issues in the society [38]. However, use of vehicle tracking sensors is helping to avoid or at least reduce false information transmission [39]. Consequently, implementing a trust judgment method and dual authentication schemes are helping to minimize the security related issues in the tracking systems [40]. Some studies have optimized multi-objective computing offloading tasks from vehicles using Edge Computing Devices in the Internet of connected vehicles' environment. This will support to conserve the energy and reduce the transmission delays [41, 42]. Machine learning and artificial intelligence technologies promise to enhance every aspect of cyber-physical systems (CPS), the link between the physical world and the internet [43–45], such as planning, maintenance, and optimization. CPSs can continuously enhance based on knowledge and adaptation to varying situations.

Considering all research findings and facts, it can be concluded that there is seamless connectivity between users, cloud, and IoT-driven systems to track the moving objects. However, this study has identified numerous limitations relating to moving objects. When the communication distances between two objects are less than 1-m, it influences to the achievement of optimum speed, acceleration, network delay, transmission time to the cloud, and responsiveness of the tracking system. Implication of governance rules and regulations for the use of tracking devices may help to improve the polite and liable technology in future. Some studies have suggested the use of user-friendly mobile applications. However, those applications have limited functionality for the users. Developing a user-friendly environment can improve the commercial use of this technology. The overall conclusion of the study has emphasized that more experiments should be conducted to enhance the quality of the IoT-driven tracking systems.

7.5   Conclusion

This chapter has provided a survey of the most recent IoT devices, sensors, and actuators used in smart vehicle tracking systems. The authors have studied 30 peer-reviewed papers published during 2012–2018 and listed the attributes of these during the information gathering process. Also, the favorable circumstances and difficulties of utilizing such smart vehicle tracking devices were examined along with the proposed solution by the authors. The analysis of the collected data has demonstrated that a greater part of the IoT devices and systems had been utilized to distinguish the security effect and anti-theft issues in vehicle tracking systems. Tracking of lost vehicles, position of school buses, public vehicles, passengers, and vehicle number plates are just a few common examples of that. Despite the tracking for security purpose, the results revealed that location, speed, and time are the most significant sensing attribute utilized in IoT vehicle tracking devices. However, most of the tracking systems are functioning with GPS sensors and GPS/GSM sensors.

Future research can provide careful consideration toward designing IoT-based sensors, actuators, and microcontrollers which have more and better features relating to tracking and environmental conditions. Finally, this study gives a reasonable view on the road facilities research and development authority, communication technologist, and field specialist regarding the subsequent stages that can be considered designing the newer versions of the IoT-based smart vehicle tracking systems.

Author Contribution and Acknowledgements U.F and M.N.H conceived the idea of the study and developed the analysis plan. U.F analyzed the data and developed the initial paper. M.N.H helped to prepare the figures and tables and finalized the manuscript. R.S completed the final editing of the manuscript. All authors read the finalized manuscript.

References

1. Bandyopadhyay D, Sen J (2011) Internet of things: applications and challenges in technology and standardization. Wirel Pers Commun 58(1):49–69
2. Piyare R, Lee SR (2013) Towards internet of things (IoTs): integration of wireless sensor network to cloud services for data collection and sharing. arXiv preprint arXiv:1310.2095
3. Raj JT, Sankar J (2017) IoT based smart school bus monitoring and notification system. In: 2017 IEEE region 10 humanitarian technology conference (R10-HTC). IEEE, pp 89–92, Dec 2017
4. Uddin MS, Ahmed MM, Alam JB, Islam M (2017) Smart anti-theft vehicle tracking system for Bangladesh based on internet of things. In: 2017 4th international conference on advances in electrical engineering (ICAEE). IEEE, pp 624–628, Sept 2017
5. Almishari S, Ababtein N, Dash P, Naik K (2017) An energy efficient real-time vehicle tracking system. In: 2017 IEEE Pacific rim conference on communications, computers and signal processing (PACRIM), Victoria, BC, Canada
6. Lee S, Tewolde G, Kwon J (2014) Design and implementation of vehicle tracking system using GPS/GSM/GPRS technology and smartphone application. In: 2014 IEEE world forum on internet of things (WF-IoT). IEEE, pp 353–358, Mar 2014
7. Desai M, Phadke A (2017) Internet of things based vehicle monitoring system. In: 2017 fourteenth international conference on wireless and optical communications networks (WOCN). IEEE, pp 1–3, Feb 2017
8. Charoenporn T, Sunate T, Pianprasit P, Kesphanich S, Bunpeng A, On-uean A (2016) Selection model for communication performance of the bus tracking system. In: 2016 international computer science and engineering conference (ICSEC). IEEE, pp 1–5, Dec 2016
9. Koubaa A, Qureshi B (2018) Dronetrack: cloud-based real-time object tracking using unmanned aerial vehicles over the internet. IEEE Access 6:13810–13824
10. Pendor RB, Tasgaonkar PP (2016) An IoT framework for intelligent vehicle monitoring system. In: 2016 international conference on communication and signal processing (ICCSP). IEEE, pp 1694–1696, Apr 2016
11. Jisha RC, Jyothindranath A, Kumary LS (2017) IoT based school bus tracking and arrival time prediction. In: 2017 international conference on advances in computing, communications and informatics (ICACCI). IEEE, pp 509–514, Sept 2017
12. Chang TH, Lee DH, Lin CY, Hao S, Lin SC (2018) A practical dynamic positioning and tracking on the vehicular ad-hoc network. In: 2018 IEEE 4th world forum on internet of things (WF-IoT). IEEE, pp 338–341, Feb 2018

13. Lohokare J, Dani R, Sontakke S, Adhao R (2017) Scalable tracking system for public buses using IoT technologies. In: 2017 international conference on emerging trends & innovation in ICT (ICEI). IEEE, pp 104–109, Feb 2017
14. Mangla N, Sivananda G, Kashyap A (2017) A GPS-GSM predicated vehicle tracking system, monitored in a mobile app based on Google Maps. In: 2017 international conference on energy, communication, data analytics and soft computing (ICECDS). IEEE, pp 2916–2919, Aug 2017
15. Weimin L, Aiyun Z, Hongwei L, Menglin Q, Ruoqi W (2012) Dangerous goods dynamic monitoring and controlling system based on IOT and RFID. In: 2012 24th Chinese control and decision conference (CCDC). IEEE, pp 4171–4175, May 2012
16. Anusha A, Ahmed SM (2017) Vehicle tracking and monitoring system to enhance the safety and security driving using IOT. In: 2017 international conference on recent trends in electrical, electronics and computing technologies (ICRTEECT). IEEE, pp 49–53, July 2017
17. Rahman TA, Rahim SKA (2013) RFID vehicle plate number (e-plate) for tracking and management system. In: 2013 international conference on parallel and distributed systems. IEEE, pp 611–616, Dec 2013
18. Tang H, Shi J, Lei K (2016) A smart low-consumption IoT framework for location tracking and its real application. In: 2016 6th international conference on electronics information and emergency communication (ICEIEC). IEEE, pp 306–309, June 2016
19. Pratama AYN, Zainudin A, Yuliana M (2017) Implementation of IoT-based passengers monitoring for smart school application. In: 2017 international electronics symposium on engineering technology and applications (IES-ETA). IEEE, pp 33–38, Sept 2017
20. Balbin JR, Garcia RG, Latina MAE, Allanigue MAS, Ammen JKD, Bague AVB, Jimenez JM (2017) Vehicle door latch with tracking and alert system using global positioning system technology and IoT based hardware control for visibility and security of assets. In: 2017 IEEE 9th international conference on humanoid, nanotechnology, information technology, communication and control, environment and management (HNICEM). IEEE, pp 1–5, Dec 2017
21. Yu M, Zhang D, Cheng Y, Wang M (2011) An RFID electronic tag based automatic vehicle identification system for traffic IOT applications. In: 2011 Chinese control and decision conference (CCDC). IEEE, pp 4192–4197, May 2011
22. Bojan TM, Kumar UR, Bojan VM (2014) Designing vehicle tracking system-an open source approach. In: 2014 IEEE international conference on vehicular electronics and safety. IEEE, pp 135–140, Dec 2014
23. Bojan TM, Kumar UR, Bojan VM (2014) An internet of things based intelligent transportation system. In: 2014 IEEE international conference on vehicular electronics and safety. IEEE, pp 174–179, Dec 2014
24. Zambada J, Quintero R, Isijara R, Galeana R, Santillan L (2015) An IoT based scholar bus monitoring system. In: 2015 IEEE first international smart cities conference (ISC2). IEEE, pp 1–6, Oct 2015
25. Özbaysar E, Borandağ E (2018) Vehicle plate tracking system. In: 2018 26th signal processing and communications applications conference (SIU). IEEE, pp 1–4, May 2018
26. Khan MA, Khan SF (2018) IoT based framework for Vehicle Over-speed detection. In: 2018 1st international conference on computer applications & information security (ICCAIS). IEEE, pp 1–4, Apr 2018
27. Chellaswamy C, Rahul C, Kumar P, Santhanaraman S (2017) IoT based rail track joints monitoring system using cloud computing technology. In: 2017 2nd international conference on computational systems and information technology for sustainable solution (CSITSS). IEEE, pp 1–6, Dec 2017
28. Sutar SH, Koul R, Suryavanshi R (2016) Integration of smart phone and IOT for development of smart public transportation system. In: 2016 international conference on internet of things and applications (IOTA). IEEE, pp 73–78, Jan 2016

29. Chellaswamy C, Dhanalakshmi A, Chinnammal V, Malarvizhi C (2017) An IoT-based frontal collision avoidance system for railways. In: 2017 IEEE international conference on power, control, signals and instrumentation engineering (ICPCSI). IEEE, pp 1082–1087, Sept 2017

30. Seiber C, Nowlin D, Landowski B, Tolentino ME (2018) Tracking hazardous aerial plumes using IoT-enabled drone swarms. In: 2018 IEEE 4th world forum on internet of things (WF-IoT). IEEE, pp 377–382, Feb 2018

31. Srinivasan R, Sharmili A, Saravanan S, Jayaprakash D (2016) Smart vehicles with everything. In: 2016 2nd international conference on contemporary computing and informatics (IC3I). IEEE, pp 400–403, Dec 2016

32. da Costa TMF, da Silva VT, dos Santos GL, Duarte Filho NL, da Costa Botelho SS, de Oliveira VM (2017) Hotlog: an IoT-based embedded system for intelligent tracking in shipyards. In: IECON 2017-43rd annual conference of the IEEE industrial electronics society. IEEE.8. Additional Readings, pp 3455–3459, Oct 2017

33. Hong K, Lillethun D, Ramachandran U, Ottenwälder B, Koldehofe B (2013) Mobile fog: a programming model for large-scale applications on the internet of things. In: Proceedings of the second ACM SIGCOMM workshop on Mobile cloud computing. ACM, pp 15–20, Aug 2013

34. Hu Q, Wu C, Zhao X, Chen X, Ji Y, Yoshinaga T (2017) Vehicular multi-access edge computing with licensed sub-6 GHz, IEEE 802.11 p and mmWave. IEEE Access 6:1995–2004

35. Sherly J, Somasundareswari D (2015) Internet of things based smart transportation systems. Int Res J Eng Technol 2(7):1207–1210

36. Lee S, Tewolde G, Kwon J (2014) Design and implementation of vehicle tracking system using GPS/GSM/GPRS technology and smartphone application. In: 2014 IEEE world forum on internet of things (WF-IoT). IEEE, pp 353–358, Mar 2014

37. Blackman S, Popoli R (1999) Design and analysis of modern tracking systems (Artech House Radar Library). Artech House

38. Mach P, Becvar Z (2017) Mobile edge computing: a survey on architecture and computation offloading. IEEE Commun Surv Tutor 19(3):1628–1656

39. An K, Chiu YC, Hu X, Chen X (2017) A network partitioning algorithmic approach for macroscopic fundamental diagram-based hierarchical traffic network management. IEEE Trans Intell Transp Syst 19(4):1130–1139

40. Zhang S, Choo KKR, Liu Q, Wang G (2018) Enhancing privacy through uniform grid and caching in location-based services. Future Gener Comput Syst 86:881–892

41. Wang X, Xiong C, Pei Q, Qu Y (2018) Expression preserved face privacy protection based on multi-mode discriminant analysis. CMC: Comput Mater Contin 57(1):107–121

42. Premsankar G, Ghaddar B, Di Francesco M, Verago R (2018) Efficient placement of edge computing devices for vehicular applications in smart cities. In: NOMS 2018–2018 IEEE/ IFIP network operations and management symposium. IEEE, pp 1–9, Apr 2018

43. Gupta A, Mohammad A, Syed A, Halgamuge MN (2016) A comparative study of classification algorithms using data mining: crime and accidents in Denver City the USA. Education 7(7):374–381

44. Wanigasooriya C, Halgamuge MN, Mohamad A (2017) The analyzes of anticancer drug sensitivity of lung cancer cell lines by using machine learning clustering techniques. Int J Adv Comput Sci Appl (IJACSA) 8(9)

45. Singh A, Lakshmiganthan R (2017) Impact of different data types on classifier performance of random forest, Naive Bayes, and k-nearest neighbours algorithms

Chapter 8
REView: A Unified Telemetry Platform for Electric Vehicles and Charging Infrastructure

Kai Li Lim, Stuart Speidel and Thomas Bräunl

Abstract Charging stations networks and connected vehicles play a pivotal role in the advent of smart cities and smart grids. A cornerstone of these infrastructures is often a platform or a service that handles the copious amounts of data generated, processes and saves it for monitoring and analyses purposes. In this contribution, we present a software platform, that we named as REView, that automatically collects, analyses, and reviews live and recorded data from electric vehicles (EVs) as well as from EV supply equipment (EVSE or "charging stations"). It also provides a unified monitoring platform for infrastructures that are both modular and scalable. For analysis purposes, the data described in this chapter has been collected from the Western Australian Electric Vehicle Trial and the WA Charging Station Trial. A secure web portal was also designed with different viewing perspectives for electric vehicle users, charging station users and charging station operators. REView includes presentation of informative statistics about a user's driving efficiency and the energy use of an EV; and then compares the collected data with the average of all other users' similar data. It further includes a smartphone application for live monitoring and producing itemized billing. In this chapter, we discuss the development of REView, including mechanisms to generate and collect the information. Finally, we show and discuss various aspects of the visualized data itself, including charging time, charging duration, energy used, as well as utilization metrics of the charging infrastructure. We promote an open source approach to charging station software development. Our work also illustrates a single-software backend to handle multiple stations from different manufacturers, promoting competition and streamlining the integration of charging technologies into other devices. The results obtained from this network and platform have ultimately

K. L. Lim (✉) · S. Speidel · T. Bräunl
The Renewable Energy Vehicle Project (REV), Department of Electrical,
Electronic and Computer Engineering, The University of Western Australia, Perth, Australia
e-mail: kaili.lim@uwa.edu.au

S. Speidel
e-mail: sjmspeidel@gmail.com

T. Bräunl
e-mail: thomas.braunl@uwa.edu.au

© Springer Nature Switzerland AG 2020
Z. Mahmood (ed.), *Connected Vehicles in the Internet of Things*,
https://doi.org/10.1007/978-3-030-36167-9_8

167

enabled us to perform quantitative investigations towards the driving and charging behaviours, as well as the overall electric vehicle trends around Perth in Australia.

Keywords Electric vehicle · Charging station · Tracking · Monitoring · Web portal · Charging statistics · Internet of Vehicles · Data visualisation · Data analytics · Analytics infrastructrue/architecture

8.1 Introduction

The REV Project at The University of Western Australia (UWA) operates one of the largest Electric Vehicle (EV) charging station networks in Western Australia, where it is also the largest network operated by an academic institution in the country. This network includes 23 units of 7 kW AC chargers (Level 2) and one 50 kW DC fast charger (Level 3). All of the energy generated for on-campus charging is provided through a 20 kW-peak solar photovoltaic (PV) array. Additionally, our successful run of the Western Australian Electric Vehicle Trial (from 2011 to 2014) has enabled us to establish a vehicle fleet-tracking platform for EVs. Being able to collect individual user data for EVs and data relating to EV Supply Equipment (EVSE), and relay them back to the user in both itemized and statistical form, provides a substantial added value for individual mobility.

Our telemetry platform, named as REView, is a web-based software package that receives, processes and stores incoming telemetry data from connected infrastructures and more importantly, it facilitates data sharing from the various streams of incoming data, thereby enabling it to directly influence system automation without external intervention. This data is then visualized on REView's front end where it performs automatic statistical evaluations and presents results to users in a meaningful and informative way, letting each user know about his or her own individual mobility costs, as well as their ranking alongside other EV users of this software platform. We find that by illuminating the user's patterns and contrasting them with data from other users, we can motivate individuals to reduce their energy consumption and carbon emissions, as well as, in general, educate people about zero emission transportation. The approach of competing users against each other is known as gamification and has already been used in related areas [1].

The statistics generated allow drivers and fleet managers to monitor their vehicles' efficiency and use of energy. It allows charging network operators to monitor the effectiveness and usage of their stations, and it allows station users to monitor their energy usage and costs. This information is made available in several ways, including via live mobile phone web applications, desktop web applications, data exporting and printing.

REView was developed as part of two different projects:

1. The Western Australian Electric Vehicle Trial, Australia's first EV trial, consisting of eleven locally converted EVs based on Ford Focus cars owned by various businesses and government agencies [2].

2. The installation of the Western Australian charging station network, a set of 23 AC and one DC electric vehicle charging outlets, which are made available to the public.

REView has helped in analyzing driving and charging behaviours of EV drivers [3–5] and statistics generated from this system have been used in setting up an acceptance study among EV drivers [6]. It is currently providing real-time monitoring of vehicles, charging stations and solar installations around Western Australia, with new statistics generated every half hour. The software is presented as a full stack project to be entirely web based, utilizing a combination of Python servers, batch scripting of Python for statistical processing, PHP server side and PostgreSQL back end with JavaScript, CSS and HTML for the user interface. As these open source languages (along with several open source libraries) make up the system, it has the ability to be used extensively and be available to any educational or not-for-profit organizations. We hope this will promote further research into charging station infrastructures and show that it is possible to fill the void between a research organization's need for data collection and the government/corporate sponsors need for at least satisfactory investment return.

That said, the motivation for proposing REView mainly stems from the lack of availability of an open and scalable telemetry and data monitoring platform that comprehensively incorporates multiple sources of data across various infrastructures. Throughout the development of this software, there were several lessons learned from stakeholders' requirements, acceptance of features by station users, and general possibilities and options for charging station software. We believe that the features that are developed in this system help to form a baseline for future software developments in vehicle tracking and charging station monitoring. As many of these infrastructures become more integrated, telemetry platforms are required to become more centralized as an effective yet holistic solution to convey its information to the users. However, we have found that many existing solutions are exclusive, only supporting certain and often proprietary products and are usually impossible to integrate with other infrastructures. Through the EV Trial, we found that commercial software which is sold bundled with charging stations, often lacks vital functions and can be very awkward to operate. Also, in all cases we have seen, such software is limited to the associated company's charging hardware, and does not support interoperability with other stations. With many different charging station manufacturers in the market, this makes management, analysis and billing cumbersome.

The solution proposed by REView is to store and organize telemetry data. The significance of its contribution is the ability to handle, process, store and visualize this data effectively, which can occur on demand and in real time. The platform is scalable whereby it is able to process data across different manufacturers and infrastructures. Monetization options are also included in REView, as monthly bills are automatically generated for users, station operators and administrators to quantify the running cost and usage of both the charging stations and the vehicles. In addition to monitoring and visualizations, we also use this collected data to analyze charging behaviours and the local EV adoption rate. The combination of these features aims to present a centralized solution to inform and encourage EV market penetration.

We organize the remainder of this chapter as follows. Section 8.2 introduces the background into the development of the REView system. Section 8.3 presents the overview of the system design. Section 8.4 describes the telemetry design for charging infrastructures. Sections 8.5 and 8.6 describe the telemetry design for vehicle fleets and data management regarding energy generation. Section 8.7 features usage and billing from telemetry data, whereas Sect. 8.8 introduces REView mobile application. Section 8.9 evaluates data interpretation and gathered results. Finally, Sect. 8.10 draws the concluding remarks.

8.2 Background

In this section, we present the background research that resulted in the development of REView. These topics cover EV adoption, environmental impacts and telematic platforms, which are elaborated in the following subsections.

8.2.1 Adoption of Electric Vehicles and Charging Stations

When starting the Western Australian (WA) EV Trial and the charging station trial in 2010, there were no OEM-built (commercially built) EVs available in Australia and there were no charging stations in Western Australia. Since then, several manufacturers have released new electric vehicles and plug-in hybrid vehicles into the Australian market, including Mitsubishi, Nissan, Hyundai, Holden, Renault, Jaguar, BMW, Porsche and Tesla Motors. Globally, the number of Electric Vehicles has grown from 700,000 in 2014 to over three million in 2017 [7]. Electric vehicles are no longer a pipe dream, but a reality, and every year we will see more on our roads. For fleet managers, tracking and logging of energy usage, as well as localization and utilization, are valuable tools to reduce carbon emissions and expenses.

Meanwhile, the number of charging station manufacturers has increased significantly, and many governments around the world are subsidizing their installation to meet the desire to reduce dependence on oil. While charging station manufacturers currently have incompatible customer identification methods and competing management software, the Open Charge Point Protocol (OCPP) [8] has been introduced as a possible new standard for EVSE communication. OCPP is an open, uniform communications protocol that can be used across all charging stations. Already, many manufacturers are supporting this protocol—now being the most popular protocol for new stations. This means that external companies can access the data and control of the stations via an application program interface (API), no matter which manufacturer.

The number of EVs in Perth, Western Australia, has grown exponentially from 15 in 2010 to over 700 in 2018. Various consulting firms and governments have forecast the growth of EV sales in Australia. In 2009 the Department of Environment and Climate Change commissioned the consulting firm AECOM to study the economic viability of electric vehicles [9] and they projected that supply constraints would limit the sales of EVs (including hybrids) in Australia until 2020; and in each of their three projections over 60% of new cars would be plug-in hybrids or pure EVs by 2040. AECOM released another report in 2011 for the Victorian Department of Transport where, through the use of a vehicle choice model, they concluded that sales of mild hybrid vehicles in Victoria will be more predominant in the short term, (up to five years), plug in EVs in the medium term (five to 10 years) and pure battery EVs in the long term (more than 15 years) [10]. They also found evidence that high levels of charging infrastructure will significantly increase the adoption of EVs.

In 2012, ABMARC performed a survey of motorists in Australia with a conservative estimate of EV uptake. They concluded that, without a breakthrough in battery technology, the adoption of EVs by 2020 would likely be 0.4% of new car sales [11]. However, plug-in hybrid electric vehicles would constitute a much large proportion of 6.4% of the new vehicle market.

The Energy Supply Association of Australia reviewed several different forecasts for Australia, showing that they all had several factors in common that controlled EV uptake, with a major factor being available EV charging infrastructure [12].

8.2.2   Measuring the Environmental Impact

The environmental impact of running any vehicle needs to be analyzed from its source. The environmental benefit in terms of CO_2 emissions of EVs relies quite heavily on the way the electricity is generated. In 2018, 21% of all energy generated in Australia came from renewable sources. This is made up as 35.2% from hydro, 33.5% from wind, 19.6% from small-scale solar PV, 7.1% from bioenergy, 3.9% from large-scale solar PV, and 0.8% from medium-scale solar PV [13].

The Union of Concerned Scientists released a report in 2012 in the US that related the source of electricity generation directly to the environmental benefit of a Nissan Leaf EV [14]. Their report showed that, in some regions, the difference in carbon emissions in electricity generation varied as much as three times, where a nuclear and renewable energy mix of generation is compared to a coal and gas driven power generation. This meant that EVs in areas with high electricity emissions were comparable to highly efficient petrol vehicles (17% of all Americans live in these areas).

A report by SMEC in 2008 for the Department of Transport states that in Western Australia, the amount of $kgCO_2$ Emissions per kWh is 0.936 [15]. From

Table 8.1 Efficiency and theoretical emissions of electric vehicles

Model	Efficiency (Wh/km)	Range (km)	CO_2 (g/km)
Hyundai Ioniq EV	115	280	104
Hyundai Kona EV	131	557	118
Renault ZOE	133	403	120
BMW i3	137	335	123
Tesla Model S (70)	185	455	166
Tesla Model X (60)	208	363	187
Jaguar I-Pace	230	446	207

this information and the efficiencies of the models from the Australian Department of Infrastructure and Regional Development [16], one could calculate theoretical CO_2 emissions per km of the EVs available in WA. Table 8.1 lists such information according to the EVs that are currently available for purchase in Australia. However, this would assume that these cars are charged entirely from the average grid without any renewables, which is clearly not the case. Note that, in Table 8.1, Model, kWh/km and range information is taken from the greenvehicleguide.gov.au website [17]; CO_2 emission in the 4th column is calculated from SMEC 2008.

Many early EV adopters also have solar PV generation at home and are able to charge their cars completely emission free. Also, just assuming the local energy mix for EV charging is too simplistic, as it does not take into account the time of charging. It should be noted that:

- Public charging stations are mostly used during sunshine hours with a usage pattern quite similar to solar PV. This means that EV charging can make use of excess solar energy during the middle of the day
- Home charging can be shifted to convenient hours of the night after the evening peak. At these times, EV charging can make use of excess wind energy and use the otherwise wasted baseload energy of coal-fired power stations, which run through the night.

Furthermore, the focus on CO_2 values ignores more harmful emissions, such as carbon monoxide and particulate matter (PM10 and PM2.5), which can be better controlled in power stations than in combustion engine cars.

In 2013, the Australian National Transport Commission released a report discussing the carbon dioxide emissions of new Australian vehicles [18]. They found that the average gCO_2 per km was 199 g/km, meaning that EVs in the worst-case scenario generate less emissions than the average new petrol car. To reduce or remove CO_2 emissions for EVs, they must be charged (or arguably offset) from a renewable energy resource.

It is important to note that air quality in metropolitan areas will improve through the use of EVs, even when charged from a "dirty grid". EVs produce zero local emissions and power stations are typically located in less populated areas outside a city.

8.2.3 Telemetry Platforms and Networks

While we note the scarcity of a cohesive solution to telemetry monitoring for public charging stations, vehicles and solar PV systems, exclusive solutions for these systems are well documented and established. Charging station networks have since relied on the expanding EV market for their own expansions in installation base and coverage area. Commercial global networks such as Chargepoint [19] and Tesla's Superchargers [20] have tens of thousands of installations to track usage and billing. Additionally, governmental networks such as China's GB-T aims to achieve more than 200,000 installations nationwide by the end of 2018 [21]. Locally in Australia, networks such as ChargeStar [22] operate a system where stations owners can participate as part of their network to bill customers. Billings from these networks are often outsourced to external companies such as Go Electric Stations [23].

Telemetry for connected vehicles became popular when automotive manufacturers installed CAN buses onto their products, thereby allowing third-party access to vehicle data. By connecting cellular-enabled GPS tracking devices to a fleet of vehicles, a connected fleet telematics system (FTS) can thus be established to monitor these vehicles. These systems are often controlled through an application service provider (ASP) such as OpenGTS [24], Traccar [25] and GPSGate [26].

There are several examples of commercial GPS tracking software packages for fleet vehicles. Commercial vehicle tracking is used in many different industries including mining, trades, utilities, transportation, and government agencies. There are a number of products available in Australia, such as EZY2C [27], Fleetmatics [28] and Linxio [29]. These systems claim to provide solutions that will reduce fuel costs, improve productivity, reduce labor costs, and increase accurate reporting. These services install GPS tracking devices into the fleet vehicles to monitor them remotely. The major drawback of such commercial systems is that they do not record energy usage or the status of charging, air conditioning, heating and headlights but rather exclusively rely on a GPS unit. Additionally, they do not include information from other devices, such as charging infrastructure. All these systems are aimed at the petrol fleet market.

Similarly, PV monitoring systems are also readily available. Many solar PV inverters such as SolarEdge [30] have Wi-Fi connectivity, which allows for remote monitoring on PCs and smartphones. Grid-connected PV systems can utilize the smart grid to manage solar feeds and power allocations. Locally, Synergy [31] offers such a system, which enables customers to monitor their PV system and solar feeds, as well as managing solar tariff returns.

In the academic scene, works that incorporate telemetry monitoring for connected vehicles [32–34], charging stations [35–37] and solar PVs [38, 39] are not uncommon. Many institutions around the world have started research into tracking and monitoring of electric vehicles and charging stations, using their own GPS systems and charging infrastructure. In the North East of England, Blythe performed a study, tracking 15 electric vehicles and a charging station network [40].

They concluded with stating their ability to use the data from tracked vehicles to derive the state of the charging station network. From there, they are able to predict possible future problem areas for electricity power generation.

While individual applications for charging network management and connected fleet or vehicles do exist, these are typically closed sourced and proprietary. Although, we may not refute the availability of proprietary or undisclosed software, at the time of writing, we are unable to find any work relevant to a telemetry monitoring system that combines vehicle tracking and charging station usage.

8.3 System Design: Overview

REView's overall architecture can be summarized as shown in Fig. 8.1. It consisting of seven components viz: an EV server, charging station server, a solar downloader, data processing scripts, a database, a web server and its web interface.

The charging stations and vehicles are all fitted with machine-to-machine (M2M) modems, which are perpetually connected to Telstra's 3G Universal Mobile Telecommunications System (UMTS) network on the 850 MHz frequency. These modems then connect to our local server to transmit data over the Transmission Control Protocol (TCP) across two dedicated ports: one for the charging stations

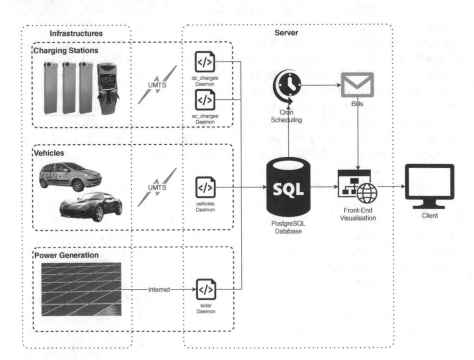

Fig. 8.1 Network architecture of REView

and another for the vehicles, as the charging stations and vehicles package data through different protocols. The server handles these transmissions through a *SocketServer* framework on Python 2.6.

The Python daemons that run the *SocketServer* function, receive and parse the incoming telemetry data from the charging stations and vehicles, before appending these data onto a PostgreSQL database. This process also verifies data integrity and consistency by filtering any duplicates and non-events. This PostgreSQL database is installed onto the same server, which stores all data relating to telemetry, charging stations, vehicles and users. The platform's Apache front end can then access this data for visualization and analysis.

The REView website features several pages including vehicle tracking, vehicle statistics, charging station status, charging station statistics, billing, heat maps, journey lists, charging lists, mobile tracking, and more. Depending on the type of user (station operator, station user or EV tracker), some pages are restricted or hidden. The website is a secure HTML 5 site with live information, interactive maps, graphs and customizable time scales. The supported browsers are Chrome, Microsoft Edge, Firefox and Safari, allowing access from computers, tablets and smartphones. At the time of writing, we are running PostgreSQL 8.4.20 and Apache 2.2.15 on a Red Hat Enterprise Linux (REHL) 6.10 server platform.

8.4   Charging Infrastructures

Our charging station network was established in 2011 as part of our research initiatives into the Western Australian EV landscape, which enables the collection of charging data to quantify charging trends and behaviours. This began with the installation of the Level 2 AC stations in the Perth metro area, followed by the installation of the 50 kW DC fast-charging station at UWA in 2014. We offer these charging stations free of charge to EV owners in return for research and data collection. The AC and DC stations transmit data through different protocols but are monitored for the same information. In other words, we collect information pertaining to charge times, duration and energy consumption from the stations for data visualization and modelling. The following subsections detail the functions of these charging stations.

8.4.1   DC Charging

The 50 kW DC fast-charging station at UWA is a Tritium Veefil-RT [41] (see Fig. 8.2), which supports charging over the CHAdeMO [42] and SAE Combined Charging System (CCS) Type 2 (IEC 62196-3) [43] standards. This station performs telemetry through the Open Charge Point Protocol (OCPP) 1.6 over a 3G UMTS network. Data from the station is first pushed onto Tritium's server before it is pushed back to our local server, this allows Tritium to collect and consolidate data from its charging stations, and to streamline their maintenance and support on the station. The simplified class diagram in Fig. 8.3 summarizes REView's functions on the DC station.

Fig. 8.2 UWA's Tritium Veefil-RT DC fast charging station

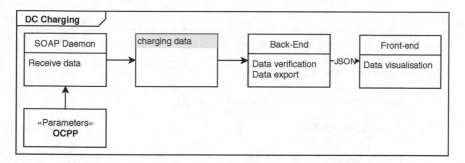

Fig. 8.3 Simplified class diagram of REView's DC station management

8.4.1.1   Communication Protocols

Open Charge Point Protocol (OCPP) data from Tritium's servers are transmitted through the Simple Object Access Protocol (SOAP). We run a PHP SOAP server with its service functionality described in a Web Service Description Language (WSDL) file, following the service list as described by Elaad.nl [44], to which we have ensured its compatibility with the Veefil charging stations. The charging station constantly transmits timestamped SOAP messages relating to heartbeats, start/stop charging events, status notifications, user authorization and energy meter values.

By referencing the WSDL file, our SOAP server receives and logs incoming data, subsequently appending it to the *dccharge* table. Each charging station user is assigned a unique identification (ID) for he charging vehicle, and each station broadcasts a unique station ID that corresponds to its installed location. For each charging event, the charging station measures energy consumption in Wh units, along with its start and end times. Finally, the station also broadcasts the charging standard used (CHAdeMO or CCS2) for each charging event.

From the series of SOAP messages, the server appends charge data following Algorithm 1.

Algorithm 1 DC Charges database append

Procedure dcappend (incoming header)

        open Connection to database and table *dccharge*

        read incoming header

        if header is *StartTransaction* **then**

                append to *dccharge* with values *TransactionID, UserID, StationID, StartTimestamp, StartUnits, ConnectorID*

        if header is *StopTransaction* **then**

                update *dccharge* with values *UserID, StopTransaction, StopUnits, TransactionData*, where same *TransactionID*

                update *status* = ChargeComplete

        close Connection to database

The combination of this data allows to perform time series analysis and modelling with regards to charge frequencies, duration, energy consumption and charge standard used. Additionally, we have instigated measures to maximize user participation through the station's proximity to the city center, while providing the service free of charge. These measures provide us with reliable data to model EV trends and fast-charging behaviours around Perth.

8.4.1.2   User Authentication

To improve data integrity, our DC station supports user authentication based on near-field communication (NFC). Users are given the flexibility to use any compatible NFC card in their possession, including those for public transportation, work or school. Figure 8.4 shows a non-exhaustive example of NFC cards that the station accepts. The charging station reads these cards for a four-byte unique identification number (UID), which is presented as eight hexadecimal numbers. The UID is then crosschecked with its database entry for charging authentication. In addition to the tag number, users are required to register with UWA for their names, email, vehicle model and registration number; whereby these fields are stored in a separate table in the database.

Fig. 8.4 RFID cards supported by the DC station: a Transperth Smart-Rider card, a UWA Student ID and a MIFARE RFID card used by station users

8.4.1.3   Data Visualization

Data visualization for DC stations is performed through a series of SQL queries which are parsed through a series of JSON encode/decode processes over PHP. REView visualizes such data in the form of time series graphs and pie charts, as well as a page delimited table with the option for a *.csv* file export.

A back-end PHP script performs data parsing by connecting to the PostgreSQL database, which we use to send SQL queries. For the visualization of charts, these arrays are appended with their headers, and then combined with chart parameters such as graph types, chart title, axes titles and legends; which are subsequently encoded into a JSON representation. For a table visualization, the JSON representation will also include the total number of charging events to enable the page delimiter to determine the number of pages for the table.

Likewise, a front-end PHP script *gets* the JSON representations from the back-end script and decodes it. This script interfaces with the web browser and is therefore also programmed with HTML and JavaScript as a webpage. Users of this webpage can specify visualization periods with a start and end date, which is then used as part of the SQL statement to generate the query. The front-end script then parses the decoded JSON representation to determine all chart parameters, and subsequently plots them in a table using the Google Visualization API [45] as illustrated in Fig. 8.5.

On the other hand, the table visualization front end, as illustrated in Fig. 8.6, draws an HTML table and fills it with the array obtained from a query that returns data from all DC charging events with the columns storing start and end times,

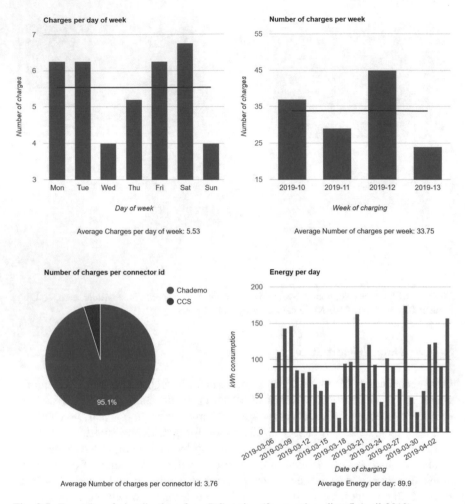

Fig. 8.5 Examples of visualizations for a DC station (for month ending 5 April 2019)

charge duration, energy consumption and type of connector used. Each page is limited to 50 entries as defined in the query. This page supports *.csv data exports, whereby the file will be downloaded with the same table headers.

8.4.2 AC Charging

Our AC charging station network consist of 11 dual outlets (see Fig. 8.7) and one single outlet Elektromotive Elektrobay charging stations, totaling 23 stations. These charging stations act as individual clients and connect to a central server node in a many-to-one configuration. Each charging station is powered through a 7 kW

Count	ID	Start Time	End Time	Duration	Energy (kWh)	Connector
1	13974	2019-04-05 12:34:04	2019-04-05 12:55:04	00:21:00	13.978	CHAdeMO
2	13973	2019-04-05 12:08:01	2019-04-05 12:34:00	00:25:59	16.715	CHAdeMO
3	13972	2019-04-05 02:59:00	2019-04-05 04:02:02	01:03:02	38.824	CHAdeMO
4	13971	2019-04-04 22:01:02	2019-04-04 22:27:04	00:26:02	4.069	CHAdeMO
5	13970	2019-04-04 16:59:01	2019-04-04 17:04:05	00:05:04	1.051	CHAdeMO
6	13969	2019-04-04 15:32:04	2019-04-04 15:58:00	00:25:56	17.593	CHAdeMO
7	13968	2019-04-04 14:36:05	2019-04-04 15:30:01	00:53:56	35.025	CHAdeMO
8	13967	2019-04-04 13:13:00	2019-04-04 14:10:05	00:57:05	41.922	CHAdeMO
9	13966	2019-04-04 12:01:03	2019-04-04 13:09:03	01:08:00	49.971	CHAdeMO
10	13965	2019-04-04 09:29:02	2019-04-04 09:41:02	00:12:00	7.606	CHAdeMO

Fig. 8.6 Screen grab for DC charges table on REView (showing last ten charges)

three-phase AC supply, either wall mounted or floor mounted. The charging outlets support the IEC 62196-2 (Type 2/Mennekes) standard, which is compatible with all recent EVs sold in Australia. The stations are water resistant and fitted with overcurrent protection and RCD switches. The functions of REView on our network are as summarized in the simplified class diagram in Fig. 8.8, and are elaborated in the respective subsections.

8.4.2.1 Communication Protocols

The AC station network achieves telemetry through the RS-232 serial standard, wherein for each station, a DE-9 cable connects to a Four-Faith F2414 [46] High-Speed Downlink Packet Access (HSDPA) M2M modem (see Fig. 8.9), transmitting charging data directly to our local server. We configured these modems using AT Commands, which instruct them to connect to our server through its IP address and port number, using TCP for data transmission. These modems establish a perpetual cellular connection upon power on, enabling them to transmit data from the stations on demand. Unlike the DC station, each AC station can be configured directly using the Elektromotive EB Connect software to set station parameters such as authorization control, energy metering, charge limits, charge event and energy tracking. It also gives the administrator the ability to remotely login to the station or disconnect a user or reset a station. To allow this software to connect to the station, the server can open a Secure Shell (SSH) tunnel between the station and the administrator PC. This can be used to either remotely configure the station's modem or the charging station itself.

Fig. 8.7 UWA's dual outlet
Elektromotive Elektrobay AC
charging station

Telemetry data from the AC charging stations follow Elektromotive's propri-etary protocol that transmits a series of concatenated, timestamped hexadecimal strings. In order to append the database, we run a Python-based data parsing script as a daemon that first splits the string into parts that correspond to the database variables, and then converts these parts into intelligible ASCII texts. This script logs all incoming telemetry data from the charging stations, which for every charging station, transmit at five-minute intervals. It also checks and sets the stations real time clock, and requests information from the stations internal database.

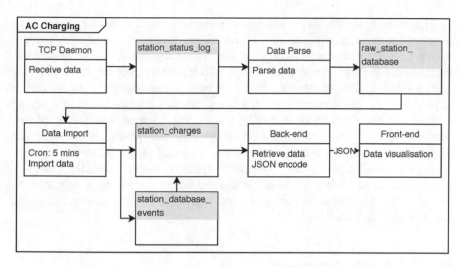

Fig. 8.8 Simplified class diagram for REView AC station management

Fig. 8.9 A Four-Faith F2414 M2M modem used in AC charging stations

The charging station modems are configured to connect to the server, allowing them to use dynamically allocated IP addresses, which are generally a cheaper option to using more convenient static IPs. To distinguish the charging stations, each modem is programmed to transmit its Modem ID as a header for all outgoing TCP messages. The Modem ID is a four-byte alphanumerical variable that is appended into every database entry to uniquely identify the station and its charging outlet (left/right side). In other words, a dual outlet station will consist of two modems, one for each charging outlet. The server then parses all messages and groups them according to their charging stations based on this Modem ID.

Each station keeps a record of several types of events, including charging, disconnect, power failure and reset. When the number of recorded events at the server is less than that at the station, the excess records are downloaded and stored for later statistical analysis.

8.4.2.2 Telemetry Parameters

In addition to the Modem ID, other relevant parameters are further detailed in the following paragraphs.

Clock

Each AC charging station keeps a clock which timestamps telemetry messages as they are produced. The stations' clock differs from the server's clock whereby the server's clock is used to timestamp the telemetry messages as they are received. This clock redundancy is particularly useful at times when the UMTS network is unreliable, in which case the charging station will store these messages in memory before transmitting them in series as soon as the UMTS network re-establishes. While the charging stations' clocks support over-the-air synchronization, this process occurs intermittently on demand, which implies that the server's clock, being consistently connected to the Internet, is more accurate. To accommodate this issue, the data parsing script checks for any time discrepancy between the server and the station, setting the message timestamp to the server's time if it is less than 12 h apart.

Station Attributes

A *flag* variable transmits station attributes through an array of 16 binary objects that correspond to various station statuses. These array elements, in fixed sequence are: "data erased", "station restarted", "no return timer activated", "no power drain", "excess power drain", "power failure", "door jammed", "door forced open", "record trip", "mains status", "door status", "charge time exceeded", "transaction start", "transaction end", "remote event", and "force end charge cycle". The values presented in this array are important for data collection and station diagnostics. For example, we use a combination of these station attribute flags to decipher station logs and append them as charging events into the database.

Station Status

The combination of short telemetry intervals with its station attributes enables REView to determine the status of each station, where they are classified as "In use", "Not in use" or "Unknown", and subsequently visualized on the front end for users to check on the station bays' occupancy. While the "in use" and "not in use" states can be ascertained from the telemetry logs, an "unknown" state is set for a station when more than 15 min have elapsed since its last telemetry message was received, considering that each station is programmed to transmit messages every five minutes. A station that is "in use" also displays its current charging time and allows registered users to track their charging status on their phones via a mobile website.

8.4.2.3 User Authentication

AC charging station users have been supplied with RFID tags for identification to allow monitoring and future billing. This also reduces the risk of cable theft, as only the correct tag can release the charging cable. Other identification methods used elsewhere include smartphone login, credit card swipe and in-vehicle identification, but these require higher security standards (in case of credit card readers) and constant Internet connection, which makes these methods more expensive.

User authentication on AC charging station network is established through Elektromotive-supplied RFID tokens (see Fig. 8.10). Each charging station is fitted with an RFID tag reader that conforms to the MIFARE DESFire [47] standard. These RFID tokens also communicate through the 13.56 MHz band and are compatible with the DC charging station. Users intending to charge their vehicles at charging station network are required to register their details and give consent to

Fig. 8.10 An RFID token used for AC charging station network (Pictured here is the tag for User ID REV182)

having their charging data collected for research purposes. We issue the
Electromotive RFID tokens to all registered users, and each RFID tag number
uniquely identifies the user through the storing of the tag's user ID.

8.4.2.4 Database

The database stores charging data from the charging station network in a normal-
ized manner across three redundant tables, as shown in Fig. 8.8, namely
raw_station_database, *station_database_events* and *station_charges*. This redundancy not only protects the data from accidental deletions, it also enables us to
experiment and modify filter and merging rules for the tables while preserving data
integrity.

For the longer than average charge duration of Level 2 stations (AC stations
where users leave their vehicle for charging while they go to work), we have set up
REView to feature per-hour and per-day-of-week data recordings for charging
duration and energy consumption. For instance, a size 24 array represents each hour
of the day; a size 7 array represents each day of the week. By segregating these
analyses into a time series, we can visualize and model the stations' utilization as
time progresses. The longer charge duration also implies that vehicles are left
plugged into the charging stations even after they are fully charged. These vehicles
are therefore subjected to a trickle charging state where their batteries are charged at
their self-discharging rate, where the charging stations are **maintaining** the charge
at an **idle** state. To differentiate and analyze this charge maintaining state, we
classify all charging instances that draw less than 1 kWh per hour as one that
maintains charge, and that of more than 1 kWh per hour as one that is actively
charging.

We present Algorithms 2 and 3 for the importation of data between tables, each
representing a different parsing script.

Algorithm 2 *station_status_log* to *raw_station_database*

Procedure acappend1 (incoming telemetry message)

> open Connection to database and table *station_status_log*
>
> read incoming *NumRecords, ModemID*
>
> fetch lastEntry from *station_status_log*
>
> **while** incoming.*NumRecords* < lastEntry.*NumRecords* **do**
>
>> *amount* = incoming.*NumRecords* – lastEntry.*NumRecords*
>>
>> fetch past *amount* record from *station_status_log*
>>
>> **for** each *amount* record **do**
>>
>> append into *raw_station_database* with values *StationID, Flags,*
>>
>> *ServerTime, StationTime, MeterReading, EnergyReading, Index,*
>>
>> *UserID, ModemID*
>
> close Connection to table *station_status_log*

Algorithm 3 raw_station_database to station_charges

 procedure ACAPPEND2(incoming header)

  open Connection to database and table *station_database_events*

  for each *ModemID* in *station_database_events* **do**

   lastEvent = max(*endtime*) from *station_database_events* at *ModemID*

   get first *event*

   for each *event* where *StartTime* < *lastEvent* **do**

    if not *stationRestarted* and *transactionStart* and *serverTime* < *currentTime*

    then

     if not *chargeStart* **then**

      chargeStart = *event*

     end if

     lastEvent = *event*

    end if

    if chargesStart and (transactionStart and transactionEnd) or

    (chargeRestart and not transactionStart) **then**

     kWh = *EndMeterReading* - *StartMeterReading*

     append into *station_database_events* with values *stationID*,

     StartTime, EndTime, kWh, UserID, rightSide, ModemID

    end if

   end for

   open Connection to table *station_charges*

   for each appended *event* **do**

    append into *station_charges*

   end for

  end for

  close Connection to all tables and database

 end procedure

We improve data integrity by setting charging limitations on the charging stations. These include a 12-h charge duration limit, and an energy threshold between 3 Wh and 14.4 kWh. The charge duration limit ensures that users who forget to tag off after a charge, or subject their vehicles to extended periods of trickle charge, are not accounted for data collection. Similarly, current limits protect the charging stations from overloading and the charging event is properly terminated as soon as the vehicle's battery is fully charged. On the database, we have programmed the data import script to filter for extremes and negative values. These filters ensure that:

- Energy values are between 0 and 200 kWh. The upper limit of 200 is, at the time of writing, higher than the battery capacity of any commercially available EV.
- Charging durations are between 0 and 12 h, to ensure that the charge duration limits are coherent with those set at the charging stations.

Any data entry that does not satisfy the filter boundaries are not imported into the *station_charges* table.

8.4.2.5   Data Visualization

Unlike the DC charging station's architecture, data from AC charging station network is stored entirely on the same local server and does not involve any communication between servers. Having all data eventually consolidated into a single table also enables us to condense, query and retrieve all data that is required for visualization into a single SQL query that combines data from charging events with its tag owner. The result is encoded as JSON representation that is subsequently sent to the front end that calls it.

Data visualization at the front end is similar to the DC charging station's whereby it generates a series of charts and a page delimited table. However, the added complexity of a charging station network, along with the increased variety of collectable data from it implies a differentiation from our descriptions in Sect. 8.4.1.3. The Google Visualization API is again used to render charts for the AC stations. The difference, however, is that data parsing is performed directly on a single SQL query result, as compared to sending multiple queries for the DC station. A JavaScript function is written to parse the decoded JSON representation and group them into multiple charts based on their headers. Users can select a specific time period for visualization periods, as well as specific charging stations. In addition to time series charts describing energy usage and charge duration, having per-station and per-user data also enables cost patterns and energy usage to be visualized for each user and station in the form of pie charts. An example of these visualizations is given as Fig. 8.11.

Furthermore, using this query approach along with a visualization period selection enables the front end to include a summary table for each charging station and its network. This table summarizes, for any selected period, its total energy consumption, estimated running cost (peak and off peak), number of charges, total plugged-in time, total charging and maintaining charge time, average charge duration, average charging and maintaining charge duration, percentage of time in use and power used for peak, shoulder and off-peak times. Each of the statistics generated is useful to charging station operators. The table shows the following factors:

- Which electricity plan is more useful, displaying the cost of a flat rate of electricity price (e.g. 21.87¢ per kWh) and a tiered rate (peak/shoulder/off-peak,

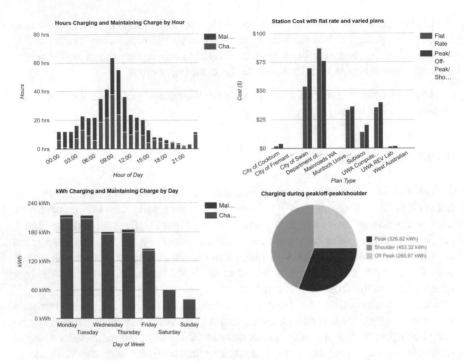

Fig. 8.11 Visualization examples for AC stations (for month ending 5 April 2019)

e.g. 42.15¢, 21.44¢ and 11.32¢ per kWh respectively), which charges more during peak times and less during off-peak periods.

- Time spent charging the vehicle (drawing more than 1 kW) and the time spent plugged in and not charging as a percentage. This shows if a station is more used for charging or if a location is more used as a parking spot.
- Time spent on a transaction (how long the vehicle is plugged in on average).
- Amount of time the station is actually in use versus its total time installed. Showing how often the stations are utilized as an average over all locations.

8.5 Vehicle Monitoring

The electric vehicle tracking functionality of REView was first established and presented as part of The REV Project's WA Electric Vehicle trial in 2011, whereby a fleet of 11 electric-converted Ford Focus cars were fitted with mobile GPS tracking units, which track the vehicles' movement patterns, battery level, charging, headlight, heater, air conditioning and ignition status. Approximately 2.7 million data points were collected over the trial period, which includes 5,000 independent journeys averaging 9.3 km per trip and 1,600 charges. Since then, our vehicle tracking platform has expanded to include our electric Jet Ski, electric boat and autonomous driving projects. The incorporating of EV tracking onto REView enables us to better analyze usage behaviours of EVs and to model future commute habits. Results stemming from our vehicle tracking system are published in [5]. Vehicle tracking on REView is performed according to the class diagram in Fig. 8.12, and is detailed in the subsequent subsections.

8.5.1 Communication Protocols

Our fleet network was initially fitted with Astra Telematics' AT110 [48] tracking devices, which were then upgraded to the AT240 (version 8.5) [49] to support the 3G UMTS network. These tracking devices communicate with the server via an internal UMTS modem, using a SIM card with machine-to-machine (M2M) capabilities.

The AT240 device, as shown in Fig. 8.13, is capable of vehicle tracking through a Global Navigation Satellite System (GNSS) over the *ublox* EVA-M8M module [50], and is IP67 waterproof rated, making it suitable for our watercraft projects. Each tracking device is based on a Cortex M3 microcontroller [51], which is powered by a 510 mAh battery that is able to sustain three days of hourly updates, and a three-axis accelerometer to detect motion and driving behaviour. Input/output options include six digital inputs and five digital outputs, as well as two inputs connected to an analogue-to-digital converter.

The five digital input lines of the tracking devices are connected to the air conditioning, ignition, headlights, radio and, heater statuses. The analogue input line is connected to a battery level logging device that outputs the battery level percentage as an analogue voltage. The battery meter counts the energy flowing in and out of the main electric vehicle battery pack using a current sensor. Serial

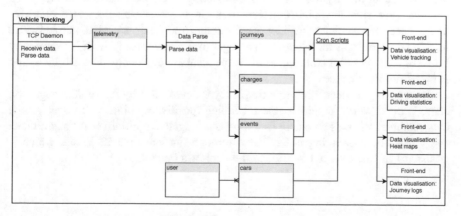

Fig. 8.12 Simplified class diagram for REView vehicle tracking system

Fig. 8.13 Astra Telematics AT240 tracking device (used in vehicle fleet network)

communications are facilitated through the RS232 standard. In the case of our road vehicles, these connect through OBD-II over a CAN bus, which transmits the vehicle's journey information to the tracking device. This journey information supplements the sensor readings from the device which are then packaged and transmitted over to our server.

The charging system monitoring is done with a Python daemon running the Threading Socket server library. This library listens for TCP connections on a defined port and creates a new thread for each one, which can handle the processing of the data received. The processing is done by parsing the incoming message from the byte stream into Python variables, connecting to the database and inserting the new data points. Each vehicle connects to our local server via a dedicated port over TCP, where the vehicles are identified through the devices' IMEI number. The tracking devices transmit telemetry messages to the server using Astra's proprietary Protocol A. This script connects to the same database and deciphers each incoming packet from the vehicles.

The tracking devices' reporting frequency is determined by the vehicle's ignition state. In other words, a vehicle that is running (positive ignition) will transmit data every minute; whereas a vehicle that is parked (negative ignition) will transmit data every half hour. Charging is facilitated through the vehicle's 12 V rail, which is connected in line over a 1 A fuse to the vehicle's battery.

8.5.2 Database

Our server's vehicle tracking daemon receives and parses all incoming telemetry data from the vehicles, which is subsequently appended into the *telemetry* table in the PostgreSQL database. The following information is recorded in the database for each data point: latitude, longitude, time logged on device, time received at server, vehicle speed, vehicle heading, altitude, journey max speed, journey max acceleration, journey distance, journey idle time, ignition status, alarm line status (unused), air-conditioning status, headlights status, heater status, charging status, and car battery level. The GPS positions and line inputs are uploaded onto the server either at every minute or at every ten meters, whichever occurs first.

The server runs a separate parsing script, scheduled as a cron job, to further process and classify the variables into three separate and redundant tables, as follows.

1. The *journeys* table records data pertaining to journeys made with the condition that the device records an increase in distance travelled. Entries are appended at one-minute intervals.
2. The *charges* table is appended from any telemetry data that whenever the vehicle is charging, are appended at one-minute intervals.
3. The *events* table logs and timestamps any appends pertaining to charges and journeys, summarizing and batching each data parsing event according to the vehicle's IMEI number.

This is followed by ten cron-scheduled Python scripts that calculate and update the imported entries for these tables, activating every 30 min. These scripts perform the following activities:

- Generate vehicle journeys, charging events, idle events, missing data events.
- Generate charging station events.
- Combine similar charging station and vehicle charging events based on user tag, time and location.
- Compress data, such as air conditioning, heater, headlights into a data point for a journey.
- Compress charging data into charging/maintaining charge, divide into by-hour and by-week arrays.
- Generate heat maps.

As separate scripts are used for different functions, adding new functionality or statistics can be easily done by adding the additional script. This limits the need for modifying existing software and reduces integration problems and helps in isolating errors. The following paragraphs detail notable information that the data import relates to:

Charging Places

By assuming that each charging event will take place at either the trial participant's workplace, home or a public charging station, each participant nominates a work and home location for charging. Charging places are hence classified as charging stations, workplaces and homes with their data stored in a separate table in the database.

Data Loss

As each tracking device is programmed to transmit data at every minute when the vehicle is running, and every 30 min when the vehicle is parked or idle, any subsequent data packets that arrive later than 30 min will be identified as a data loss, these are recorded into a *data_loss* table.

Time Series Modelling

Data from the *telemetry* table are parsed into day/week time series representation through scripts 8 and 10. This is done by selecting and grouping telemetry data from day and week segments. Travelling distances and energy consumption is summed as they are appended into the *charges* and *journeys* tables. Time series modelling is vital for the visualization of our vehicles' data as they are disclosed to trial participants to gauge and understand their driving habits.

8.5.3 Data Visualization

REView visualizes vehicle data across four categories: vehicle tracking, driving statistics, heat maps and journey logs, as detailed in the following subsections. All visualizations are presented on REView's front end, which is presented through a PHP script that connects to any of the three related tables on the database through SQL queries. Users wanting to view their vehicle's data on REView are required to log in with their credentials, which is linked to their registered vehicles. In this context, we establish a one-to-many relationship between the *User* and *Cars* tables in the database. We use a one-month sampling period to represent the figures in this section.

8.5.3.1   Vehicle Tracking

The vehicle tracking page looks up a list of vehicles that corresponds to the user viewing it. The user can select an individual vehicle or all vehicles in a fleet and the time period to display. The graph is interactive, allowing the user to drag and zoom in on the time scale.

This page is a result of PHP scripts that supply information to JavaScript code using several free-to-use libraries including jQuery [52] for communication with the server, Google Maps for the map and Dygraph [53] for the interactive graph. The Dygraph library is open source and was modified for use in the website.

To display the GPS data, the map has the ability to show individual interactive points that can be clicked on for additional information or an image that is generated by a PHP script on the server. The number of interactive points is limited to 150, as too many points can cause instability in the browser. However, the image overlaid over the map can contain any number of points, which allows users to see data over longer time periods. Generating an image at the server is also useful, because the information sent from the server to the user is significantly less. The server caches all images generated and generates differently scaled images for different map zoom levels.

For performance and stability reasons, the graph below the map (in Fig. 8.14) is limited to displaying 2,000 points at a time. To reduce the load on the server, the graph caches data and only requests additional information when necessary. When a user pans the graph to the left or right, only the missing information is requested. The granularity of the data is also important, and the server is designed to send sub-divided data when the time period selected has more than the 2,000-point maximum. When the user zooms on a section of the graph, the server is asked for sub-divided or raw information for this smaller time period.

A query reads all IMEIs associated to that user and retrieves a list of related vehicles from the *Cars* table, which is subsequently presented as a selectable radio button list. For each selected vehicle, the vehicle tracking page visualizes the vehicle's position and its trajectory from the selected time period, which is displayed using the Google Maps JavaScript API. The vehicle's current location is given as a color-coded pin that corresponds to the selected vehicle, whereas its trajectory is plotted through an array of markers obtained from the vehicle's historical location data from the *telemetry* table.

Additionally, the vehicle tracking page also includes a chart that allows users to check, by defining a sampling period, the vehicle's driving behaviour as a time series, which is also exportable as a *.csv* file. REView plots multiple line graphs on this chart, which includes battery levels, vehicle speed, travelled distance, as well as air conditioning, headlights and heater usage. Figure 8.14 shows an example of this visualization.

8.5.3.2   Driving Statistics

Visualization for driving stations is presented in two modes: one as an overview for all vehicles, and another for a detailed analysis of driving patterns. The overview page consolidates statistics across all vehicles, and for a specific range of dates with examples, as shown in Fig. 8.15.

Leaderboard

The leaderboard table ranks the vehicles according to their distance travelled, and compares them alongside the other vehicles. The table also lists the vehicles' driving times, total number of journeys, average journey distance and average journey time. An example of the leaderboard is shown in Fig. 8.16.

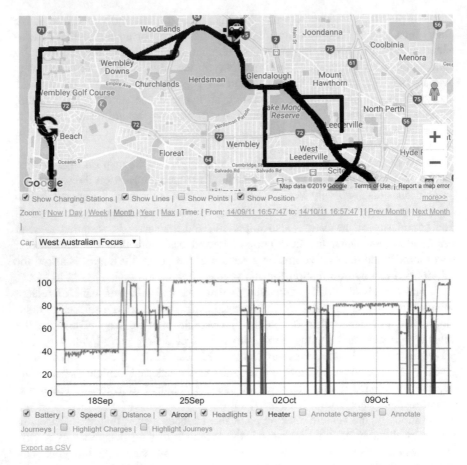

Fig. 8.14 Partial screen print of vehicle tracking page (showing historical trajectories on Google Map, and driving behaviour)

Chart Legends

Legends used in the various charts are described according to the following:

- Vehicle name: Name of vehicle e.g. UWA Lotus, UWA Getz etc.
- Location type: Type of charging station location e.g. home, business or public station.
- Vehicle state: Vehicle status e.g. idle, driving or plugged in.
- Plug type: Charging current e.g. 32 A, 15 A or 10 A.
- Distance: Daily distance travelled in km, with an average plot.

Likewise, driving statistics for the individual vehicles are presented across three sections: distance, charging and journeys, as shown in Fig. 8.17. For each statistic, results from each vehicle are compared against the average results from the rest of

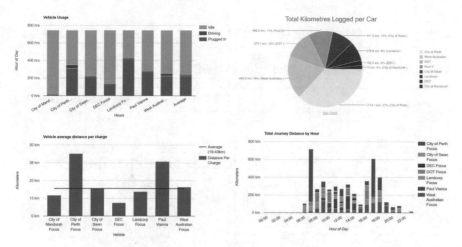

Fig. 8.15 Examples of charts showing driving data

Leaderboard Table

Rank	Vehicle	Distance Travelled	Time driving	Total Journeys	Average Journey Distance	Average Journey Time
1st		1714.1 km	1 day 19:10:22	110	15.58 km	00:23:32
2nd		845.5 km	1 day 01:52:12	142	5.95 km	00:10:55
3rd		570.1 km	16:57:48	51	11.18 km	00:19:57
4th		498.2 km	14:17:07	65	7.66 km	00:13:11
5th		447.2 km	14:09:58	83	5.39 km	00:10:14
6th		378.8 km	12:22:07	39	9.71 km	00:19:01
7th		102.5 km	03:10.48	12	8.54 km	00:15:54
8th		73 km	01:24:39	4	18.25 km	00:21:09

Fig. 8.16 A sample leaderboard displaying driving statistics for trial vehicles for an average month. Vehicle names pixelated to preserve privacy

the vehicles (community). This, along with the leaderboard, is a gamification feature, which we find useful to keep monthly reports interesting for the users. These are mostly presented in bar charts of grouped pairs to illustrate the statistical difference between that vehicle and the community.

8.5.3.3   Heat Maps

REView generates heat maps of the tracked vehicles through the aggregation of their GPS coordinates and location type geofencing. They are automatically

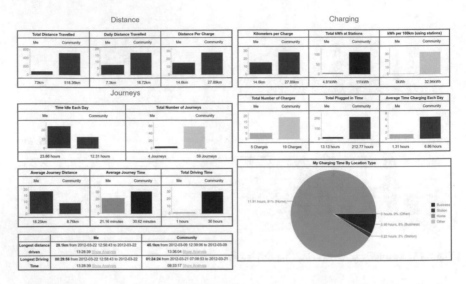

Fig. 8.17 Individual driving statistics (presented as distance, journeys and charging sections)

generated and can show areas where vehicles drive, park and charge within certain time periods. The server periodically runs a heat map generating Python script as a cron job, which searches through the journeys, charges, telemetry and any low charges from the *journeys* table, eventually generating a *.kml* file that contains the location of the heat maps.

The Python script imports Jeffrey Guy's heatmap library [54], which utilizes Python Imaging Library (PIL) to generate a heat map *.kml* file along with a *.png* file containing an image overlay that corresponds to the heat maps. Areas with increasing overlapping entries from the database progress from a navy to a red hue, as shown in Fig. 8.18. These heat maps can be generated according to user's selection, which includes the type of heat map (places parked, charged or driven), time frame (year or month) and time period.

8.5.3.4 Journey Logs

The completion of any drive routine from a tracked vehicle results in the tabulation and display of its journey log onto REView as shown in Fig. 8.19.

The cost of equivalent petrol, electricity and carbon emissions are calculated based on the distance travelled, which is 10.43 cents, 3.47 cents and 168 g per kilometers respectively. These values were taken from the Carbon Dioxide Emissions from New Australian Vehicles 2013 information paper [18] by the Australian National Transport Commission along with current electricity tariffs as an average. The monetary savings are calculated as a difference between petrol and electricity cost. Importing options for *.csv* data is also available.

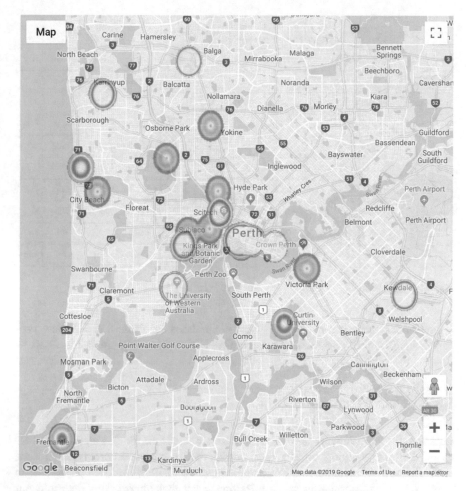

Fig. 8.18 Heat map of a vehicle fleet tracked over a month for possible charging locations utilized during the day

8.6   EV Charging Power Generation

UWA's AC chargers are powered by a solar PV array installed on the roof of the Human Movement building. This is a dual inverter PV plant that is rated at 20 kWp and collectively generates 114.1 kWh per day, which is greater than the combined usage of all our AC chargers. This plant uses two Sunny STP 10000TL-10 [55] 10 kW three-phase inverters with a maximum efficiency of 98.1% at a nominal 600 V. Monitoring on the solar plant is done through SMA Sunny WebBox [56], which provides a web interface for configuration, including that for telemetry data transmission to our local server at up to one-minute intervals. The average

Count	ID	Car	Distance (km)	Time	Battery Used (%)	Maximum Speed (km/h)	Idle Time (minutes)	Start Time	End Time	Emissions Saved (kg CO2)	Electricity Cost ($)	Petrol Equivalent ($)	Savings ($)
4701	27505		39 km	00:36:22	0 %	112 km/h	3	2014-06-05 09:17:12	2014-06-05 09:53:34	6.591 kg	$1.35	$4.07	$2.72
4702	27506		39 km	00:36:22	0 %	112 km/h	3	2014-06-05 09:17:12	2014-06-05 09:53:34	6.591 kg	$1.35	$4.07	$2.72
4703	27503		14.9 km	00:22:43		80 km/h	6	2014-06-05 09:02:59	2014-06-05 09:25:42	2.5181 kg	$0.52	$1.55	$1.03
4704	27504		14.9 km	00:22:43		80 km/h	6	2014-06-05 09:02:59	2014-06-05 09:25:42	2.5181 kg	$0.52	$1.55	$1.03
4705	27501		13.3 km	00:20:29		80 km/h	2	2014-06-05 08:28:13	2014-06-05 08:48:42	2.2477 kg	$0.46	$1.39	$0.93
4706	27502		13.3 km	00:20:29		80 km/h	2	2014-06-05 08:28:13	2014-06-05 08:48:42	2.2477 kg	$0.46	$1.39	$0.93
4707	27500		0.2 km	00:12:25	0 %	30 km/h	11	2014-06-05 06:20:21	2014-06-05 06:32:46	0.0338 kg	$0.01	$0.02	$0.01
4708	27499		0.2 km	00:12:25	0 %	30 km/h	11	2014-06-05 06:20:21	2014-06-05 06:32:46	0.0338 kg	$0.01	$0.02	$0.01
4709	27498		0.1 km	00:01:40	0 %	20 km/h	0	2014-06-04 21:41:15	2014-06-04 21:42:55	0.0169 kg	$0	$0.01	$0.01
4710	27497		0.1 km	00:01:40	0 %	20 km/h	0	2014-06-04 21:41:15	2014-06-04 21:42:55	0.0169 kg	$0	$0.01	$0.01

Fig. 8.19 Table showing journey logs of a tracked vehicle

instantaneous measurement of the plant is tabulated in Table 8.2. REView processes data from this infrastructure according to the class diagram in Fig. 8.20.

8.6.1 Data Visualization

Data from Sunny WebBox is pushed daily to our server via FTP as a series of *.csv spreadsheets, containing timestamped entries at five-minute intervals that record each inverter's instantaneous measurements, similar to values presented in Table 8.2. We subsequently wrote a Python script that imports this data into our server's PostgreSQL database as the *sunny_solar* table, which runs on a five-minute cron schedule. This script reads the last event in the table and imports every successive data from the spreadsheet, synchronizing the data between the two

Table 8.2 Average instantaneous measurements of the solar PV plant

DC current	1 A
DC voltage	510 V
DC power	1000 W (2 inverters)
AC grid frequency	50 Hz
AC grid power	1600 W
AC phase current	1.1 A
AC phase voltage	225 V
AC phase power	500 W
AC day yield	56.67 Wh (average)

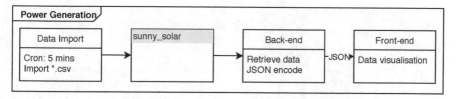

Fig. 8.20 Simplified class diagram showing REView energy generation system

entities. The data from the solar system includes time stamps, power generated, voltages at the panels and grid and operation health flags. PV systems are connected to and scanned every 15 min for data download.

This data in the *sunny_solar* table is, in turn, visualized as a series of line graphs on REView's front end, which can be selected through a day, week or month period. Visualizations are given in the solar plant's power and energy generation and consumption, as accumulated from the selected period. We achieve this visualization by importing the JpGraph [57] library, whereby a back-end PHP script calls all related SQL queries which then encodes it as a JSON representation for the front end, as Fig. 8.21 illustrates.

8.7   Usage Billing

The billing page allows an EV user to view his or her monthly mobility cost. It also lets station operators to view the utilization and energy usage of their stations. In both cases, summaries as well as itemized bills are generated. All bills are automatically generated with several informative graphs, including distance travelled, distance per charge and kWh per km of the individual versus the community; a time-of-use energy graph is also generated.

To provide users and charging station owners with a summarized update, bills are generated at every first Monday of the month through a scheduled cron job, which is then emailed to related users who opt into this service, where they are classified as either vehicle owners or station operators. For this purpose, we have configured a CSS script that enables the printing of the contents directly through HTML as a Tax Invoice. This is presented as a combination of charts and tables that are dependent on the type of bill presented: Vehicle/User (User Itemized Billing), Charging Station (Station Operator Billing) or Summary (Network Overview). Each bill type is generated through individual PHP back ends that encode SQL queries as a JSON representation as detailed in the subsections.

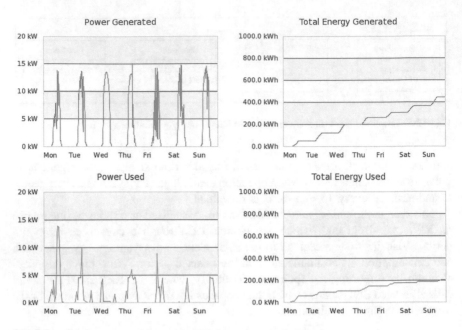

Fig. 8.21 Graphs generated from solar PV data over a typical week

8.7.1 Itemized Billing

An itemized bill enables users to individually monitor and track their charging behaviours, which is displayed as a line graph followed by three tables (see Fig. 8.22). If the vehicle is part of our tracked fleet in Sect. 8.5, its driving statistics (total distance travelled, distance per charge and kWh per km) will be presented above the line graph and compared against the rest of the community. REView obtains these data by running a query for each graph or table for a given bill. This query is applied for the *station_charges* table according to the accessing user and its selected sampling timeframe, which returns the charging timestamps, duration, location and per-hour energy usage to the front end.

The line graph illustrates the cumulative charging energy consumption per hour for the vehicle according to the charging places (as mentioned in Sect. 8.5.3), which is individually labelled as different colored lines. The data used is summed through the per-hour energy usage data obtained in the query as mentioned in the first paragraph. To achieve this, the system uses an incremental function to individually increment the energy consumption for each location type at each hour of the day.

The mobility cost table is the first one presented following the line graph, which tabulates the cost of charging based on the user's charging location. Headers are given as: charging locations, number of events per location, energy consumption in kWh, tariff units, and totals with GST which is calculated as 10% of the total price. We use a standard tariff of 25 cents per kWh for charging stations and 21 cents per

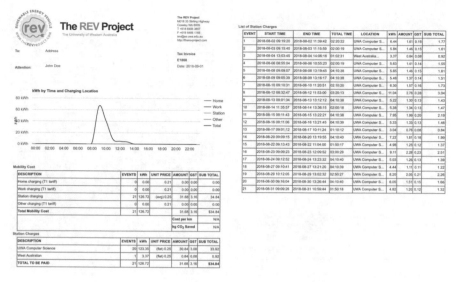

Fig. 8.22 Itemized bill generated for a station user (for August 2018)

kWh for other locations. When presented for a tracked fleet vehicle, it also calculates the cost per kilometers travelled and the amount of CO_2 saved.

The second table presents station charges using the same headers found in the first table, which itemizes the charging cost according to the charging stations used, detailing the individual station charging events found there.

The third table further itemizes the second one by listing each charging event of that vehicle on charging stations, listing each event using headers that correspond to its starting and ending timestamps, duration, station location, energy consumption, and amount totals with GST.

8.7.2 Station Operator Billing

Bills are also sent to station operators to summarize the monthly usage of their charging stations. Station operator bills consist of a community comparison, a line graph and an itemized table (see Fig. 8.23). Data is obtained from the *station_charges* table through a query that returns the charging user, starting and ending timestamps, energy consumption, per-hour charging and maintaining energy consumption, charging duration and connector side (left or right).

Community comparisons are drawn through the billing station against the average charging parameters of the other charging stations, where these parameters are given as their energy consumption in kWh, number of transactions and plugged in time. These are illustrated as a pair of color bar charts for each parameter.

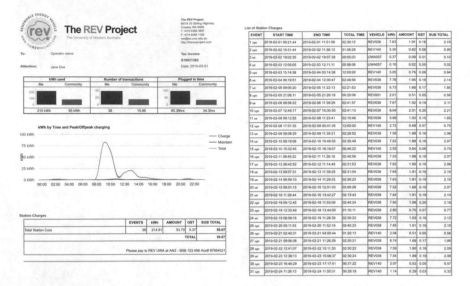

Fig. 8.23 Itemized bill generated for station operators (for February 2019)

The line graph plots the per-hour energy consumption of the charging station on an average day, separated by its charging, maintaining and total plots. The total plot is the instantaneous sum of the "charge" and "maintain" energy consumption per hour.

Finally, the *station_charges* table begins by listing the total operating cost for the selected billing period, which lists the number of charging events and total energy consumption for that period, followed by the total energy costs on a 25 cents per kWh tariff. Another table itemizes each charging event chronologically under the following headers: connector side, starting and ending timestamps, charging duration, vehicle tag, energy consumption and charging price with GST.

8.7.3 Network Overview

REView displays an overview of the charging station network and vehicles' usages, which is accessible only to project administrators. This bill is generated for a selected period through a drop-down menu that evokes a query that produces a usage summary for our entire AC charging station network. Data on the network overview bill is presented as a line graph that is followed by two tables (see Fig. 8.24). Data for this bill is obtained through a query that returns individual events under the following headers: station name, vehicle tag, energy consumption and per-hour energy usage (charging and maintaining charge).

The line graph plots the cumulative per-hour energy usage across all stations for each hour of the day. This graph consolidates data under the per-hour energy usage header from the query, adds the energy usage for charging and maintaining charge across each hour of the day, and subsequently presents it as a sum of energy consumption for each event in the selected period.

Each of the two tables aggregate charging events, one for the charging stations, and the other for the vehicles. These tables present, for each charging station or vehicle, its number of charging events and total energy consumption in kWh. Additionally, this bill also includes cost (expenses), revenue (return) and profit generated from each event. [At the time of writing, our expenses and return tariffs are equally set as we are operating on a non-profit model]. Data from the query is processed through a PHP back end that calculates values for an aggregated number of events and energy usage, as well as its associated costs. For each charging station and vehicle, an incremental function finds and increments the number of events and its associated energy usage, and subsequently its operating costs, revenues and profits, before encoding the entire table as a JSON representation.

8.8   Mobile Applications

REView has two mobile phone applications for EV drivers and station users as illustrated in Fig. 8.25. The first allows users to view their vehicle status on their mobile phone, showing location, status and battery level. The second allows station users to see if their vehicle is still drawing power or if their EV is fully charged. It also allows users to check remotely if a station is occupied or free, allowing EV drivers to plan their trip ahead. These applications help ease "range anxiety" where drivers fear their vehicle will not have enough energy left in the battery to make it to a destination.

Designing mobile web pages instead of apps makes sure they can be used for every smartphone or tablet model. The pages are developed as lightweight web pages using HTML 5 and JavaScript, which communicate periodically with the server for data updates. Each of the charging stations are listed on a page, with their availability indicated by a blue or green icon.

8.9   Results

In this section, we discuss results from the WA Electric Vehicle Trial, represented in REView graphs. We also present usage analyses and forecasts for the charging infrastructures.

The REV Project
The University of Western Australia

The REV Project
M018 35 Stirling Highway
Crawley WA 6009
T +618 6488 3897
F +618 6488 1168
rev@ee.uwa.edu.au
http://therevproject.com

Billing Summary Sheet   Date: 2019-03-01

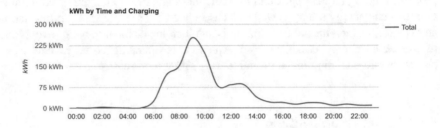

Stations

STATION	EVENTS	kWh	EXPENSES	RETURN	PROFIT
Subiaco	23	142.4	35.6	35.6	$0
Murdoch University CREST	38	214.81	53.7	53.7	$0
City of Swan	12	89.18	22.29	22.29	$0
UWA Computer Science	18	104.82	26.2	26.2	$0
City of Cockburn	6	5.93	1.48	1.48	$0
UWA REV Lab	14	36.1	9.03	9.03	$0
TOTAL	111	593.23	148.31	148.31	$0

Vehicles

VEHICLE	EVENTS	kWh	EXPENSES	RETURN	PROFIT
REV052	20	91.99	23	23	$0
REV038	27	192.91	48.23	48.23	$0
REV140	8	19.42	4.86	4.86	$0
UWA057	15	18.91	4.73	4.73	$0
REV170	1	17.85	4.46	4.46	$0
UWA 026	12	89.18	22.29	22.29	$0
REV001	1	2.02	0.51	0.51	$0
REV104	1	4.86	1.22	1.22	$0
UWA 020	11	45.89	11.47	11.47	$0
REV006	7	60.7	15.18	15.18	$0
REV005	1	10.26	2.57	2.57	$0
REV057	1	30.4	7.6	7.6	$0
REV047	5	7.93	1.98	1.98	$0
REV018	1	0.9	0.22	0.22	$0
TOTAL	111	593.23	148.31	148.31	$0

Fig. 8.24 Network overview bill (for March 2019)

Fig. 8.25 REView Smartphone applications execution

8.9.1 Overall Energy Usage

Figure 8.26 shows the energy consumed by charging EVs by hour of day and location. This information can be used for analyzing EV grid impacts and the usage of renewable energy. The locations are defined as followed:

- Home: A residential area.
- Business: A commercial or industrial area.
- Station: An EV charging station.
- Unknown: An undefined area.

It is found that the peak of the energy supplied for charging vehicles (averaged over all locations) is during the morning hours around 9–10 am. This means, EVs are commuting from home to work and use a charging facility at work (most likely free of charge). It is worth noting that the majority of energy supplied is during sunshine hours especially for station charging and business charging. For unknown and home charging, there is a much smaller peak at around 6 pm, as vehicles are returning home and charging there. This also suggest that the majority of unknown locations are unlabeled home locations.

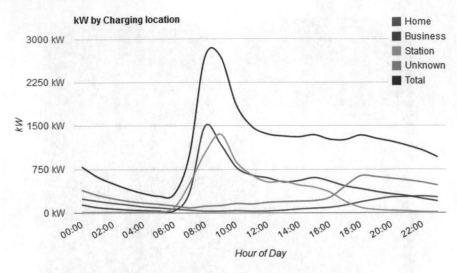

Fig. 8.26 Power drawn by hour of day for EV charging at various location types

From this information and solar information gathered (see Fig. 8.28), we can show that typical charging scenarios can be offset almost ideally by solar technology. Most of the charging occurs during the day, which differs fundamentally from the scenario propagated by some energy suppliers, which shows that all EVs charge around 6 pm when they return to home.

8.9.2 Usage of Charging Infrastructure

The statistics in this subsection are taken from REView Stations page showing summary of all charging stations as part of the WA Charging Station Network, from the beginning in June 2012 until March 2019.

In Figs. 8.27 and 8.28, we show the difference between charging and maintaining the charge. It is common that an EV charger will draw a large amount of power until the battery pack is full, at which point the charger will continue to draw power but at a significantly lower rate. When drawing power at the lower rate, the EV can be doing several things including maintaining the charge of the battery pack, pre-conditioning the interior of the vehicle with heating or cooling or maintaining the temperature of the battery pack to improve driving efficiency. To distinguish between charging at maintaining charge, we define a vehicle to be "charging" if it is drawing more than 1 kW of power; otherwise we define it as "maintaining".

From Fig. 8.27, it is clear that throughout the day the majority of energy consumed from the stations is done during a charging cycle. The energy for charging varies heavily depending on the time of day with the majority of energy being used

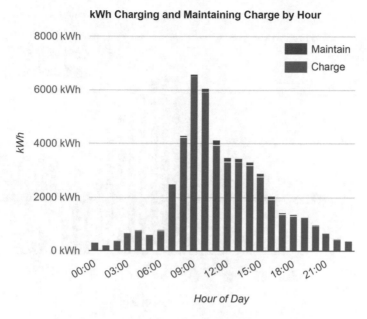

Fig. 8.27 Energy drawn (kW) by hour of day stacked with power drawn for charging vs. power drawn for maintaining charge

throughout the day, reducing steadily into the evening and bottoming out around midnight. However, the maintaining charge energy consumption is similar in every hour throughout the day and night. This is because the electric vehicles are sometimes parked at the charging station overnight, and possibly over days when the EVs are not being used. The maintaining charge consumption remains steady throughout the time the vehicle is idle.

In Fig. 8.28, we show the amount of time spent for charging and maintaining charge. From the discrepancy between the time required for charging and the time actually spent plugged in at the charging station, it can be seen that the charging stations in many cases are being misused as free parking locations for EVs.

Table 8.3 shows the summary of the charging station usage. From the information collected and automatically analyzed, we can draw several conclusions. The flat rate plan of buying electricity is cheaper than the peak-shoulder-off peak plan. Only 46% of the time spent at a station is used actually charging, while for the remaining 54% of the time, the vehicle sits idle and blocks a charging station. This could allow for vehicle-to-grid technologies, however, as shown in [58], V2G applications are not cost effective with current battery technology, as the addition wear and tear from extra charge cycles by far outweighs the marginal energy cost. The stations themselves were only in use 3.2% of the time logged, leaving a large proportion of the outlets idle.

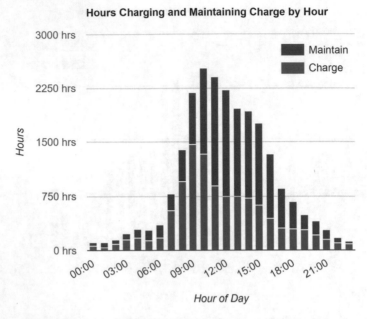

Fig. 8.28 Amount of time spent (hours) at charging station with stacked charging time and maintaining charge time

8.9.3 Solar PV Monitoring

In Fig. 8.29, we show the average hourly power output of the 20 kWp solar system at UWA for a typical summer day. The solar system begins generating energy at 6 am and shuts down at 6 pm, with the energy output peaking at 12 pm. The PV

Table 8.3 Charging station statistics June 2012–March 2019 (81 months)

Total kWh	48788.343 kWh
Estimated cost (21.87¢ per kWh)	$10670.01
Estimated cost (peak/shoulder/off)	$13891.92
Number of transactions	6917
Plugged in time	957 days, 23:29:36
Charging time	444 days, 2:46:05 (46.36%)
Maintaining charge time	513 days, 20:43:31 (53.64%)
Avg transaction time	3:19:26
Avg charging transaction time	1:32:27 (46.36%)
Avg maintaining time	1:46:58 (53.64%)
Percentage time in use	3.21%
Power used in peak	21785.22 kWh (44.65%, $9182.47)
Power used in shoulder	21965.21 kWh (45.02%, $4709.34)
Power used in off-peak	5037.91 kWh (10.33%, $570.29)

Fig. 8.29 Average power output of 20 kW of UWA's solar PV system (on a given day in February)

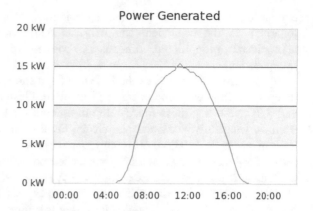

system generates approximately 80 kWh per day of operation. So, this system is generating around 30 MWh per year. In comparison, the 23 AC charging stations use only 5.7 MWh per year on average (see Table 8.3). This shows that one large solar PV installation can effectively power a number of EV charging stations.

8.9.4 Heat Maps for EV Tracking

With reference to Fig. 8.18, we generated a heat map of the charging locations for tracked EVs from 2010 to July 2014. By looking at the charge events that took place during the day between 7 am and 6 pm, we can identify possible public locations in the Perth Metro area. The heat map shows several heavily utilized areas, including residential and business locations. One hot spot in Landsdale WA is the location of an EV conversion company that services most of the tracked EVs. From the heat map, we can determine that this is a place where a charging station would be highly frequented. The heat map also shows hot spots around most existing stations such as at The University of Western Australia.

8.9.5 Charging Infrastructure Usage Forecast

The historical data of the charging stations enables us to forecast the usage on these stations, and consequently predict the EV uptake rate in Western Australia. To achieve this, we have sampled data pertaining to charge frequencies C and energy consumption E for the AC charging network and the DC charging station. Data points were sampled from 1 December 2014 to 1 March 2019 (51 months) for DC charging, and from 1 June 2012 to 1 February 2019 (80 months) for AC charging.

For both data sets, we conducted an augmented Dickey-Fuller test (ADF), which resulted in the rejection of the null hypothesis. This prompted us to fit the forecast over a logarithmic linear model using a log-level regression. We use a natural logarithm as the coefficients on its scale that can be interpreted directly as approximate proportional differences, which we describe in our model interpretations.

By modelling the usage of the DC charging station, we present our forecasts of its charging frequency and energy consumption, illustrated as graphs in Figs. 8.30 and 8.31. Note that these models do not account for any usage saturation for the stations, where given the average charge duration of 3 h on an AC station and 24 min on the DC station, charging frequencies will saturate at 1800 charges per month for the DC station, and 240 charges per month for an AC station, assuming a back-to-back use case scenario, which is beyond the scope of the forecasts.

The coefficients for the models are given as Eq. 8.1 for charge frequencies, and Eq. 8.2 for energy consumption, with time t measured per day:

$$\ln(C_{\text{DC}}) = 0.0007116t - 10.9732 \tag{8.1}$$

$$\ln(E_{\text{DC}}) = 0.0005839t - 9.4866 \tag{8.2}$$

We can therefore infer from the coefficients, that the charging frequency will increase by 0.07% per day; the average energy consumption will increase by 0.05% per day. This results in a 26.0% annual increase in charge frequency, and a 21.3% annual increase in energy consumption per charge.

Similarly, we model our AC charging station network's usage across the whole network as it compensates for the lower charging frequency on AC charging

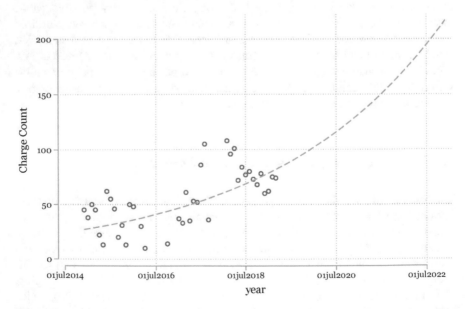

Fig. 8.30 Regression model forecasting the per-month charge frequency on DC charging station

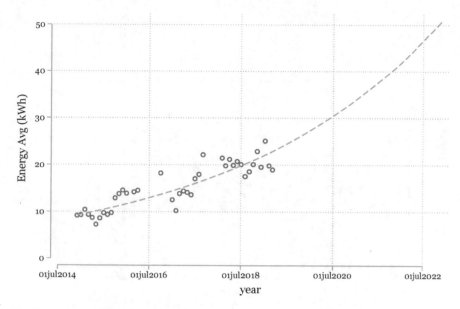

Fig. 8.31 Regression model forecasting the per-month charging energy consumption per charge on DC charging station

stations due to its longer charging duration. Here, we present models for charging frequency and charging energy consumption as Figs. 8.32 and 8.33.

Likewise, the coefficients for the AC charging network's models are given as Eq. 8.3 for charge frequencies, and Eq. 8.4 for energy consumption.

$$\ln(C_{AC}) = 0.0002787t - 1.362995 \tag{8.3}$$

$$\ln(E_{AC}) = 0.0000206t - 1.592417 \tag{8.4}$$

From Eq. 8.3, we interpret the coefficients to suggest that the charging frequency for the AC network will increase by 0.03% per day, or 10.2% per annum. This equates to average energy consumption per charge at an AC station increase by 0.002% per day, or 0.75% per annum.

From the above discussion, we can deduce that DC charging is the more preferred charging method for local EVs, and that the increasing energy consumption per charge could suggest the increasing battery capacity and range for newer EVs. As the number of EV uptakes in WA increases, so does the charging frequency at the DC station. Note that we have observed different charging behaviours across AC and DC charging stations whereby our DC charging station registers more unique users, and many of the charging events registered large energy consumptions, as shown in Fig. 8.34. This also implies that most users are charging their vehicle from a low charge state. On the contrary, however, most AC charging users are routine users, whereby the charging bays are occupied for extended periods

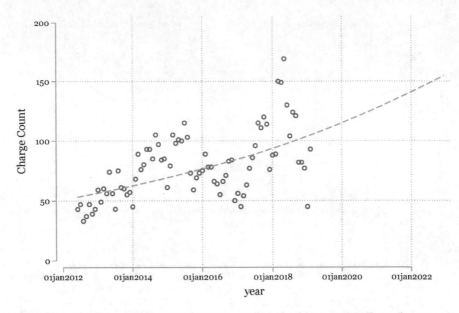

Fig. 8.32 Regression model forecasting the per-month charge frequency on AC charging network

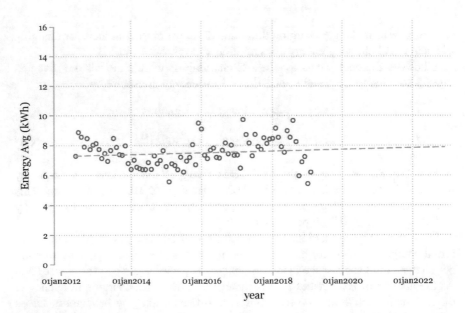

Fig. 8.33 Regression model forecasting the per-month charging energy consumption per charge on AC charging network

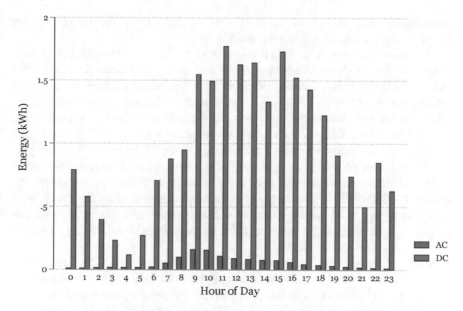

Fig. 8.34 Per-hour energy usage comparison between AC and DC charging stations

while the drivers are at work, for example. This explains the low increase in energy consumption across AC charges, as charging events are often top-up charges that recharge the vehicle after a daily commute, and where the vehicle is unlikely to be at a low charge state. However, the increase in charge frequency on the AC network also implies the increasing EV uptake in WA.

8.10   Conclusion

In this chapter, we have presented our telematics platform, REView, for connected electric vehicles and its infrastructure. The suggested system comprises live data information portals for customers as well as for fleet operators and charging network operators. It provides statistical information on time and location of charge events and includes a time-of-use billing system. It interfaces with charging stations, vehicle-based data loggers and solar PV systems. This required configuration and testing for each of the different devices in parallel with server and database development.

The software was written for server-based and client-based processing and data display. Each of the different aspects of this project (server, data processing, and interface) was developed in tandem to ensure integration. The software was designed in a modular way with separate scripts for individual features, making unit testing easier, reducing integration problems and isolating failures. All of the programming languages used in the system are interpreted, which means that design changes could be made very quickly.

The main contribution of REView is that it consolidates incoming data from connected vehicle fleets, charging stations and power generators into a unified platform to improve the efficiency of information presentation. These are presented entirely as a web-based solution for increased accessibility and are supplemented by additional features such as billing which supports the monetization of the system.

From the data collected and analyzed, we can deduce that solar technology is an effective way for offsetting energy required for charging EVs at public charging stations and place-of-work. For home charging, energy is mostly required outside of solar generation hours and would need to be provided by a domestic energy storage system. A 20 kW solar PV system was more than enough to offset the energy used by EV charging at 23 public charging stations. The results produced from this system have, throughout the years, enabled us to perform various analyses on the EV landscape of Western Australia.

Moving forward, with the increase of data volume and infrastructures, future plans for REView involve its eventual compliance with the arrival of the Internet of Vehicles (IoV) standards. With this proposal, we are planning to integrate REView as a cloud application while utilizing the Platform as a Service (PaaS) environment. This, hopefully, will enable us to streamline further developments on REView and to increase its modularity through more efficient workflows, while improving scalability.

Acknowledgements The authors would like to thank all partners of the WA Electric Vehicle Trial, as well as the ARC as sponsoring body. We would like to thank all donors and sponsors of the REV Project, especially Galaxy Resources, Synergy, the WA Department of Transport, and UWA. For the REView project, we would like to especially thank Telstra Australia, who provided us with M2M SIM cards for vehicles and charging stations.

References

1. Magana VC, Munoz-Organero M (2015) GAFU: using a gamification tool to save fuel. IEEE Intell Transp Syst Mag 7:58–70. https://doi.org/10.1109/MITS.2015.2408152
2. Mader T, Bräunl T (2013) Western Australian electric vehicle trial. The University of Western Australia, Perth, Australia
3. Speidel S, Bräunl T (2014) Driving and charging patterns of electric vehicles for energy usage. Renew Sustain Energy Rev 40:97–110. https://doi.org/10.1016/j.rser.2014.07.177
4. Jabeen F, Olaru D, Smith B et al (2013) Electric vehicle battery charging behaviour: findings from a driver survey. Brisbane
5. Speidel S, Jabeen F, Olaru D et al (2012) Analysis of Western Australian electric vehicle and charging station trials
6. Jabeen F, Olaru D, Smith B et al (2012) Acceptability of electric vehicles: findings from a driver survey. Perth, Australia
7. International Energy Agency (2018) Global EV Outlook 2018. Paris

8. Rodríguez-Serrano Á, Torralba A, Rodríguez-Valencia E, Tarifa-Galisteo J (2013) A communication system from EV to EV Service Provider based on OCPP over a wireless network. In: IECON 2013—39th annual conference of the IEEE industrial electronics society, pp 5434–5438

9. Australia AECOM (2009) Economic viability of electric vehicles. AECOM Australia Pty Ltd., Sydney

10. Kinghorn R, Kua D (2011) Forecast uptake and economic evaluation of electric vehicles in Victoria. AECOM Australia Pty Ltd., Melbourne

11. Roberts N (2012) Media release: ABMARC releases key findings from their electric & hybrid vehicles report

12. Energy Supply Association of Australia (2013) Sparking an electric vehicle debate in Australia. Energy Supply Association of Australia, Melbourne

13. Clean Energy Council (2019) Clean energy Australia report 2019. Clean Energy Council, Melbourne

14. Anair D, Mahmassani A (2012) State of charge. Union of Concerned Scientists 10

15. van Namen K, Tieu A, Olden P (2011) Green house gas emissions from households in Western Australia. SMEC Australia Ltd., Perth

16. Department of Infrastructure and Regional Development (2016) Vehicle emissions standards for cleaner air

17. Department of Infrastructure and Regional Development (2018) Green vehicle guide. https://www.greenvehicleguide.gov.au/. Accessed 2 Feb 2018

18. National Transport Commission (2014) Carbon dioxide emissions from new Australian vehicles 2013. National Transport Commission

19. ChargePoint Inc. (2019). https://www.chargepoint.com. Accessed 2 Apr 2019

20. Tesla Supercharger (2019). https://www.tesla.com/en_AU/supercharger. Accessed 2 Feb 2018

21. China Electric Vehicle Charging Infrastructure Promotion Alliance (2017) Zhongguo diandong qiche chongdian jichu sheshi fazhan niandu baogao 2016–2017 ban [China electric vehicle charging infrastructure development annual report 2016–2017 edition]. National Energy Administration, Beijing, China

22. ChargeStar (2019). https://www.chargestar.com.au/. Accessed 10 Apr 2019

23. Go Electric Stations S.r.l.s (2019). https://goelectricstations.com/. Accessed 2 Apr 2019

24. GeoTelematic Solutions, Inc. (2019) OpenGTS. http://www.opengts.org/. Accessed 2 Apr 2019

25. Traccar Ltd. (2019). https://www.traccar.org/. Accessed 2 Apr 2019

26. GpsGate AB (2019). https://gpsgate.com/. Accessed 2 Apr 2019

27. NETSTAR EZY2c GPS Tracking (2019). https://www.ezy2c.com.au/. Accessed 23 Apr 2019

28. Verizon Fleetmatics (2019). https://www.verizonconnect.com/au/fleetmatics/. Accessed 23 Apr 2019

29. Linxio (2019) In: linxio.com. https://linxio.com/. Accessed 23 Apr 2019

30. SolarEdge Technologies Inc. (2019). https://www.solaredge.com/aus/. Accessed 2 Apr 2019

31. Synergy Solar & Battery (2019) In: Synergy. https://www.synergy.net.au:443/Solar-and-battery. Accessed 2 Apr 2019

32. Rybak IB, Wood D, Murray J et al (2017) Systems and methods for extraction and telemetry of vehicle operational data from an internal automotive network

33. Amadeo M, Campolo C, Molinaro A (2016) Information-centric networking for connected vehicles: a survey and future perspectives. IEEE Commun Mag 54:98–104. https://doi.org/10.1109/MCOM.2016.7402268

34. Siegel JE, Erb DC, Sarma SE (2018) A survey of the connected vehicle landscape—architectures, enabling technologies, applications, and development areas. IEEE Trans Intell Transp Syst 19:2391–2406. https://doi.org/10.1109/TITS.2017.2749459

35. Amjad M, Ahmad A, Rehmani MH, Umer T (2018) A review of EVs charging: from the perspective of energy optimization, optimization approaches, and charging techniques. Transp Res Part D Transp Environ 62:386–417. https://doi.org/10.1016/j.trd.2018.03.006

36. Shaukat N, Khan B, Ali SM et al (2018) A survey on electric vehicle transportation within smart grid system. Renew Sustain Energy Rev 81:1329–1349. https://doi.org/10.1016/j.rser.2017.05.092

37. Zhou Y, Kumar R, Tang S (2018) Incentive-based distributed scheduling of electric vehicle charging under uncertainty. IEEE Trans Power Syst 1–1. https://doi.org/10.1109/TPWRS.2018.2868501

38. Manghani H, Prasanth Ram J, Rajasekar N (2018) An internet of things to maximum power point tracking approach of solar PV array. In: SenGupta S, Zobaa AF, Sherpa KS, Bhoi AK (eds) Advances in smart grid and renewable energy. Springer Singapore, pp 401–409

39. Touati F, Al-Hitmi MA, Chowdhury NA et al (2016) Investigation of solar PV performance under Doha weather using a customized measurement and monitoring system. Renew Energy 89:564–577. https://doi.org/10.1016/j.renene.2015.12.046

40. Hill GA, Blythe PT, Suresh V (2012) Tracking and managing real world electric vehicle power usage and supply. In: 9th IET data fusion & target tracking conference (DF&TT 2012): algorithms & applications. IET, London, UK, pp 15–15

41. Tritium Pty Ltd. (2019) VEEFIL-RT 50KW DC FAST CHARGER. https://www.tritium.com.au/product/productitem?url=veefil-rt-50kw-dc-fast-charger. Accessed 3 Apr 2019

42. Anegawa T (2010) Development of quick charging system for electric vehicle. In: Proceedings of world energy congress. Tokyo Electric Power Company

43. Commission IE (2014) IEC 62196-3: 2014. plugs, socket-outlets, vehicle connectors and vehicle inlets–conductive charging of electric vehicles–part 3

44. ElaadNL (2019). https://www.elaad.nl/. Accessed 3 Apr 2019

45. Google Developers (2019). https://developers.google.com/chart/interactive/docs/reference. Accessed 3 Apr 2019

46. Xiamen Four-Faith Communication Technology Co., Ltd. (2019) F2414 UMTS/WCDMA/HSDPA/HSUPA IP MODEM(DTU). https://en.four-faith.com/f2414-wcdma-ip-modem.html. Accessed 3 Apr 2019

47. NXP Semiconductors (2019) MIFARE DESFire EV1. https://www.mifare.net/en/products/chip-card-ics/mifare-desfire/mifare-desfire-ev1/. Accessed 3 Apr 2019

48. Astra Telematics Limited (2019) AT110 GPS-GPRS vehicle tracking device. https://gps-telematics.co.uk/products/at110-gps-gprs-fleet-management-applications/

49. Astra Telematics Limited (2019) AT240 3G UMTS/GNSS IP67 waterproof vehicle tracking device. https://gps-telematics.co.uk/products/at240-waterproof-vehicle-tracking-device/

50. u-blox AG (2016) EVA-M8 series. https://www.u-blox.com/en/product/eva-m8-series

51. Yiu J (2015) ARMv8-M architecture technical overview. ARM Limited. https://community.arm.com/cfs-file/__key/telligent-evolution-components-attachments/01-2142-00-00-00-00-66-90/Whitepaper-_2D00_-ARMv8_2D00_M-Architecture-Technical-Overview.pdf

52. jQuery JavaScript Library (2019). https://jquery.com/

53. Vanderkam D (2019) Interactive visualizations of time series using JavaScript and the HTML canvas. https://github.com/danvk/dygraphs

54. Guy JJ (2019) Python module to create heatmaps. https://github.com/jjguy/heatmap

55. SMA Solar Technology AG (2017) Sunny Tripower 5000TL-12000TL. https://files.sma.de/dl/17781/STP12000TL-DAU1723-V10web.pdf

56. SMA Solar Technology AG (2017) Sunny Webbox. http://files.sma.de/dl/4253/SWebBox-BA-US-en-34.pdf

57. Asial Corporation (2019) JpGraph. https://jpgraph.net/. Accessed 4 Apr 2019
58. Mullan J, Harries D, Bräunl T, Whitely S (2012) The technical, economic and commercial viability of the vehicle-to-grid concept. Energy Policy 48:394–406. https://doi.org/10.1016/j.enpol.2012.05.042

Part III
Security and Privacy in the IoT

Chapter 9
Security and Privacy Challenges in Vehicular Ad Hoc Networks

Muath Obaidat, Matluba Khodjaeva, Jennifer Holst
and Mohamed Ben Zid

Abstract The steady increase in the number of vehicles on the roads comes with an increased number of accidents and fatalities. The manufacturers' interest in providing services to the driver (customer), along with safety applications, have contributed to connecting vehicles in networks on the fly (i.e., Ad Hoc networks), to provide certain services and information to the driver. These Ad Hoc networks consist of mobile vehicles that are located in a certain geographical zone and within a certain radius of each other, and communicate with each other or with road side units (RSUs) over the wireless medium. In addition, these mobile vehicles share some common characteristic, e.g., driving direction. These networks are well known as Vehicular Ad Hoc Networks (VANETs). From the extensive adoption and development of Internet of Things (IoT), and the integration of and convergence between VANETs and IoT, has emerged a new type of network known as Internet of Vehicles (IoV). VANETs, IoV and Intelligent Transportation Systems (ITS) have witnessed an explosive growth over the past two decades. This growth and the wide gamut of applications and services, these systems and networks have also increased the threats and attacks against these networks as well as raised many security and privacy concerns. In this book chapter, we describe the need for security in VANETs in terms of security requirements, current challenges in securing VANETs and present current security issues related to these networks. In particular, we deliver a comprehensive and up-to-date summary of threats and attacks in VANETs. Additionally, we present proposed solutions and countermeasures to mitigate these threats and attacks in order to secure and defend VANETs against them.

M. Obaidat (✉) · M. Khodjaeva · J. Holst · M. Ben Zid
Center for Cybercrime Studies, John Jay College of Criminal Justice
of the City University of New York, New York, USA
e-mail: mobaidat@jjay.cuny.edu

M. Khodjaeva
e-mail: mkhodjaeva@jjay.cuny.edu

J. Holst
e-mail: jholst@jjay.cuny.edu

M. Ben Zid
e-mail: mbenzid@jjay.cuny.edu

© Springer Nature Switzerland AG 2020
Z. Mahmood (ed.), *Connected Vehicles in the Internet of Things*,
https://doi.org/10.1007/978-3-030-36167-9_9

223

Keywords VANET · Security · Safety · Attack · Privacy · Threat ·
Countermeasure · RSU · Availability · Authentication · Integrity ·
Non-repudiation · Topology · Mobility · Key · Encryption

9.1 Introduction

Mobile Ad Hoc Network (MANET) is a mobile wireless Ad Hoc network that is
self-configured, infrastructure-less, autonomous and multi-hop [1–3]. A Vehicle Ad
Hoc Network (VANET) is a special type of MANET, where nodes are mobile
vehicles that communicate over the wireless medium [1, 2]. The topology is very
dynamic and the wireless link characteristics may change significantly over time as
mobile vehicles (i.e., nodes) join and leave the network frequently [1, 2]. Hence,
trust between vehicular nodes exists for a very short space of time. In addition,
VANETs differ from MANETs in many aspects. Two of the differences relate to the
number of vehicles that cannot be predicted in advance, and vehicle density that is
often unknown, random and asymmetrical [2].

The advancement of technology has led to the exponential growth of the Internet
of Things (IoT). To serve the purpose of safety and provide more services neces-
sitates connecting vehicles to the Internet, as in the IoT. This has led VANETs to
evolve to become a more sophisticated network with more services and applications
known as Internet of Vehicles (IoV) [4]. With its connectivity to the Internet, IoV
has exposed itself to new threat and attack vectors more than VANETs.

One of the main distinctions between VANETs and IoV is the geographic
coverage. Besides, VANETs are smaller in scale than IoV. IoV is a global network
that integrates vehicles, the Internet and the relevant infrastructure into one network
with a broad range of services and applications [4, 5]. In addition, VANETs are
constrained by the number of vehicles joining or leaving the network and the
mobility of them. VANETs are restricted in scope and functionality in an urban
environment due to the obstacles in the terrain such as tall buildings, traffic jams
and irregularity and complexity of the road network. However, IoV does not suffer
from these constraints and is not affected by such limitations [5]. Even though IoV
provides more functionality, services and applications, and is witnessing increased
interest, it comes with more security and privacy challenges and issues [4]. This is
due to the heterogeneity of IoV networks, that is a consequence of its wide variety
of technologies, standards, services and applications [4].

VANETs have become an important area of research in Intelligent
Transportation Systems (ITS) since the initial and main application of such networks is the safety of drivers on the road. According to the U.S. Department of
Transportation Federal Highway Administration, the number of vehicles on the
road in 2017 was approximately 272.48 million [6]. The number of traffic related
fatalities for the same year was approximately 37 thousand [7]. In 2017, the World
Health Organization (WHO) reported that more than 1.25 million human lives were
lost due to car accidents, while 20–50 million people were left with some injury or

disability [8, 9]. It was cited in [2] that the time between the broadcast of an early warning message and the driver taking action is approximately 0.75–1.5 s.

VANETs have three main application categories: road safety, comfort and infotainment, and traffic management via enhancing the effectiveness of transportation systems [10, 11]. The main goal of safety applications is to increase safety on the road, decrease probability of accidents and save lives [11, 12] by disseminating safety and warning messages among vehicular nodes in the networks [2]. Some of the safety examples are intersection collision warning, lane change assistance, and response to hazardous conditions. Some of the traffic management applications aim at congestion avoidance, speed limit notifications and navigation guidelines. Comfort and infotainment applications provide drivers with services to enhance their driving experience. These services that usually require connection to the Internet, include video streaming, weather information and gaming, among others [10, 11].

The fact that VANETs' main goal is safety on the road and saving of lives makes it necessary to secure these networks. Just like MANETs, VANETs are vulnerable to many kinds of threats and attack vectors. This is even more crucial when we address VANETs in the context of first responders and law enforcement agencies.

Since VANET is the core of Intelligent Transport Systems (ITS), and hence of the IoV, this chapter focuses on addressing the security and privacy concerns in vehicular networks. It provides an overview of the state of the art in VANETs, security challenges and an up-to-date presentation of threats and attacks against vehicular networks, along with proposed countermeasures to mitigate them. Most of what this chapter covers applies to IoV as well.

The rest of the chapter is organized as follows. In Sect. 9.2, we provide an overview, cover the architecture and components of VANETs, and address some of the unique characteristics of VANETs. Section 9.3 presents security and privacy in vehicular networks, including the need for security, security requirements, challenges in implementing security in VANETs, and adversary models. Classification of threats and attacks against vehicular networks is presented in Sect. 9.4. Mitigation and countermeasures are discussed in Sect. 9.5. Section 9.6 concludes the chapter.

9.2 VANET Overview

VANET is a type of a mobile Ad Hoc network where vehicles use the wireless medium to establish communications amongst themselves. Short- and long-range radio technologies are used to establish V2V (vehicle-to-vehicle, i.e., inter-vehicle) communication and V2I (vehicle-to-infrastructure, i.e., intra-vehicle) communication [4, 10, 13]. VANETs have many applications such as safety (e.g., collision avoidance, accident prevention), entertainment, traffic management (e.g., escape congested zones, reroute traffic in case of an accident) advertisement and weathercast [2, 4, 10, 13]. Due to the wide range of applications that VANET provides, this setup is known as Intelligent Transport System (ITS) [10, 13]. Communications in VANETs can be peer-to-peer (P2P) or multi-hop [2, 10, 13]. Even though VANETs are considered a subset of MANETs, they differ in some aspects [10, 13]. Some of these differences are frequently changing topology, high possibility of network segmentation, and high mobility, while power consumption is not an issue in VANETs [2, 10, 13].

We now review the different lines of communication within VANETs, introduce and describe the fundamental components and interfaces in a VANET, and note the characteristics that distinguish VANETs from other Ad Hoc networks.

9.2.1 VANET Architecture Components

VANETs consist of the following main modules: vehicles, Road Side Units (RSU) and the Trusted Authority (TA). The communication in VANETs illustrated in Fig. 9.1, take the form of vehicle-to-vehicle (V2V), vehicle-to-infrastructure (V2I), infrastructure-to-infrastructure (I2I) [14], hybrid communication which is a combination of V2V and V2I [15–18], and vehicle-to-anything (V2X) [8]. To enable such communication, the main components that make up a VANET include the application unit (AU), on board unit (OBU), and the road side unit (RSU). RSUs are usually connected to the network access which in turn is connected to the Internet. The following, as shown in Fig. 9.2, are the fundamental components of VANETs with a brief description of each:

- Application Unit (AU): The AU uses an application and is in charge of mobility and networking tasks. This unit might be coupled along with the OBU in one physical device or it could communicate with the OBU via wired or wireless medium. This unit uses the OBU to communicate with the network as the OBU is in control of that function [14, 16, 19].
- On Board Unit (OBU): This interface connects to other OBUs and other network components. It is the unit that is used to exchange information with other components (i.e., OBUs and RSUs) in VANET architecture using the IEEE 802.11p radio technology. The OBU consists of sensors (e.g. GPS), a network device (i.e., wireless communication element), a central control module (CCM), and an interface component [19]. The CCM consists of a user interface, a resource command processor (RCP), memory used for storage and information retrieval that includes read/write operations, and a transceiver [10, 14, 19, 20]. Data is usually collected and processed at the OBU [21]. In addition to providing wireless radio communications, OBU has several other functions including geographical and Ad Hoc routing, data security, network congestion control mechanism, message dissemination, and IP mobility [14].

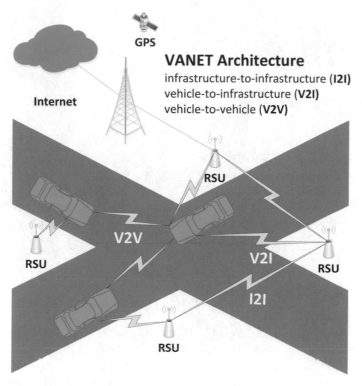

Fig. 9.1 VANET general architecture

- Road Side Unit (RSU): This is a fixed infrastructure unit that is usually located
  on the road side or in vital locations such as intersections or active areas (e.g.,
  parking lots) [10, 20]. The RSU is a wireless access in vehicular environments
  (WAVE) or dedicated short-range communications (DSRC) device. RSU
  includes a short-range wireless communications device based on the IEEE
  802.11p wireless technology to enable communications with vehicles on the
  road [10, 14, 17, 19]. Also, it has the capability to communicate with other
  infrastructure units in the network [10, 20]. The role of the RSU is similar to the
  role of an access point (AP) in wireless Ad Hoc networks [10, 21]. So, it
  disseminates and forwards (routes) traffic.
  RSUs deliver information such as traffic status, infotainment, and safety messages from central authorities [15]. These units perform the following functions:
  (i) disseminate and redistribute messages to other RSUs and OBUs to extend the
  communication range beyond a single Ad Hoc network; (ii) function as a
  gateway to OBUs to the Internet; and (iii) act as a data source and run safety
  applications using infrastructure-to-vehicle communications [10, 14–16, 20]. In
  addition, all RSUs (within a certain geographical zone) are interconnected to
  each other [20].

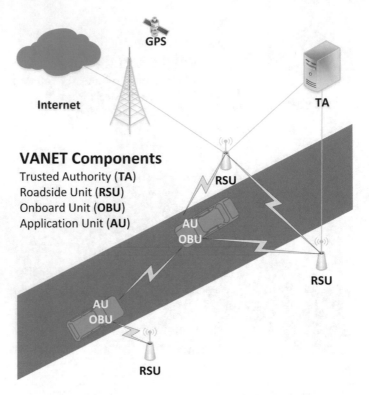

Fig. 9.2 VANET components

- Trusted Authority (TA): This is the unit responsible for controlling and managing all RSUs. It has high computational and storage capacity as it authenticates all vehicles and invalidates vehicles that transmit false messages. A TA has many security functions, such as verification of digital signatures and certificates [10, 20, 21] among others.

9.2.2 VANET Characteristics

VANET shares many common characteristics with other Ad Hoc networks. Nonetheless it has its own characteristics that makes it different from other Ad Hoc networks and in particular MANETs. This sub-section is not meant to be an exhaustive list, rather it only covers characteristics that pertain to VANETs. The following are the unique characteristics of VANETs:

- Power: Because vehicles are equipped with rechargeable batteries they have longer life expectancy than MANETs. Therefore, there is no issue of power constraints in VANETs [8, 10, 14, 22, 23].
- Mobility: Vehicles are constrained by the road topology, and are obligated to comply with traffic lights and road signs, and react to other vehicles in the VANET. Therefore, there is a higher degree of predictability in their mobility pattern in comparison to other Ad Hoc networks where mobility is completely random and unpredictable [8, 10, 14, 16, 20, 22, 23].
- Network Density: Vehicle density on the road depends on different factors, the services the roads provide (e.g. local vs. highway), the time of the day (e.g., peak traffic periods vs. non-peak periods), and the geographical location of the road (e.g., city vs. suburban) [8, 14, 22].
- Topology: VANETs are very dynamic due to high mobility and vehicles joining and leaving the network. Driver behavior is another factor in rapid changing topology. The change in wireless link quality contributes to dynamic topology as this is affected by the radio communication range [10, 14, 16, 20, 23, 24].
- Scale: VANETs could be very large in scale such as in urban dense environments and busy highways [14, 22]. This has a direct impact on communication and the infrastructure, since there is a lot of traffic generated by different users in VANETs [22, 24].
- Computational Ability: Unlike other types of Ad Hoc networks, nodes (i.e., vehicles) in VANETs have sufficient computational resources such as Global Positioning System (GPS), processors, memory capacity, and smart antennas that enables vehicles to report accurate information regarding their position, speed, direction and other necessary messages to exchange [10, 14, 20].
- Delay: There are some delay constraints on time critical applications such as early warning messages, collision warnings, pre-crash warnings and network breakdown [25]. This is to guarantee that a message is received and acted upon by the driver in a timely manner [16, 20, 22, 24].

9.3 Security and Privacy in Vehicular Networks

The security aim in wireless networks is usually to achieve the goals of information security, availability, confidentiality and integrity. On the other hand, security in VANETs is different from other Ad Hoc networks as the concern significantly affects and deals with human lives.

In this section, we present the need for security in VANETs with respect to the safety concerns, privacy of the driver and privacy of vehicle data contained within the network. We enumerate security requirements in terms of confidentiality, integrity, availability and other important principles. Challenges in implementing security, given the inherent nature and constraints of VANETs, are also reviewed. Finally, we provide a taxonomy of adversary types and models that describe the methods and motives behind their attacks.

9.3.1 Need for Security

VANETs are extensively used in Intelligent Transportation Systems (ITS) to support various relevant applications. One of the key concerns in VANETs is safety, security and privacy.

Providing security in VANETs is a challenging task due to their high mobility, dynamic topology and frequent wireless links breakages as discussed in Sect. 9.2.2. Even though there is a common basic security model for VANETs, different applications may require additional security measures. For example, an infotainment application needs a different security approach than a safety application that disseminates an early warning message. This increases the cost of secure communications in VANETs. The fact that a VANET is a self-organized network with no infrastructure makes it more vulnerable to attacks.

In addition, the nature of wireless medium makes these networks vulnerable to threats and attacks that can exploit the broadcast nature of the radio technology used in VANETs [8, 26]. Sensitive information is being communicated over the wireless channel in VANETs. This data must be secured as it draws adversaries' attention. The absence of security in VANETs exposes them to different types of attacks that can have serious consequences for the safety of drivers on the road including loss of life.

Users' privacy must be protected, as information exchanged between vehicular nodes contains private information about the driver and the vehicle. In addition, the control authority should be able to identify the driver's liability while at the same time preserving the driver's privacy [27]. So, privacy aware and secure communication must be an integral component of VANETs [28]. Forged information broadcasted by a malicious node (vehicle) could have serious consequences for drivers, passengers and pedestrians lives. Also, if a legitimate safety message sent by a VANET component is modified, delayed or dropped by an attacker, then it can have severe consequences for human lives as well [29]. This means that only legitimate and true information should be communicated over VANETs [28].

As vehicles become smarter, security becomes an ever more challenging issue for VANETs. It is reported that 100% of 2016 vehicle models contain wireless technology and 60% of vehicles for the same year have some Internet access [30, 31]. This leads to a VANET of vulnerabilities as vehicles can join and leave the network without any security measures [31]. Smart vehicles contain tens of Electrical Control Units (ECUs) as part of their in-vehicle network. If these ECUs are compromised they can jeopardize drivers on the road [31]. A security attack on VANETs can have an enormous impact on human lives and assets [8].

All previously discussed reasons contribute to the need for better security in VANETs. Nevertheless, security and privacy requirements must be fulfilled in order to provide secure communications in VANETs [26, 28].

9.3.2 Security and Privacy Requirements

Safety of data and communication is the main requirement of VANETs, so it is essential to have a secure VANET system. Messages communicated can save lives or cause fatalities, as in the case of a malicious node injecting false information. Before discussing threats and attack vectors against VANETs, it is paramount to address security and privacy requirements in VANETs. Failure to satisfy these requirements may lead to serious exploits of vulnerabilities in VANETs. A framework that satisfies security and privacy must fulfil the core requirements in order to have a secure VANET. These requirements are defined in various perspectives as discussed below.

Integrity and Data Trust

This is an important factor when communication takes place between two parties. A message that is sent by a sender must be received without alteration at the receiver side [27, 32]. The content of the message should not be modified, deleted or dropped [4, 10, 13, 17, 20, 24, 33–35]. If a fabricated message is received and accepted by any node on the network, integrity is violated [32]. For example, a warning message that has been modified can lead to serious consequences on the road. A detection scheme must be implemented in case of any malicious activity on the network [13].

Authentication and Identification

In order to allow only the legitimate traffic communication over the wireless link in VANETs, all vehicles (nodes) must be authenticated before being allowed to join and transmit any data. This authentication is to ensure that a malicious vehicle is denied access to join the network and messages sent by such nodes are blocked before they are communicated [4, 10, 13, 20, 21, 24, 25, 33–37]. This is to prevent a malicious node from duplicating a genuine node's identity or impersonating another vehicle's identity [37]. This is critical in VANETs as vehicles respond to a message from other vehicles in the network. For example, if an early warning message is deleted by a malicious node, this can have serious consequences, as other drivers will not respond accordingly [31]. Authentication can be achieved via using a certificate by the originator and using a pseudonym by the receiver to verify sender's identity [27]. Authentication can prevent many attacks such as the Sybil attack, which is discussed below, by giving a unique ID to each vehicle in the network [32]. In addition, a strong authentication mechanism is very important in providing legal forensic evidence to law enforcement agencies [32]. Only authenticated and authorized vehicles and RSUs should use and benefit from services the VANET provides [33].

Availability

Users (vehicles) should be able to send and receive messages at all times even when a VANET is under an attack such as a D/DoS or jamming attack [25, 27] or under any malicious activity [27, 32]. For example, in a very congested area the authentication server might fail due to unintentional DoS [4, 10, 20, 31, 33–35]. Part of availability is to have a backup in case there is a break in the network [38]. Availability requires abundant bandwidth and high connectivity [37]. Also, VANET design must be resilient with a high degree of fault tolerance and have the ability to resume normal operations after an incident [32, 39]. The importance of availability arises from the fact that some messages are delay sensitive and must be transmitted in real time, otherwise they will lose their value (e.g., message about road conditions) and might even be harmful (e.g., hazardous reporting message) to the users in the network [32].

Privacy

Drivers' privacy must be preserved even when they are liable for an act on the road. A malicious node (vehicle) in VANET should not be able to get access to neither driver nor a vehicle's private information [27]. Vehicle information must be preserved during communication (i.e., sending and receiving messages). The actual identity of the vehicle, driver and location should not be disclosed to any unauthorized parties. Therefore, actual identity information should not be known from the message [4, 10, 13, 20, 25, 33, 35, 36, 40]. A certain level of anonymity must be preserved while certain authorities should be able to obtain the driver's identity and vehicle information in case of any liability issue. It is worth mentioning here that privacy is different from confidentiality; privacy is not to allow a malicious node to access driver information, while confidentiality is to hide the data communicated over the network [33].

Confidentiality

Data transmitted between two vehicles or between a vehicle and an RSU must be protected from being leaked and eavesdropped upon [10, 32, 41]. Other users on the network should not be able to understand the message communicated between two parties on the network [32, 41]. Confidentiality is achieved via implementing encryption schemes which protect drivers' confidential data such as identities [32]. Encryption mechanisms could be symmetric or asymmetric techniques.

Non-repudiation

This is to ensure that vehicles in the network cannot deny their responsibility for certain acts such as sending a message [10, 13, 19, 20, 25, 27, 32, 34, 37]. The TA is the only authority that should have access to this information [24, 36]. Non-repudiation enables identifying attackers by providing forensic evidence regarding the origin of the message, which is useful in tracking communication exchange in case of an accident [37]. The objective of non-repudiation is to have evidence and make it available and denial-proof in case of certain action or in disputing a claim [32]. All vehicle information, such as speed and time stamp, is stored in a Tamper Proof Device and only authorized authorities are able to recover such information [32].

Access Control

This is about implementing access rights and privileges for the users of the network. A user in VANET should not be able to hear the communication between other vehicles that communicate sensitive information such as law enforcement [32]. Each user must have access only to the services and applications that are needed to perform their task and not gain access to other services [32]. This is to prevent vehicles on the road from accessing applications or services that they should not have access to. Access control policies should be implemented to determine the rights and privileges of each entity on the network, which includes RSU and vehicular nodes [32].

Traceability and Revocation

These are required for an authorized entity to be able to trace a malicious vehicle in the network [21]. The authority is responsible for revoking the abuser vehicle's certificate in the network [21]. The network must have the mechanism to track a vehicle, based on its identity [33].

Data Verification

Only true and legitimate messages should be allowed on the network. This is to exclude fabricated messages and to reduce the risk of threats and attacks that try to inject false information into the network [27, 33, 34, 36].

9.3.3   Challenges in Implementing Security in VANETs

This section presents some challenges with regards to implementing security mechanisms in VANETs.

Security and Privacy Balance

This is one of the most challenging requirements in VANET communication. VANETs must be secure and trust must be established between senders and receivers without violating users' privacy [14]. Messages must be transmitted securely and at the same time authenticated. This should be accomplished without knowing the actual identity of the vehicle or being able to track vehicles from these messages [16, 24, 27]. Also, accountability of any action must be clearly identified, particularly in case of an accident [42]. However, privacy concerns must be addressed and handled properly, such as when biometrics are used to access and control vehicles [42].

Scalability

VANETs are generally very large Ad Hoc networks in comparison to other types such as MANETs. Nonetheless, there are no standards for them, as each region has its own VANET [37]. For example, the standard for DSRC in the U.S. is different from that of its peer in Europe [37]. So, there is a challenge in finding common agreement between different regions [27]. This increases complexity and decreases performance [10, 13]. Communicating an early warning message to the maximum number of drivers in a certain zone in order for them to react and take action in a timely manner in a network where connections are sporadic is hard to accomplish [16, 22, 33, 37, 43]. Also, scalability introduces complexity in distributing encryption keys [10, 13]. Scalability must be satisfied even in dense areas and highly congested zones without compromising VANET security or performance.

Real Time Constraint

Safety applications (e.g., early warning messages) and vehicular assistance applications are time critical and delay sensitive. They should have higher priority and be transmitted and delivered in the order of less than a few hundred milliseconds [2, 37, 42]. This imposes restrictions on the cryptographic and authentication mechanisms implemented in VANETs, as they should be fast, reliable, resilient, not complex, and minimize the number of broadcasts [23, 34, 37, 42].

Low Error Tolerance

Some of the protocols implemented in VANETs are probabilistic protocols [23, 34]. This means that an error can have hazardous consequences [23]. In VANETs there is very little tolerance for errors, as VANETs deal with human lives. Also, drivers must take action in response to an incident in the order of few hundred milliseconds in critical situations [16].

Key Distribution

Majority of security protocols use cryptography as a technique to encrypt and decrypt messages, and use keys as a means of implementation [27]. The keys that are used at different ends of the communication could be symmetric keys or asymmetric keys [23]. With many manufacturers installing their own keys, this imposes a challenge in distributing keys in VANETs [16, 34]. In addition, it is very challenging to regulate who is the Certificate Authority (CA) as there are many players such as the government and vehicular manufacturers [39].

Mobility

VANETs are characterized by high mobility. Therefore, execution time for security protocols and encryption mechanisms must be fast and efficient, as nodes (vehicles) in VANETs do not remain within the same network for a long time, which causes frequent wireless link breakage [34]. Messages not authenticated within a certain time frame can lead to an inefficient response to an incident that requires a fast reaction [43]. Also, high mobility makes it very difficult to perform a handshake to establish secure communication [27]. High mobility can lead to reliability issues. After an accident occurs, VANET has the capability to identify the message source, so it helps in liability matters as well as in the case of an attack [27, 37]. On the other hand, privacy of users must be preserved through providing anonymity of the message source and protecting private information such as the driver's biometric data [37].

Reliability

Because the link's lifetime is very short, it is a very challenging task to ensure that the message is received by the intended nodes and acknowledged [34].

Volatility

Due to high mobility in VANETs, the lifetime of links between vehicles is very short, so implementing security measures to verify users identities is a challenge [10, 13]. Using a password-based mechanism to establish secure vehicular communication is very hard to accomplish because of the absence of relatively long-lived connections [37].

Heterogeneity

VANETs consist of vehicles from different manufacturers and, therefore, have many systems and subsystems that are not standardized. In addition, the numerous applications used in VANETs and their requirements increase the degree of heterogeneity, adding another vector of complexity to secure the network [42]. Efficiency and reliability are the requirements that address heterogeneity to provide security in VANETs [42].

9.3.4   Adversaries and Adversary Models

Many attack vectors exist in wireless Ad Hoc networks; VANETs are no exception. A strong authentication mechanism should be implemented in order to prevent many attacks. This sub-section presents types of adversaries and a list of adversary models known up to now.

Based on the adversaries' objectives and activities, researchers have come up with a list of recognized names, as mentioned below:

- Selfish and Greedy Driver: This is a driver who carries out an attack for personal benefit or interest. The adversary sends a false warning message that deceives other drivers on the road to make them choose an alternate route (i.e., redirects traffic) so the adversary can have the road clear just for itself. Another scenario is where the adversary sends a false message about an accident to create congestion on the road [27, 44].
- Eavesdropper: This adversary tries to gather information about the VANET, its resources, and drivers. The information is then used to launch other attacks on the system [27].
- Prankster: This adversary's attacks are solely for personal enjoyment and satisfaction. It exploits a vulnerability in the system to launch an attack such as a DoS (Denial of Service) attack or to send a false message to create congestion on the road [27, 44].

- Industrial Insider: This adversary is often an employee or someone within the organization who tampers with hardware components during a firmware update or key distribution [27].
- Malicious Adversary: This type of adversary uses applications in a VANET to cause harm and damage to the network. It usually has knowledge of the network and the network's resources [27, 44].

Classification of adversary models is based on the method, membership, motivation and scope of attack [8].

- Insider versus Outsider: An insider vehicle (i.e., a node) is an authenticated vehicle that passed all authentication measures before gaining access and joining the network, meaning it is part of the network. On the other hand, an outsider is a vehicle that is not part of the network and has not passed the authentication mechanism required for joining the network [17, 21–23, 27, 45, 46]. As in any network, the insider threat is more dangerous than the threat from the outside.
- Active versus Passive: In a passive attack, the adversary has no authorization and is not doing any harm other than sniffing or listening to the communication over the wireless channel. An active attacker can actually do harm by modifying a message or injecting a harmful message on the network to accomplish some gain [17, 21, 23, 27, 32, 45, 46]. As described, it is clear that authentication cannot prevent passive attackers from eavesdropping on communication, while it can prevent active attacker from harming the network.
- Malicious versus Rational: While rational attackers have a personal interest in gaining something out of their attacks, malicious attackers have the intention of causing harm to the system or other users in the VANET, without such a personal interest. Authentication does not distinguish between these two varieties of attackers [17, 22, 23, 45, 46].
- Local versus Extended: A local attacker targets vehicles and RSUs that are relatively close in vicinity, which limits its attack to a local zone. Extended (global) attackers extend their attack beyond their local zone by targeting vehicles and RSUs that are sporadically disseminated throughout the network. So, the extended attacker has larger view of the network [21, 23, 27, 45].
- Static versus Adaptive: The static attacker carries out its plan regardless of the outcome of the attack (i.e. they are not sure whether the attack succeeds or not). Adaptive attackers, on the other hand, have a wealth of information and knowledge of the target network configuration and parameters, and act accordingly [21]. This means the adaptive adversary knows the results and consequences of the launched attack.

9.4 Threats and Attacks in Vehicular Networks

Vehicular networks are vulnerable to different threats and attacks, as already discussed in previous sections. However, Internet of Vehicles (IoV) is more vulnerable than VANETs since IoV is connected to the Internet and the cloud, which increases the number of threats as well as the probability and the severity of an attack launched against it. On the other hand, VANETs are not connected to the Internet and are limited in their scope. So, the number of threats and attacks is lower in VANETs.

This section presents a taxonomy of different attacks on VANETs. The taxonomy is with respect to some of the security requirements discussed in Sect. 9.3.2. These attacks on vulnerabilities of VANETs can be classified into attacks on the fundamental requirements such as authenticity, availability, integrity and data trust, confidentiality, and non-repudiation. Different forms of attacks are discussed in the following sub-sections, as they appear in the literature and other research studies as cited below.

9.4.1 Attacks on Authenticity and Identification

Authentication verifies identity using login credentials (e.g. login and password) [47]. This is an essential security requirement in VANET as it grants nodes access to join the network and avail of its services through a verification process. Authentication provides the first layer of security to protect users on the network from intruders (i.e., malicious users) who try to gain access to the network through spoofed ID or any other means. The goal of the attack is to allow unauthorized users to join the network and cause harm or damage to it. This could be through altering or forging messages, and injecting fabricated messages to disrupt and affect communication within a VANET. Some common forms of attacks are briefly discussed below.

- Message Tampering Attack: Communication between two entities must be secured. In this type of attack, as illustrated in Fig. 9.3, a genuine message from an authenticated user is captured by a malicious node. The content of the message is tampered with then forwarded to the destination [33, 48]. This attack violates the integrity principle [32]. A data verification mechanism can help mitigate such an attack by checking and verifying the content of the message [32].
- Impersonation Attack: An adversary uses the identity of a trustworthy user to utilize network resources and services or to gain access to sensitive information and send out fraudulent messages on the network [32, 33, 41, 42, 48]. The victim node whose ID is used to send these fraudulent messages might be excluded from using network services [32].

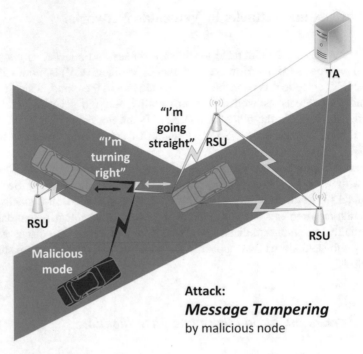

Fig. 9.3 Message tampering attack

- Replication Attack: In this attack, the adversary replicates the identity of a genuine node (i.e. vehicle) on the network. This attack adds nodes in the network that enable the malicious node to send fabricated messages [33].
- GPS Spoofing Attack: The integrity of the GPS signal is compromised in this attack. Information about the locations of all vehicles in the VANET is stored in a satellite [42, 49]. The GPS signal is vital in VANETs as it provides information about the location and the time for a vehicle. The adversary tries to modify the GPS signal and change the information randomly by providing a stronger signal than the genuine GPS signal [42, 48]. This causes the compromised vehicles in the VANET to synchronize with the new rogue GPS time [33]. See Fig. 9.4 for an illustration of this attack.
- Tunneling Attack: An adversary creates a communication channel between two parties in the network. Usually, the two parties are far apart in the network but look like they are neighbors. This tunnel-like communication between them is used to launch other attacks against the VANET to disrupt its functionality [33, 49]. Some of these attacks are the traffic analysis attack and the selective forwarding attack [42, 48].
- Free Riding Attack: A vehicle uses the services offered by the network and then leaves the network [33]. Two types of this attack exist: free riding without authentication, and free riding attack with fake authentication. In the first attack

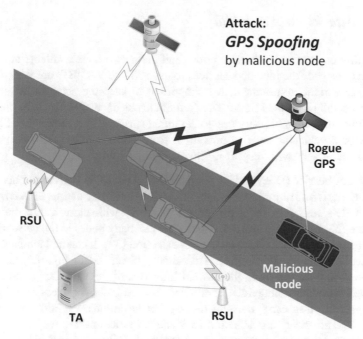

Fig. 9.4 GPS spoofing attack

(i.e., without authentication), a vehicle that is not part of the authentication process uses the authentication as a result from another vehicle. In the second attack (i.e., fake authentication), the attacker pretends it is part of the authentication process by generating fake authentication; this is subtler than the first attack [50].

- Sybil Attack: An adversary injects false information into a VANET and might be able to gain access to the entire network. The adversary acts as multiple nodes on the network by claiming multiple identities [39, 42, 48]. This eventually misleads other vehicles to make wrong decisions based on the false information received. Also, it affects the consistency throughout the VANET [33].
- Wormhole Attack: An adversary compromises two malicious nodes (i.e. vehicles) in the network and creates a tunnel between them. Three types of wormhole attack exist: (i) open, where malicious nodes (vehicles) have knowledge of all malicious nodes along the route and are part of the reply request packet (RRP) header; (ii) closed, where the message is communicated between the two ends of the wormhole; and (iii) half-open, where one side is able to alter the message [33].

9.4.2 Attacks on Availability

Availability ensures that vehicular nodes and the network are available at all times and able to provide the intended services to users. Since VANETs are very dynamic and a time variant environment, it is important to keep the network in functional status, especially in critical times. The goal of attacks on availability is to prevent an entity from providing its services to legitimate users. This means disabling the network or a vehicular node to prevent it from performing its normal functions. Some of the attacks with respect to the availability aspect are:

- Denial of Service (DoS) and Distributed DoS (DDoS): In this case, a node is not able to perform its normal tasks due to a malicious user or users overwhelming the node's resources (e.g., processor, memory) with more traffic than it can handle [33, 48, 51]. This attack usually affects many nodes in the network and is challenging to recover from even after detection [32]. It can be launched against vehicular nodes or RSUs [51]. This attack, illustrated in Fig. 9.5, has consequences beyond disabling the normal functions of the vehicular node or RSU. D/DoS can cause dangerous conditions, for example, suppression of the dissemination of an early warning message of hazardous conditions on the road [42]. The adversary can launch a DoS attack in one of three ways: (i) consume the vehicular nodes' resources and keep them fully used and busy; (ii) cause signal jams to disrupt communication; and (iii) drop the message [42].

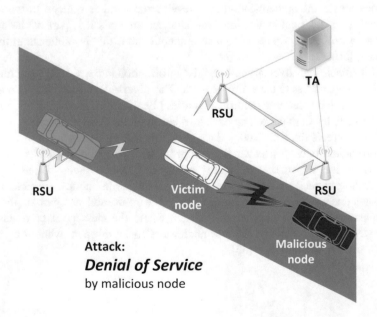

Fig. 9.5 Denial of Service (DoS) attack

- Black hole Attack: The adversary vehicular node does not participate in forwarding messages that pass through them causing a black hole in the network [42]. Another case is when a compromised node is instructed to drop messages instead of routing (i.e., forwarding) them. This increases latency and prevents some nodes from receiving a message in a timely manner, affecting their response time to an incident [4, 49].
- Spamming Attack: In this attack, the adversary transmits spam messages on the network such as advertisement or gaming [42]. These messages are forwarded by other receiving vehicular nodes on the VANET. This depletes precious resources of the VANET such as bandwidth [33, 42]. Ultimately, this attack increases latency for transmitted messages and can have serious consequences for VANET users [42, 48, 49].
- Jellyfish Attack: There are two types of this attack: (i) reordering attack where a malicious vehicular node reorders messages received before forwarding them, hence out of order acknowledgements are received, requiring the network to retransmit the messages [33]; and (ii) periodic dropping attack, where messages are dropped sporadically at random times from a communication session, causing wrong routing information at the vehicular node since congestion is acknowledged as the current state [33]. This makes a vehicular node drop messages from the communication session which leads to longer retransmission times. In Jellyfish attacks, the main aim of the adversary is to introduce a state of congestion so that the vehicular nodes change their routing tables accordingly [33].
- Gray Hole Attack: Messages are dropped by the vehicular node. This is a subtle attack to detect as there is no distinction between malicious and legitimate vehicular nodes as both are involved in forwarding messages during a communication session.
- Malicious Software (Malware Attack): The adversary is an insider who uses software to carry out the attack [48]. In particular, the attack targets the OBUs and the RSUs. This causes disruption in the network's normal operations [48, 49]. This attack usually infects VANETs when there is a software update to the intra-vehicular software component or RSU [51].
- Greedy Behavior Attacks: In this attack the adversary is looking for personal benefit. For example, the adversary might generate a false message (i.e., congestion or accident) along a road. This is to divert traffic so that the road becomes clear with minimal traffic [33].
- Jamming Attack: VANETs use the wireless shared medium as a means of communication. This makes them vulnerable to different attacks such as signal jamming or interference at the physical layer [33]. The adversary corrupts the radio wave signal by overpowering the original signal until communication is lost via the original signal [49].

9.4.3   Attacks on Confidentiality and Privacy

Data confidentiality must be preserved during a communication session between any two parties in a VANET. It assures that message content is not disclosed to any user on the network other than the intended recipient and can be read only by a legitimate authorized user [39]. VANET is characterized by high mobility, so it is more challenging to provide routing that guarantees security than in any other Ad Hoc networks [48]. Some relevant threats are briefly mentioned below.

- Man-in-the-middle attack: The adversary positions itself between the sender and receiver (V2V or V2I), where they are not aware of the adversary in the middle. This enables the attacker to be able to see what is communicated and gain access to modify, delete or drop a message or multiple messages [33, 49, 51].
- Eavesdropping attack: The attack is performed by getting the session key and listening to the communication session [33]. The aim of the adversary is to gain access to confidential information [32, 51].
- Traffic analysis attack: The purpose of this reconnaissance attack is to gather information about the target victim, and analyze and categorize traffic flows to later carry out an attack against the target [33, 49].
- Home attack: The adversary hijacks control of the vehicle from another user on the network. This enables the adversary to control all systems in the vehicle [52].
- Identity disclosing: The adversary tries to get the actual identity of the vehicular node's driver and gain access to sensitive information that can be further used against the driver, or to abuse and harm network resources.
- Location tracking: A vehicular node should not be tracked by any other vehicular node in the VANET. Only the trusted authority should be able to trace drivers on the network for protection and accountability purposes.

9.4.4   Attacks on Non-repudiation (Accountability)

In a repudiation attack, a malicious vehicular node denies sending or receiving a message in the case of sender or receiver respectively. In the latter case, it demands that the sender retransmit the message [51]. This consumes network resources (e.g., bandwidth) and increases latency throughout the VANET [33]. However, in the case of impersonation, it is difficult to recognize the adversary [32]; it is also difficult for the TA to track the adversary [49]. This attack lets the adversary deny accountability for their actions [42]. Also, a compromised vehicular node might be used to drop messages instead of forwarding them to their destination.

9.4.5   Attacks on Integrity and Data Trust

Some of the relevant threats in this category are discussed below.

- Replay attack: A malicious vehicular node receives a beacon from another legitimate node in the network. The malicious node stores the information in the beacon and intentionally replays it at a future time when the information is invalid or outdated, which can lead to harmful consequences on the road [33, 53]. The transmission of the replayed message is through a new connection [42]. For example, an old beacon contains the speed of a car with a high value, though later the speed changed to a lower value. The outdated information in the beacon when replayed periodically can insinuate to the following drivers that the car's speed is still high even though in reality it is not, which can lead to a collision or a series of collisions [53].
- Masquerading (spoofing) attack: The adversary spoofs the identity of another vehicle in the network and uses it to inject fabricated information into the network [33]. This is also considered an attack against authenticity [49]. For example, an adversary may pretend to be an emergency vehicle to clear the path for itself [49].
- Illusion attack: In this attack, illustrated in Fig. 9.6, an adversary injects counterfeit information into the network such as a report of an accident or congestion on the road. This deceives other drivers on the road and leads them to make the wrong decision based on fake, misleading information [33, 41]. This can have devastating consequences on the safety of the VANET users, create congestion, consume network resources and reduce network efficiency [41].
- Message tampering: Here the aim is to send wrong information into the network by changing the content of a message [4, 42]. For example, the adversary might send a false congestion message about a route to make other vehicles on the road change their route [49].
- Timing attack: In this attack the adversary receives messages, adds latency to them, then sends them out. This latency causes the messages to arrive later in time which defeats the purpose of safety on the road in the case of an early warning message [47, 52].

9.5   Mitigation and Countermeasures

This section discusses proposed solutions for threats and attacks presented in the previous section. Before going into the descriptions of these solutions, it is necessary to keep in mind VANETs' fundamental constraints such as message transmission delay of less than 100 ms, managing certificates, key distribution, and low error tolerance.

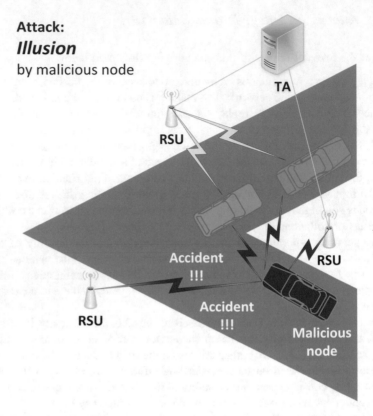

Fig. 9.6 Illusion attack

Most of the previously presented threats and attacks on VANETs can be mitigated by implementing cryptographic techniques, digital signatures and secure routing protocols or by using an intrusion detection system. Moreover, these techniques satisfy the security and privacy requirements in VANETs.

9.5.1 Intrusion Detection Systems

An Intrusion Detection System (IDS) is a fundamental element in securing networks and in particular VANETs. IDSs protect against attacks by collecting and analyzing data from various network components that define security rules which then serve as a baseline of the network behavior. Any behavior that violates these predefined security rules is flagged as anomalous behavior (i.e., a potential attack). These IDSs protect against both internal and external threats. Based on the detection approach, IDSs can be categorized into signature-based and anomaly-based detection systems [54], as follows.

- In a signature-based IDS, a database is created from the signatures of well-known attacks and the network state is checked against the database. If the network state matches any signature stored in the database it is flagged and a set of instructions is executed. It is clear that this detection systems works fine for predefined and known signature attacks. However, for a new, non-stored signature of attack, the detection system does not work, so the network is exploited via a vulnerability that was not protected from the new threat [4, 54, 55].
- An anomaly-based IDS is designed to have the baseline normal behavior of the network environment. Any traffic on the network that does not follow the predefined existing baseline behavior is flagged as abnormal and action is taken against the potential threat or attack. It takes a long time for the network to have enough information to define the baseline, which needs updating as the network evolves and new technologies are implemented [4, 54–56].

Nonetheless, IDSs are very flexible where new rules (i.e., security strategy) can be added on demand as needed at any time. However, this can be a costly and time-consuming process.

9.5.2 ID-Based Security Systems

An ID-based cryptography mechanism is used to maintain privacy by implementing a pseudonym-based technique. The scheme works by utilizing known information that can be used to determine the user's identity to verify digital signatures. The information could be one or a combination of user name, network address, or email address. The key advantage of ID-based security systems is the use of a pseudonym that can be changed as needed. Also, to preserve privacy, a user can have more than one pseudonym. The ID-based scheme has advantage over the Public Key Infrastructure (PKI) mechanism as it does not require storage, or fetching and validating the public key certificate by a trusted third party. This translates into less storage capacity, time and bandwidth required than PKI. The main drawback of this technique is user privacy assurance [4, 33, 55, 56].

9.5.3 Public Key/Asymmetric-Based Schemes

Key-based security solutions are implemented in various types of wireless networks and in VANETs as well. Digital signatures are implemented using public key cryptography (asymmetric cryptography). Asymmetric cryptography uses two mathematically different keys for communication called public and private keys. In VANETs, each node has a private key that no one else in the network has. On the sender's side, all messages that belong to the same vehicular node are signed using the private key before transmission. On the other side of the communication, the receiving vehicular node uses the private key to verify the authenticity of the message using the digital signature against the sender's public key. Hence, digital signatures in communication are used in data authentication and non-repudiation. So, the public key must be authenticated by vehicular nodes that are communicating [4, 33, 55, 56].

The Certificate Authority (CA) is the entity responsible for distributing the certificate to validate and authenticate public keys for each vehicular node in the network. Since VANETs are infrastructure-less and Public Key Infrastructure (PKI) requires infrastructure, it is a challenge to implement PKI in VANETs. In addition, PKI cannot guarantee effective authentication, fair revocation, and location privacy [33, 55, 56].

Different modified versions of PKI have been proposed in literature such as decentralized PKI [57, 58], elliptic curve digital signature algorithm (ECDSA) [59] and a scheme that uses a combination of Merkle tree (MT) and elliptic curve digital signature algorithm (ECDSA) proposed in [60].

9.5.4 Symmetric-Based Schemes

Symmetric cryptography is less computationally intensive than asymmetric cryptography (i.e., higher performance) but has a more complex key management mechanism. In [61], an authentication scheme is proposed that uses a trusted RSU with high performance capabilities to assist in V2V communication. One of the earliest devised authentication schemes is Timed Efficient Stream Loss-tolerant Authentication (TESLA) [62]; the scheme introduces latency as the receiver must wait for the key that is transmitted in a different message to authenticate the message received initially [55]. In [63], a message authentication scheme that utilizes TESLA protocol and PKI is proposed. The scheme satisfies authentication and non-repudiation requirements [55].

9.5.5 Secure Routing Protocols

Routing protocols in vehicular networks can serve as a major defense mechanism. Most threats and attacks can be prevented by implementing a secure routing protocol at the network layer of the VANET architecture. This sub-section presents some of the routing protocols that provide security.

Secure Ad Hoc on-demand distance vector (SAODV), is a variant of AODV. This protocol can provide a secure route discovery phase that is able to satisfy the authentication, integrity and non-repudiation requirements. It secures routing via digital signatures and a one-way hash function to validate the number of hops. SAODV protocol digitally signs each message to authenticate it. Intermediate vehicular nodes are neither able to tamper with the hashed hop count nor to change the message of the source or destination as the key fields in the route request packet are digitally signed [4, 47, 54].

Ariadne protocol is a variant of the Dynamic Source Routing (DSR) protocol employing symmetric key cryptography. It uses the TESLA scheme as authentication mechanism that implements a one-way hash function chain. Each vehicular node chooses a chain value as the TESLA-key to compute the MAC attached to the routing message between nodes using a shared key. The protocol introduces latency due to the authentication and using a one-way hash function chain [4, 47, 54].

Authenticated Routing for Ad Hoc Networks (ARAN) is another variant of the AODV protocol. In this protocol a certificate authority (CA) provides signed certificates to vehicular nodes. Every node that joins the network must request a certificate from the CA. The public key of the CA is known to all vehicular nodes. ARAN uses public key cryptography (asymmetric cryptography) to authenticate secure route discovery. ARAN has five phases: certification, authenticated route discovery, authenticated route setup, route maintenance, and key revocation. Also, the protocol ensures route authentication by adding a certificate and signature at each intermediate node along the path. A time stamp is used in ARAN to guarantee route freshness [4, 47, 64].

Elliptical Curve Digital Signature Algorithm (ECDSA) is a derivative of the digital signature algorithm (DSA). It uses digital signature, hash function and asymmetric cryptography in order to perform authentication. Elliptical curve parameters must be negotiated between source and destination. The public key is generated via DSA while signature verification is achieved via SHA algorithm [4, 47, 65].

Many other routing protocols have been proposed such as, but not limited to, Secure and Efficient Ad Hoc Distance vector protocol [27] and Robust method for Sybil Attack Detection [66].

9.6 Conclusion

The most essential requirement of ITS, IoV and VANETs is the safety on the roads. These systems have a great potential to grow and provide more applications besides making safe driving conditions on the road for drivers. Nonetheless, the connectivity of these systems to the Internet has increased the risk of compromise to these systems. The exploitation of the inherited and unique vulnerabilities of these systems by a malicious attacker can have severe consequences, such as loss of life and property damage. Hence, it is vital to secure vehicular networks. Other wireless networks were initially created and developed without security in mind; vehicular networks are no exception.

This chapter has discussed security threats, challenges and mitigating mechanisms in vehicular networks. We discussed the need for security as well as security and privacy requirements in vehicular networks, as they are unlike other wireless ad hoc networks. The challenges in implementing security in vehicular networks as they have different characteristics and requirements from other wireless Ad Hoc networks have been addressed. Also, adversaries and adversary models that threaten vehicular networks were covered. Insiders are considered a greater threat to the network as they are part of the network and have passed the authentication mechanism. In addition, the chapter gave an overview of threats and attacks against vehicular networks and possible countermeasure techniques to mitigate them.

Embedded small electronic control elements in VANETs in different types of units with different purposes increase the number of threat and attack vectors and vulnerabilities in VANETs. Also, the nature of the wireless medium is prone to exploitation. The main challenge in securing vehicular networks is the absence of security architecture standardization. This standardization is crucial to ensure that vehicular networks are resilient against most threats and attacks and for these systems to achieve their potential globally.

Even though much research has been conducted to secure vehicular networks, more work needs to be done to secure these systems against potential and malicious attacks. In particular, systems connected to the Internet are exposed to distant attackers. The dilemma of balancing security and privacy, however, remains an open issue that needs to be resolved.

References

1. Obaidat M, Ali M, Shahwan I, Obeidat S, and Obaidat MS (2013) QoS-aware multipath communications over MANETs. J Netw (JNW) 8:26–36
2. Obaidat M, Ali M, Shahwan I, Obeidat S, Toce A (2014) Dynamic suppression broadcast scheme for vehicle adhoc networks (VANET). In: 10th international conference on wireless communications, networking and mobile computing (WiCOM 2014), Beijing, pp 589–592
3. Luckshetty A, Dontal S, Tangade S, Manvi SS (2016) A survey: comparative study of applications, attacks, security and privacy in VANETs. In: 2016 international conference on communication and signal processing (ICCSP), Melmaruvathur, 2016, pp 1594–1598
4. Abu Talib M, Abbas S, Nasir Q, Mowakeh MF (2018) Systematic literature review on Internet-of-Vehicles communication security. Int J Distrib Sens Netw 14(12)
5. Alouache L, Nguyen N, Aliouat M, Chelouah R (2018) Survey on IoV routing protocols: security and network architecture. Int J Commun Syst, November 2018. Wiley
6. Statista.com (2019) U.S. vehicle registrations 1990–2017. https://www.statista.com/statistics/183505/number-of-vehicles-in-the-united-states-since-1990/. Accessed May 2019
7. Statista.com (2018) Number of traffic-related fatalities in the United States 1975–2017. https://www.statista.com/statistics/191521/traffic-related-fatalities-in-the-united-states-since-1975/. Accessed May 2019
8. Abassi R (2019) VANET security and forensics: challenges and opportunities. WIREs Forensics Sci 1(2)
9. World Health Organization (2017) Global status report on road safety 2017. www.who.int/violence_injury_prevention/road_safety_status/report/en/. Accessed May 2019
10. Mejri MN, Ben-Othman J, Hamdi M (2014) Survey on VANET security challenges and possible cryptographic solutions. Veh Commun 1(2014):53–66. Elsevier
11. Ben Hamida E, Noura H, Znaidi W (2015) Security of cooperative intelligent transport systems: standards, threats analysis and cryptographic countermeasures. Electronics 4(3): 380–423

12. Ghori MR, Zamli KZ, Quosthoni N, Hisyam M, Montaser M (2018) Vehicular ad-hoc network (VANET): review. In: 2018 IEEE international conference on innovative research and development (ICIRD), Bangkok, 2018, pp 1–6
13. Mokhtar B, Azab M (2015) Survey on security issues in vehicular ad hoc networks. Alex Eng J 54:1115–1126. Elsevier
14. Al-Sultan S, Al-Doori M, Al-Bayatti A, Zedan H (2014) A comprehensive survey on vehicular ad hoc network. J Netw Comput Appl 37(2014):380–392. Elsevier
15. Boualouache A, Senouci S, Moussaoui S (2018) A survey on pseudonym changing strategies for vehicular ad-hoc networks. IEEE Commun Surv Tutor 20(1):770–790. Firstquarter 2018
16. Begum R, Prasad K (2016) A survey on VANETs applications and its challenges. In: International conference on advanced computer science and software engineering (ICACSSE), 2016, pp 1–7
17. Ayoub T, Mazri T (2018) Security challenges in V2I architectures and proposed solutions. In: IEEE 5th international congress on information science and technology (CiSt), Marrakech, 2018, pp 594–599
18. Dai Nguyen HP, Zoltán R (2018) The current security challenges of vehicle communication in the future transportation system. In: 2018 IEEE 16th international symposium on intelligent systems and informatics (SISY), Subotica, 2018, pp 161–166
19. Kumar DA, Bansal M (2017) A review on VANET security attacks and their countermeasure. In: 2017 4th international conference on signal processing, computing and control (ISPCC), Solan, 2017, pp 580–585
20. Lu Z, Qu G, Liu Z (2019) A survey on recent advances in vehicular network security, trust, and privacy. IEEE Trans Intell Transp Syst 20(2):760–776
21. Asuquo P, Cruickshank H, Morley J, Anyigor O, Chibueze P, Lei A, Hathal W, Bao S, Sun Z (2018) Security and privacy in location-based services for vehicular and mobile communications: an overview, challenges, and countermeasures. IEEE Internet Things J 5(6):4778–4802
22. Tangade SS, Manvi SS (2013) A survey on attacks, security and trust management solutions in VANETs. In: 2013 fourth international conference on computing, communications and networking technologies (ICCCNT), Tiruchengode, 2013, pp 1–6
23. Salagar PG, Tangade SS (2015) A survey on security in VANETs. Int J Technol Res Eng 2(7)
24. Al-ani R, Zhou B, Shi Q, Sagheer A (2018) A survey on secure safety applications in VANET. In: 2018 IEEE 20th international conference on high performance computing and communications; IEEE 16th international conference on smart city; IEEE 4th international conference on data science and systems (HPCC/SmartCity/DSS), Exeter, United Kingdom, 2018, pp 1485–1490
25. Kaura N, Kad S (2016) A review on security related aspects in vehicular ad hoc networks. Procedia Comput Sci J 78:387–394
26. Sumra A, Hasbullah H, Manan JA (2011) VANET security research and development ecosystem. In: National postgraduate conference, Kuala Lumpur, 2011, pp 1–4
27. Hasrouny H, Samhat A, Bassil C, Laouiti A (2017) VANet security challenges and solutions: a survey. Veh Commun 7(2017):7–20. Elsevier
28. Rawat DB, Bista B, Yan G (2016) Securing vehicular ad-hoc networks from data falsification attacks. In: IEEE region 10 conference (TENCON), Singapore, 2016, pp 99–102
29. Elsadig MA, Fadlalla YA (2016) VANETs security issues and challenges: a survey. Indian J Sci Technol 9(28):2016
30. Markey EJ (2014) Tracking & hacking: security & privacy gaps put American drivers at risk. US Senate, 2015
31. Kelarestaghi KB, Foruhandeh M, Heaslip K, Gerdes R (2019) Intelligent transportation system security: impact-oriented risk assessment of in-vehicle networks. IEEE Intell Transp Syst Mag
32. Engoulou RG, Bellaïche M, Pierre S, Quintero A (2014) VANET security surveys. J Comput Commun 44(2014):1–13. Elsevier

33. Tanwar S, Vora J, Tyagi S, Kumar N, Obaidat MS (2018) A systematic review on security issues in vehicular ad hoc network. Secur Priv 1(5). September/October 2018
34. Jana B, Mitra S, Poray J (2016) An analysis of security threats and countermeasures in VANET. In: 2016 international conference on computer, electrical & communication engineering (ICCECE), 2016, pp 1–6
35. Sirola P, Joshi A, Purohit KC (2014) An analytical study of routing attacks in VANETs. Int J Comput Sci Eng (IJCSE) 3
36. Kaur R, Singh TP, Khajuria V (2018) Security issues in vehicular ad-hoc network (VANET). In: 2nd international conference on trends in electronics and informatics (ICOEI), Tirunelveli, 2018, pp 884–889
37. Al-Raba'nah Y, Al-Refai M (2016) Toward secure vehicular ad hoc networks an overview and comparative study. J Comput Commun 4:12–27
38. Qu F, Wu Z, Wang F, Cho W (2015) A security and privacy review of VANETs. IEEE Trans Intell Transp Syst 16(6):2985–2996
39. Sun Y, Wu L, Wu S, Li S, Zhang T, Zhang L, Xu J, Xiong Y (2015) Security and privacy in the internet of vehicles. In: International conference on identification, information, and knowledge in the internet of things (IIKI), Beijing, 2015, pp 116–121
40. Kim Y, Kim I (2013) Security issues in vehicular networks. In: The international conference on information networking 2013 (ICOIN), Bangkok, 2013, pp 468–472
41. Chaubey NK (2016) Security analysis of vehicular ad hoc networks (VANETs): a comprehensive study. Int J Secur Appl 10(5):261–274
42. Abdelgader AMS, Shu F, Zhu W, Ayoub K (2017) Security challenges and trends in vehicular communications. In: IEEE conference on systems, process and control (ICSPC), Malacca, 2017, pp 105–110
43. Vaibhava A, Shuklab D, Dasc S, Sahanad S, Johrie P (2017) Security challenges, authentication, application and trust models for vehicular ad hoc network—a survey. Int J Wirel Microw Technol 3:36–48
44. Samara G, Al-Salihy WAH, Sures R (2010) Security analysis of vehicular ad hoc networks (VANET). In: 2010 Second international conference on network applications, protocols and services, Kedah, 2010, pp 55–60
45. Riley M, Akkaya K, Fong K (2011) A survey of authentication schemes for vehicular ad hoc networks. Secur Commun Netw 4(10):1137–1152
46. Shahid MA, Jaekel A, Ezeife C, Al-Ajmi Q, Saini I (2018) Review of potential security attacks in VANET. In: Majan international conference (MIC), Muscat, 2018, pp 1–4
47. Mishra R, Singh A, Kumar R (2016) VANET security: issues, challenges and solutions. In: International conference on electrical, electronics, and optimization techniques (ICEEOT), 2016
48. Dhamgaye A, Chavhan N (2013) Survey on security challenges in VANET. Int J Comput Sci Netw (IJCSN) 2(1)
49. Azees M, Vijayakumar P, Deborah LJ (2016) Comprehensive survey on security services in vehicular ad-hoc networks. IET Intell Transp Syst J 10(6):379–388
50. Sarencheh A, Asaar M R, Salmasizadeh M, Aref MR (2017) An efficient cooperative message authentication scheme in vehicular ad-hoc networks. In: 2017 14th international ISC (Iranian Society of Cryptology) conference on information security and cryptology (ISCISC), Shiraz, 2017, pp 111–118
51. Kumar A, Sinha M (2014) Overview on vehicular ad hoc network and its security issues. In: International conference on computing for sustainable global development (INDIACom), New Delhi, 2014, pp 792–797
52. Nguyen HPD, Zoltán R (2018) The current security challenges of vehicle communication in the future transportation system. In: IEEE 16th international symposium on intelligent systems and informatics, SISY 2018, September 13–15, 2018, Subotica, Serbia
53. Amoozadeh M, Raghuramu A, Chuah C, Ghosal D, Zhang HM, Rowe J, Levitt K (2015) Security vulnerabilities of connected vehicle streams and their impact on cooperative driving. IEEE Commun Mag 53(6):126–132

54. Sun Y, Wu L, Wu S, Li S, Zhang T, Zhang L, Xu J, Xiong Y, Cui X (2017) Attacks and countermeasures in the internet of vehicles. Ann Telecommun 72:283–295
55. Le V, Hartog J, Zannone N (2018) Security and privacy for innovative automotive applications: a survey. Comput Commun 132:17–41
56. Bariah L, Shehada D, Salahat E, Yeun CY (2015) Recent advances in VANET security: a survey. In: IEEE 82nd vehicular technology conference (VTC2015-Fall), Boston, MA, 2015, pp 1–7
57. Wazid M, Das AK, Kumar N, Odelu V, Reddy AG, Park K, Park Y (2017) Design of lightweight authentication and key agreement protocol for vehicular ad hoc networks. IEEE Access 5:14966–14980
58. Kumar N, Preet Singh J, Bali RS, Misra S, Ullah S (2015) An intelligent clustering scheme for distributed intrusion detection in vehicular cloud computing. Clust Comput 18(3):1263–1283
59. Jacobs S (2015) Optimizing cryptographically based security in wireless networks. In: International conference on information systems security and privacy (ICISSP), 2015, pp 39–45
60. Smitha, A, Pai MM, Ajam N, Mouzna J (2013) An optimized adaptive algorithm for authentication of safety critical messages in Vanet. In: Proceedings of the IEEE 2013 8th international ICST conference on communications and networking in China (CHINACOM), Guilin, China, 14–16 August 2013, pp 149–154
61. Zhang C, Lin X, Lu R, Ho PH (2008) RAISE: An efficient RSU-aided message authentication scheme in vehicular communication networks. In: Proceedings of the IEEE international conference on communications, 2008. IEEE, pp 1451–1457
62. Perrig A, Canetti R, Tygar JD, Song D (2002) The TESLA broadcast authentication protocol. RSA Lab 5(2):2002
63. Studer A, Bai F, Bellur B, Perrig A (2009) Flexible, extensible, and efficient VANET authentication. J Commun Netw 11(6):574–588
64. Sanzgiri K, Dahill B, Levine BN, Shields C, Belding-Royer EM (2002) A secure routing protocol for ad hoc networks. In: Proceeding of IEEE ICNP 2002, November 2002, pp 78–87
65. Manvi SS, Kakkasageri MS, Adiga DG (2009) Message authentication in vehicular ad hoc networks: ECDSA based approach. In: International conference on future computer and communication, 2009, pp 16–20
66. Chen IC, Wang X, Han W, Zang B (2009) A robust detection of the Sybil attack in urban VANETs. In: Distributed computing systems workshop, ICDCS workshops '09. 29th IEEE international conference, 2009, pp 270–276

Chapter 10
Security Issues in Vehicular Ad Hoc Networks for Evolution Towards Internet of Vehicles

Atanu Mondal and Sulata Mitra

Abstract This chapter elaborates on the security issues of vehicular communication as a need for future evolution of Vehicular Ad Hoc NETworks (VANET) towards the Internet of Vehicles (IoV). The communication between connected vehicles is a potential issue of concern with respect to road safety, detecting traffic accidents, etc. in vehicular Ad Hoc network. In the Internet of Vehicles paradigm, each vehicle is equipped with a powerful multi-sensor platform, communication technologies, computation units, IP-based connectivity to the Internet and to other vehicles for enabling communication between connected vehicles as well as between vehicles and roads. Such communication must be protected from unauthorized message injection and message alteration. Hence a robust security solution for such networks must facilitate authentication verification of vehicles and integrity verification of the disseminated messages. In the first part of this chapter, a lightweight scheme for identification, authentication, and tracking of vehicles in hierarchical vehicular Ad Hoc network is elaborated. In the second part, a low overhead digital watermark-based vehicle revocation scheme to identify and revoke attackers of disseminated message is discussed. The combination of these two schemes ensures secure inter vehicle communication in vehicular Ad Hoc networks.

Keywords IoT · IoV · VANET · MANET · IVC · Vehicle identification · Vehicle authentication · VIN · Vehicle revocation

A. Mondal
Ramakrishna Mission Vidyamandira, Howrah, India

S. Mitra (✉)
Department of Computer Science and Technology, Indian Institute of Engineering, Science and Technology, Shibpur, Howrah, India
e-mail: bulu456@yahoo.co.in

© Springer Nature Switzerland AG 2020
Z. Mahmood (ed.), *Connected Vehicles in the Internet of Things*,
https://doi.org/10.1007/978-3-030-36167-9_10

10.1 Introduction

The Internet of Vehicles (IoV) is a convergence of mobile Internet and the Internet of Things (IoT), where vehicles function as smart moving intelligent nodes or objects within the sensing network. Over the years, the total number of vehicles in the world has seen a remarkable growth, increasing traffic density which, in turn, has resulted in more accidents. The IoV paradigm serves the applications for intelligent transportation like driver safety with reduced accidents, increasing traffic efficiency, and better infotainment. It also serves the need of smart cities for large-scale data sensing, collection, information processing, and storage. In vehicular Ad Hoc network (VANET), the connected smart vehicles with their own processor, computing, and communication power, can interact with each other to provide a safe and smart environment. The vehicles communicate and interact wirelessly, via different types of devices connected to the Internet, that lead to a specific type and customized IoT, called the IoV. It helps to achieve unified management in intelligent transportation, to improve road traffic safety, and to supervise other vehicles by providing secure and reliable information services. But the adversaries may misuse the resources by generating and communicating message with other vehicles.

The unauthentic vehicles may jeopardize the safety of other vehicles, drivers, passengers as well as the efficiency of the transportation system by generating false message like to clear the road for selfish reason. Hence identification and authentication of vehicles are essential to protect VANET from unauthorized message injection and message alteration. It is also essential to track the vehicles at various check points within a boundary or premises in high mobility VANET to manage the security and safety of vehicles in case of theft or other undesirable incidents. Moreover, the misbehaving vehicles may change the content of the message during its dissemination to misguide the other vehicles. Such misbehaving vehicles are formally known as *alteration attackers*. Hence, the protection of VANET from unauthentic vehicles and alteration attackers is a great challenge for future deployment and application of the IoV.

This chapter elaborates on secure inter vehicle communication in two parts. In the first part, a lightweight secure vehicle identification, authentication, and tracking (LIAT) scheme is elaborated to identify the unauthentic vehicles. In the second part, a low overhead watermark-based vehicle revocation (LoWVR) scheme is elaborated to identify and revoke attackers.

The existing schemes as described in [1, 2] are modified in LIAT [3]. Both the existing schemes [1, 2] are implemented in hierarchical VANET to track and to provide service to a lot of high velocity vehicles due to its increased coverage area. Both the schemes use Vehicle Identification Number (VIN) to identify a vehicle uniquely, conduct an initial registration phase for each vehicle to verify its authentication using VIN, and assign a Digital Signature (D_Sig) to each authentic vehicle. A tracking algorithm is executed to track the authentic vehicles within the coverage area of the network in both the schemes. But the communication channel between the different levels of the hierarchy is not secure in both the schemes [1, 2]. An attacker can access the messages by sensing such insecure channel. Hence the existing schemes as reported in [1, 2] are modified in LIAT [3] as M_1 and M_2 respectively, to ensure security of the communication channel among the different levels of the hierarchy.

LIAT uses symmetric key cryptography mechanism. The secret key of symmetric key cryptography is shared among the communicating nodes in encrypted form. It needs less complex computation than public key cryptography, and hence it is suitable for a high mobility network like VANET. Moreover, it needs less communication, computation, and storage overhead than existing schemes. It deals with securing the network at every step of communication within the network. The qualitative performance of LIAT is measured in terms of security analysis and overhead analysis. It is also studied through simulation considering a number of factors, viz: delay in verifying authentication per vehicle per BS, percentage of authentic vehicles per BS, packet loss per BS, service block per BS, and rate of tracking of vehicles per BS as performance metrics. Both qualitative performance and quantitative performance of [3] are compared with the existing schemes in [1, 2]. The security analysis shows that the cracking of data by the attacker is of no use as described in [3]. The overall analysis shows the significant improvement of [3] in terms of communication, computation, and storage overheads over the existing schemes. The results of simulation experiment show the improved performance in terms of the minimization of authentication delay and packet loss with minimum block of service at a BS. But in [1–3] vehicles are not capable to revoke the unauthentic and misbehaving vehicles from VANET. They depend on trusted third party for revocation of such vehicles. The trusted third parties are external aids like road side unit, BS, etc.

A Low Overhead Watermark-based Vehicle Revocation (LoWVR) [4], as discussed later in this chapter, overcomes the drawbacks detected in most of the existing schemes on securing VANET. In [4], the receiver vehicle verifies the authentication of the sender vehicle and revokes the sender vehicle if it is detected as an alteration attacker. Each vehicle maintains a unique identification number. The function of the sender vehicle after observing an event is to generate (1) message (M) and a random number (key), (2) M′ by concatenating M and a timestamp, (3) a deformed version of M′ (DVOM), (4) message digest (MD) from DVOM, (5) MD_DVOM by concatenating DVOM and MD, (6) watermark bits from unique identification and key, and (7) WM_MD_DVOM by embedding watermark bits, unique identification, key in MD_DVOM. The sender vehicle broadcasts WM_MD_DVOM.

The function of the receiver vehicle, after receiving WM_MD_DVOM, is to extract watermark bits, unique identification, key of the sender vehicle from WM_MD_DVOM for converting WM_MD_DVOM into Pad_MD_DVOM. The receiver vehicle (1) verifies authentication of sender vehicle, (2) verifies whether the sender vehicle is a alteration attacker, (3) computes M from DVOM if the sender vehicle is an authentic vehicle and also not an alteration attacker.

The unauthentic vehicles and alteration attackers are revoked from VANET by not processing the messages disseminated by them. VANET cannot afford any complex computation like conventional cryptography due to high mobility of vehicles. The security solution is required to be as robust as uncrackable within its relatively short lifespan. Hence a bit operation based small computations over short size data are proposed in LIAT to ensure light weight security. The presence of timestamp in M' makes it difficult for a vehicle to generate two different messages or more than one copy of the same message at the same time instant. Hence M', DVOM and WM_MD_DVOM are all unique. Moreover, a replay attacker may capture WM_MD_DVOM during its transmission from sender vehicle to receiver vehicle and rebroadcast it. The vehicles which are neighbor of the sender vehicle, and also the neighbor of this replay attacker, discard the second copy of WM_MD_DVOM after receiving it from the replay attacker as it is impossible for the sender vehicle to generate two identical WM_MD_DVOM. Hence, the mechanism in [4] is protected from replay attack.

Chapter Organisation

The rest of this chapter is organized as follows. Table 10.1 provides details of parameters, annotations and abbreviations used in the contribution. Section 10.2 deals with related existing literature. The existing schemes [1, 2] are elaborated in Sect. 10.3. LIAT is elaborated in Sect. 10.4 and LoWVR mechanism is discussed in Sect. 10.5. Performance analysis of LIAT along with a detailed comparative study is done in Sect. 10.6. The performance of LoWVR and its comparison with the existing schemes are discussed in Sect. 10.7. Section 10.8 provides concludes.

10.2   Related Work

Several vehicle authentications and revocation schemes have been reported in the literature, for centralized VANET.

A 3-tier architecture having trusted authority at the root level, road side unit (RSU) at the intermediate level and on-board unit (OBU) at the leaf level is proposed in [5]. The trusted authority issues a token to each authentic OBU within its coverage area. Each OBU sends this token along with the message to its parent RSU. The RSU verifies the authentication of the received token. If the received token is authentic, the RSU issues another token that contains safety message signature and a valid time period to the OBU. Now the OBU can broadcast the message to other OBUs using this safety message signature during the valid time period. The OBU leaves the safety message signature after moving into the coverage area of a new RSU. But the method of verifying the authentication of RSUs and OBUs by the trusted authority is not mentioned. The assignment of a new safety message signature by the RSU to all the OBUs within its coverage area incurs extra computation and communication overhead. Moreover, the performance of the proposed scheme is not studied by increasing the number of OBUs within the coverage area of the RSU continuously.

Table 10.1 Abbreviations and parameter list

Parameter	Description
BS	Base station
CA	Certifying authority
CA_VIN_Queue	VIN queue at CA
CRL	Certificate revocation list
COMM_OH	Communication overhead
COMP_OH	Computation overhead
Copy_DVOM	Copy of DVOM
CH	Channel
D_Sig	Digital signature
DVOM	Deformed version of message
D_VIN	Decrypted VIN
Dist_10	Reference [10] as simulated in distributed VANET
Dist_22	Reference [22] as simulated in distributed VANET
Dist_23	Reference [23] as simulated in distributed VANET
Dist_24	Reference [24] as simulated in distributed VANET
ELP	Electronic license plate
E_VIN	Encrypted VIN
F_CA	First level CA
IoV	Internet of vehicles
IoT	Internet of things
IVC	Inter vehicle communication
MD	Message digest
MD_DVOM	Concatenation of DVOM and MD
M_1	Modified version of [1]
M_2	Modified version of [2]
MC	Message content
NO_OF_BS	Number of BS
NO_OF_FCA	Number of F_CA
OBU	On-board unit
P_F_CA	Parent F_CA
R_CA	Root CA
RSA	Rivest-Shamir-Adleman
RSU	Road side unit
R_CA_VIN_Queue	VIN queue at R_CA
STO_OH	Storage overhead
Size_Pad_MD_DVOM	Size of Pad_MD_DVOM
Size_WM_MD_DVOM	Size of WM_MD_DVOM
SHA	Secure hash algorithm
TOT_V	Total number of vehicles
VANET	Vehicular Ad Hoc network

(continued)

Table 10.1 (continued)

Parameter	Description
V2I	Vehicle to infrastructure
VIN	Vehicle identification number
VIN_CA	VIN database at CA
VIN_BS	VIN database at BS
VIN_P_Time	VIN processing time
WM	Watermark

A pseudonymous authentication based conditional privacy protocol is proposed in [6] for vehicular authentication. Here, the vehicles register themselves with trusted motor vehicle department using its identity such as license plate number of the vehicle, address of the owner, etc. and get a ticket or signature. The vehicle uses this ticket to obtain token from the neighbor vehicles. The token is used by the vehicle to generate pseudonyms for anonymous message broadcasting. The sender vehicle broadcasts its pseudonym to neighbors during anonymous communication. The receiver vehicle uses this pseudonym to encrypt messages before transmission. But the attacker can also receive the pseudonym which is broadcast by the sender vehicle and can then jeopardize the security of the VANET.

A proxy-based authentication scheme is proposed in [7]. Here, each vehicle is equipped with a tamper proof device, which has three preloaded secret master keys. Each vehicle uses the keys to generate its identification and privacy key. Each vehicle also transmits the identification and privacy key to its parent RSU, which uses the privacy key as the verifier for the identification. The vehicle is authentic if the verification of identification is successful. But the authentication of a vehicle is verified using a new identification and privacy key when it enters into the coverage area of a new RSU which incurs extra communication and computation overhead.

An identity-based signature mechanism and pseudonyms are used to verify the authentication of the vehicles in [8]. Each vehicle has a packet format associated with it. A fixed identification field in the packet is reserved for the manufacturer identification. This is used as the authentication code of the vehicle. The authentication of the vehicle is verified by transmitting manufacturer identification to the RSU. But an attacker can put its own generated manufacturer identification in the fixed identification field of the packet. Since, no data set of manufacturer identification is available at RSU, the RSU cannot verify the originality of the manufacturer identification field of such a received packet.

The certifying authority certifies the public key of the vehicles along with their static attributes such as weight, color, model, etc. for vehicle authentication in [9]. The dynamic attributes like location, direction of movement, etc. of the vehicles are also considered as identification of the constantly moving vehicles which may not be done successfully using their static attributes. The dynamic attributes identify the vehicles by the laser technology assistance. But the use of static attributes increases the storage requirement as well as computational overload. On the other hand, the use of dynamic attributes incurs extra computation overhead for vehicle identification.

In [10], a vehicle sends driving license, vehicle number, log-in identification, and password to the trusted main authority for its registration. The trusted main authority provides a private key and unique id to each registered vehicle for identifying them as authentic. The assigned private key has a short lifetime. Moreover, a vehicle may revoke the private key before its expiry, and send its log-in identification and password to the trusted main authority for getting a new key which increases the registration frequency of the vehicle, key generation rate, storage overhead, and its associated communication overhead.

A biometric-key-based authentication scheme is proposed in [11]. All users receive their personal key and the personal key of the servers (RSUs) from a biometric key distribution center. Such distribution of personal key helps to make mutual authentication in between users and servers. Biometric key distribution center also generates a session ticket for all the users from their biometric information. The session key contains the name of the user and the expiration time of the session. This authentication scheme does not highlight how the personal keys are generated for the users and servers; and how it obtains the name of the users from their biometric information. Again, the personal keys are transmitted through insecure channel which in turn can easily be intercepted by the attacker. Thus, it incurs a high risk in communication paradigm in VANET.

In [12], the virtual certificate authority provides certificates to the vehicles after verifying their authentication. But the certificates are assigned to the vehicles through insecure channel. Moreover, the method of identifying the misbehaving vehicles is not elaborated.

In [13], an ID-based authentication verification and revocation scheme is proposed. Three entities such as Trusted-Third-Party, RSU, and OBU are required for the authentication and revocation of a malicious vehicle. A revocation tree is maintained by the Trusted-Third-Party and RSU to store the pseudonyms of the authentic vehicles at the root level and intermediate level of the tree. The leaf level of the tree contains the digests of pseudonyms of the malicious vehicles. The RSU searches the higher levels of the tree for the pseudonym and leaf level of the tree for the digest of pseudonym of an OBU before starting any communication; and that increases communication as well as computation overhead.

A shared ID-based revocation mechanism is proposed in [14]. The CA generates a couple of certificates for a vehicle at different instances of time, and also generates a shared key by correlating all certificates of that vehicle. It helps to keep privacy of a vehicle hidden from other vehicles. But all the certificates associated with a vehicle need to be revoked separately to make the shared key inactive in case this vehicle is identified as a misbehaving vehicle. Hence, the scheme has a large computation overhead.

10.3 Existing Schemes

In this section the existing schemes [1, 2] are elaborated. The scheme in [1] is implemented in a centralized VANET which is a hierarchy having certifying authority (CA) at the root level, base stations (BSs) at the intermediate level and vehicles at the leaf level. This scheme is extended in [2] for accommodating more vehicles in a VANET. It is implemented in a distributed VANET which is a hierarchy having CA at the root level (R_CA) and also at the first level (F_CA), BSs at the second level, and vehicles at the leaf level. The F_CAs are maintained at the first level of the hierarchy to increase the coverage area of VANET which in turn helps to accommodate more vehicles in [2] than suggested in [1]. Both CA and R_CA maintain a VIN database (VIN_CA) to store the VINs of the vehicles that are already manufactured. It is updated when a new vehicle is manufactured. Both CA and R_CA gather knowledge about the available pattern of each character in VIN of the vehicles which are already manufactured by consulting with VIN_CA after updating. Each BS also maintains a VIN database (VIN_BS) to store the encrypted VIN and digital signature pair of the authentic vehicles that enters into its coverage area in [1] rather than into the coverage area of its parent F_CA as stated in [2].

Both the schemes, as mentioned in [1, 2], consider Vehicle to Infrastructure (V2I) communication among vehicles and root through BS in the form of data. A dedicated short-range communication protocol [15] has been proposed for such fast short distance V2I communication in [1, 2]. In this protocol, the 70 MHz bandwidth at each BS is divided into 7 links, 6 of them are reserved for service and 1 link is used for control purposes. So, the bandwidth of each link is 10 MHz and it is allocated to authentic vehicle on demand at a rate of 1.28 MHz. Each 10 MHz link is divided into (10 MHz/1.28 MHz) $\approx$ 8 channels and the total number of available channels to provide service at each BS is 48. Each BS allocates one channel to each authentic vehicle within its coverage area and so can provide service to 48 vehicles using 48 channels.

In both schemes [1, 2], it is assumed that each vehicle has an electronic license plate (ELP) in which the encrypted VIN (E_VIN) of the vehicle is embedded by the vehicle manufacturer. The ELP of a vehicle broadcasts (as per IEEE P1069 and IEEE 802.11p) E_VIN after entering into the coverage area of a new BS (New_BS). The New_BS executes the identification algorithm after receiving E_VIN from vehicle, adds E_VIN in its VIN queue and searches VIN_BS for the received E_VIN. If found, the vehicle is authentic and its initial registration phase is over. The identification algorithm reads D_Sig from the record corresponding to the received E_VIN in VIN_BS and triggers the tracking algorithm by sending D_Sig. Otherwise the identification algorithm initiates the initial registration phase of this vehicle by triggering the authentication algorithm.

The authentication algorithm at New_BS calls the verification function at CA in [1] and authentication algorithm at its parent F_CA (P_F_CA) in [2] by sending E_VIN. The authentication algorithm at P_F_CA calls the verification function at R_CA by sending E_VIN.

The verification function adds E_VIN in a VIN queue and decrypts it using RSA algorithm. It verifies each character of the decrypted VIN (D_VIN) for validity. The verification function knows the possible pattern of each character in D_VIN and also has knowledge about the available pattern of each character in D_VIN from VIN_CA. If D_VIN is not valid the verification function generates a security message for the police or checking personnel to make them aware about this vehicle. Otherwise, the verification function generates D_Sig from D_VIN using SHA-1 algorithm. In [1], verification function at CA calls the insertion function at all the BSs under it by sending (E_VIN, D_Sig) pair. In [2], the verification function at R_CA calls the insertion function at P_F_CA which in turn calls the insertion function at all the BSs under it by sending (E_VIN, D_Sig) pair. Each BS under CA in [1] and under P_F_CA in [2] inserts (E_VIN, D_Sig) in their VIN_BS. During this span of time, the vehicle may reside within the coverage area of New_BS or may move to the coverage area of another BS under CA in [1] and under P_F_CA in [2]. The insertion function at the BS within which the vehicle is currently passing through (Current_BS) triggers the tracking algorithm.

The tracking algorithm at BS assigns D_Sig and allocates a channel to the vehicle. It also reassigns D_Sig to the vehicle periodically as long as the vehicle is within its coverage area.

The use of VIN for identification and authentication of a vehicle is advantageous as it is impossible to transfer VIN among vehicles, or to alter the information within it. Moreover, VIN of a vehicle remains intact even in typical environmental condition. It contains information about the manufacturer of the vehicle and description of the vehicle. So, other than identification and authentication, VIN can also be used to know the manufacturing details and the details description of a vehicle which may be required in case of accidents etc.

10.4   LIAT

As already mentioned, M_1 and M_2 are modified versions of the existing schemes as reported in [1, 2]. In this section, we elaborate M_1 and M_2 as reported in LIAT [3].

M_1 and M_2 ensure secure communication between root and BS by registering all the BSs under root in the Root-BS registration phase. The communication among each BS and vehicles within its coverage area is made secure by registering all the vehicles in the BS-Vehicle registration phase. The identification, authentication, and tracking of all the registered vehicles within the coverage area of each BS are ensured in the authentication phase.

10.4.1 Network and System Assumptions

System Model

M_1 and M_2 are implemented in the same centralized and distributed VANET as proposed in [1, 2] respectively. The root of the hierarchy is CA in M_1 and R_CA in M_2. NO_OF_BS is the number of BSs within the coverage area of CA in [1] and F_CA in [2]. NO_OF_FCA is the number of F_CAs within the coverage area of R_CA in [2]. Hence the maximum possible number of BSs under R_CA in [2] is NO_OF_FCA × NO_OF_BS. Each vehicle is identified uniquely by VIN.

The dedicated short-range communication protocol is proposed in M_1 and M_2 as in [1, 2] respectively. Each BS allocates one channel to each authentic vehicle within its coverage area so that it can provide service to 48 vehicles using 48 channels. Each channel is identified by 6 bits.

Both M_1 and M_2 have two phases of operation. The first phase is the registration phase which has two sub-phases. The Root-BS sub-phase is used for the registration of all the BSs under root. Here, the communication between the root and BSs under root is in encrypted form. The other, the BS-Vehicle sub-phase, is used for the registration of all the vehicles under BS. Here, the communication between BS and vehicles is also in encrypted form. The second phase is the authentication phase where each BS verifies the identification and authentication of all the registered vehicles under it, also tracks authentic vehicles within its coverage area. A digital signature (D_Sig) is assigned to each authentic vehicle. The messages and keys are exchanged in encrypted form among the different levels of the hierarchy during the authentication phase. The key is computed at the receiving end during decryption of the received message. This reduces key cracking probability (as discussed in Sect. 10.6.1.1) on transmission and makes the securing mechanism more robust. It ensures confidentiality and helps to maintain message integrity.

Both CA and R_CA maintain a VIN database (VIN_CA) as in [1, 2]. Each BS also maintains a VIN database (VIN_BS) to store (VIN, D_Sig) pair of the authentic vehicles that enters into the coverage area of CA in M_1 whereas into the coverage area of its parent F_CA in M_2.

Adversary Model

Security attacks in VANET may occur at all the nodes. Hence, VANET is susceptible to attackers by passive eavesdropping to active spamming and tampering of the messages. In LIAT, the attacker vehicles are not manufactured by any automakers which are registered under the law, Anti-Car Theft Act [16] passed in October 1992 by the Congress and signed by U.S.A President George H. W. Bush. Such vehicles can join the real life VANET at any time, can capture the message during its transmission, and can crack the captured message by performing brute-force operations. Such vehicles have no valid identification like VIN. The operations of normal vehicles are also not known to them. Moreover, such vehicles can occupy the same resource like authentic vehicles if they are able to prove themselves as registered vehicles as well as authentic vehicles. Hence secure and reliable message communications are required for improving driving safety and traffic management. Again, unauthentic retrieval of the stored message could be fatal to the authentic vehicles in VANET. As a consequence, efficient message management is required to keep the stored messages safe and secure at different nodes in VANET from the attacker. Hence, the objective of LIAT is to secure stored messages in nodes, and communication of messages between the nodes of the hierarchy.

Security Requirements

Since, BS verifies the identification and authentication of vehicles and tracks vehicles within its coverage area, each BS must be trustworthy which is ensured in the Root-BS registration phase. In LIAT, it is assumed that the operational capability of attacker is in between the mote-class attacker [17] and laptop-class attacker [17] of wireless network. It is also possible for the attacker to attack root, BSs, and authentic vehicles by transmitting malicious messages in VANET. Thus, it is required to provide better environment to make the vehicular communication secure among root, BSs, and vehicles.

Design Goal

The objective of LIAT is to ensure the following basic properties, as defined in [18], are considered in the above-mentioned Adversary Model:

- Authentication: This is defined as the property which ensures that the origin of an electronic message or document is correctly identified. It ensures protection against the fabrication attack
- Confidentiality: This is defined as the property where only the sender and the intended recipient(s) should be able to access the content of a message. It is achieved by encrypting the original message and sending the cipher text to the intended recipient(s)
- Integrity: This property is defined as the property where the content of the message is not changed during transmission from sender to intended recipient(s)
- Non-repudiation protection: This is the protection against non-repudiation, [18] i.e. sender sends a message, and later on refuses that said transmission
- Masquerade protection: This is the protection against masquerade where an unauthorized entity pretends to be another entity.

10.4.2   Root-BS Registration Phase

The root of the hierarchy initiates the registration phase by triggering Root-BS registration algorithm. The algorithm sends a registration packet to all the BSs under root. The Root-BS registration algorithm (for all BSs under root) receives the registration packet, and initiates the registration response phase by sending the registration response packet to the root. In this section, the registration of Bth BS (BS_B, $1 \leq B \leq NO\_OF\_BS$) under root is elaborated.

BS_B is under CA in M_1 and under C^{th} F_CA (F_CA$_C$, $1 \leq C \leq$ NO_OF_FCA) within the coverage area of R_CA in M_2.

Root-BS Registration Algorithm at Root

ID_R is the root identification of size 2 bits. The algorithm generates a timestamp ($T_{R\_present}$) of size 12 bits, uses $T_{R\_present}$ to encrypt ID_R (EID_R'' of size 14 bits) using FUNC_1, stores $T_{R\_present}$ for future communication with BS_B, and sends the registration packet in the form EID_R'' to BS_B.

FUNC_1

Input is ID_R and $T_{R\_present}$. Output is EID_R''. It pads ID_R with 10 zero bits from most significant bits, computes EID_R as ID_R XOR $T_{R\_present}$, computes EID_R' as complement of EID_R, concatenates ID_R and EID_R' to generate EID_R'' and swaps 2 most significant bits of EID_R'' with its 2 least significant bits to update EID_R''.

Root-BS Registration Algorithm at BS_B

This algorithm reads EID_R'' from the registration packet, decrypts EID_R'' to retrieve the key ($T_{R\_present}$) using FUNC_2, stores $T_{R\_present}$ for future communication with root, and initiates the registration response phase.

FUNC_2

Input is EID_R''. Output is $T_{R\_present}$. It swaps 2 most significant bits of EID_R'' with its 2 least significant bits, splits EID_R'' into ID_R and EID_R', computes EID_R as complement of EID_R', and computes $T_{R\_present}$ as EID_R XOR ID_R.

Registration Response Phase at BS_B

ID_B is the identification of BS_B of size 2 bits. The Root-BS registration algorithm at BS_B uses $T_{R\_present}$ to encrypt ID_B (EID_B'' of size 14 bits) using FUNC_3, and sends the registration response packet in the form EID_B'' to the root.

FUNC_3

Input is ID_B and $T_{R\_present}$. Output is EID_B''. It pads ID_B with 10 zero bits from most significant bits, computes EID_B as ID_B XOR $T_{R\_present}$, computes EID_B' as complement of EID_B, concatenates ID_B and EID_B' to generate EID_B'', and swaps 2 most significant bits of EID_B'' with its 2 least significant bits.

Root-BS Registration Algorithm at Root

The algorithm reads EID_B'' from the registration response packet of BS_B, decrypts EID_B'' to retrieve the key ($T_{R\_present}'$) using FUNC_4, and compares $T_{R\_present}'$ with $T_{R\_present}$. In case of match, the registration of BS_B under root is successful.

FUNC_4

Input is EID_B''. Output is $T_{R\_present}'$. It swaps 2 most significant bits of EID_B'' with its 2 least significant bits, splits EID_B'' into ID_B and EID_B', computes EID_B as complement of EID_B', and computes $T_{R\_present}'$ as EID_B XOR ID_B.

10.4.3 BS-Vehicle Registration Phase

A vehicle initiates the registration phase by triggering BS-Vehicle registration algorithm after entering into the coverage area of a new BS. The algorithm sends a registration packet to the new BS. The BS-Vehicle registration algorithm at new BS receives the registration packet and initiates the registration response phase by sending the registration response packet to the vehicle.

Let the vth vehicle (V_v, $1 \leq v \leq TOT\_V$, TOT_V is the total number of vehicles in VANET) enters into the coverage area of BS_B. In this section the registration of V_v under BS_B is considered for relevant discussion.

BS-Vehicle Registration Algorithm at V_v

VIN_v is the VIN of V_v. VIN has 17 characters and the size of each character is assumed as 8 bits (extended ASCII format). Hence the size of VIN_v is 136 bits. The algorithm generates a timestamp ($T_{v\_present}$) of size 12 bits, uses $T_{v\_present}$ to encrypt VIN_v (E_VIN_v of size 272 bits) using FUNC_5, stores $T_{v\_present}$ at V_v till V_v resides within the coverage area of BS_B, and sends the registration packet in the form E_VIN_v to BS_B.

FUNC_5

Input is VIN_v and $T_{v\_present}$. Output is E_VIN_v. It pads $T_{v\_present}$ with 124 zero bits from most significant bits, computes VIN_v' as VIN_v XOR $T_{v\_present}$, computes VIN_v'' as complement of VIN_v', concatenates VIN_v and VIN_v'' to generate VIN_v''', and swaps 72 most significant bits of VIN_v''' with its 72 least significant bits to generate E_VIN_v.

BS-Vehicle Registration Algorithm at BS_B

This algorithm reads E_VIN_v from the registration packet, adds E_VIN_v in $BS_B\_VIN\_Queue$, decrypts E_VIN_v to retrieve VIN_v and $T_{v\_present}$ using FUNC_6, and searches for a free channel. If not found, then it increases a service block counter (SE_BL_B) by 1. Otherwise it assigns a free channel (CH_v) to V_v, stores a record (R_v) in the form (VIN_v, $T_{v\_present}$, CH_v) for V_v in a list ($LIST_B$) till V_v resides within the coverage area of BS_B, uses $T_{v\_present}$ as a key for future communication with V_v, and initiates the registration response phase by sending the registration response packet to V_v through CH_v.

FUNC_6

Input is E_VIN_v. Output is VIN_v and $T_{v\_present}$. It swaps 72 most significant bits of E_VIN_v with its 72 least significant bits to generate VIN_v''', splits VIN_v''' into VIN_v

and VIN_v'', complements VIN_v'' to generate VIN_v', and computes $T_{v\_present}$ as VIN_v XOR VIN_v'.

Replay Attack

A misbehaving vehicle may disseminate the same message multiple times. Such misbehaving vehicles are formally known as replay attacker. LIAT can detect the replay attack but can't identify replay attacker. For example, let r^{th} vehicle (V_r) be a neighbour of V_v and an attacker vehicle. V_r captures the registration packet of V_v and sends the same packet to BS_B. BS_B reads E_VIN_v from the registration packet of V_r, finds E_VIN_v in $BS_B\_VIN\_Queue$, and discards the registration packet of V_r which helps to protect VANET from replay attack.

Registration Response Phase at BS_B

In this phase, the BS-Vehicle registration algorithm at BS_B uses $T_{v\_present}$ to encrypt ID_B (EID_B'' of size 14 bits) using FUNC_7 and sends the registration response packet in the form (EID_B'') to V_v.

FUNC_7

Input is ID_B and $T_{v\_present}$. Output is EID_B''. It pads ID_B with 10 zero bits from most significant bits, computes ID_B' as ID_B XOR $T_{v\_present}$, computes ID_B'' as complement of ID_B', concatenates ID_B and ID_B'' to generate ID_B''', and swaps 2 most significant bits of ID_B''' with its 2 least significant bits to generate EID_B''.

BS-Vehicle Registration Algorithm at V_v

The algorithm reads EID_B'' from the registration response packet, decrypts EID_B'' to retrieve the key ($T_{v\_present}'$) using FUNC_8 and compares $T_{v\_present}'$ with $T_{v\_present}$. In case of match the registration of V_v under BS_B is successful. The BS-Vehicle registration algorithm at BS_B initiates the authentication phase for V_v after its successful registration.

FUNC_8

Input is EID_B''. Output is $T_{v\_present}'$. It swaps 2 most significant bits of EID_B'' with its 2 least significant bits to generate ID_B''', splits ID_B''' into ID_B and ID_B'', complements ID_B'' to generate ID_B', and computes $T_{v\_present}'$ as ID_B XOR ID_B'.

10.4.4 *Authentication Phase*

In this section the identification, authentication, and tracking of V_v by BS_B are elaborated.

Identification, Authentication and Tracking Algorithm

BS_B triggers identification algorithm for V_v by sending (VIN_v, E_VIN_v) pair. The identification algorithm searches VIN_BS at BS_B (VIN_BS_B) for VIN_v. If VIN_v is found V_v to be an authentic vehicle, then execution of the authentication algorithm for V_v is not required. The identification algorithm at BS_B reads the D_Sig of V_v

(D_Sig$_v$) from the record of V$_v$ in VIN_BS$_B$ and triggers tracking algorithm by sending (VIN$_v$, D_Sig$_v$) pair to it. Otherwise, the identification algorithm at BS$_B$ generates the authentication packet containing E_VIN$_v$ of size 272 bits and triggers the authentication algorithm by sending the authentication packet.

The authentication algorithm at BS$_B$ calls the verification function at CA in M_1 and authentication algorithm at F_CA$_C$ in M_2 by sending the authentication packet. The authentication algorithm at F_CA$_C$ calls the verification function at R_CA by sending the authentication packet.

The verification function reads E_VIN$_v$ from the authentication packet, adds E_VIN$_v$ in CA_VIN_Queue in M_1 and in R_CA_VIN_Queue in M_2. The verification function decrypts E_VIN$_v$ to generate VIN$_v$ using FUNC_9.

FUNC_9

Input is E_VIN$_v$. Output is VIN$_v$. It swaps 72 most significant bits of E_VIN$_v$ with its 72 least significant bits to generate VIN$_v'''$ and splits VIN$_v'''$ into VIN$_v$ and VIN$_v''$.

The verification function verifies VIN$_v$ for its validity [19]. If VIN$_v$ is not valid, the verification function generates a security message for the police or checking personnel and BS$_B$ to make them aware about V$_v$. The authentication algorithm at BS$_B$ deletes R$_v$ from LIST$_B$ and frees CH$_v$. Otherwise, the verification function generates D_Sig$_v$ from VIN$_v$ using SHA-1 algorithm [18] and generates B$_v$ which is (VIN$_v$, D_Sig$_v$) pair. It encrypts B$_v$ using the key B$_1$ to generate EB$_v$ and encrypts B$_1$ to generate K$_1$ using FUNC_10. It also generates an insertion packet containing (EB$_v$, K$_1$) pair of size 308 bits, calls the insertion function at all the BSs under CA in M_1 and under F_CA$_C$ in M_2 by sending the insertion packet. The insertion function at F_CA$_C$ in turn calls the insertion function at all the BSs under it by sending the insertion packet.

FUNC_10

Input is B$_v$ and B$_1$. Output is EB$_v$ and K$_1$. It selects a random number (B$_1$) of size 4 bits, rotates B$_v$ by B$_1$ bits towards left to generate EB$_v$, pads B$_1$ with 8 zero bits from most significant bits, and encrypts B$_1$ as K$_1$ = B$_1$ XOR T$_{R\_present}$.

The insertion function at all the BSs under CA in M_1 and under F_CA$_C$ in M_2 reads (EB$_v$, K$_1$) pair from the insertion packet, decrypts EB$_v$ to generate B$_v$ using FUNC_11, and inserts (VIN$_v$, D_Sig$_v$) pair in their VIN_BS. Hence all the BSs under CA in M_1 and under F_CA$_C$ in M_2 receive D_Sig$_v$ after the authentication verification of V$_v$. It helps to prevent the further authentication verification of V$_v$ under CA in M_1 and F_CA$_C$ in M_2. The insertion function at BS$_B$ sends the executable version of FUNC_13 to V$_v$ through CH$_v$ and triggers tracking algorithm by sending (VIN$_v$, D_Sig$_v$) pair.

FUNC_11

Input is EB$_v$ and K$_1$. Output is B$_v$. It computes B$_1$ as K$_1$ XOR T$_{R\_present}$, rotates EB$_v$ towards right by B$_1$ bits to generate B$_v$.

However, during this span of time, V$_v$ may move to the coverage area of another BS under CA in M_1 and under F_CA$_C$ in M_2. The BS within which V$_v$ is currently passing through is the current BS of V_v (Current_BS_v). V_v triggers BS-Vehicle registration phase for its registration after entering into the coverage area of Current_BS_v. The identification algorithm at Current_BS_v triggers the tracking algorithm by sending (VIN$_v$, D_Sig$_v$) pair in the authentication phase.

The tracking algorithm for V_v is elaborated by assuming that BS_B is Current_BS_v. The tracking algorithm at BS_B searches LIST$_B$ using VIN$_v$ as the search key, reads $T_{v\_present}$ and CH_v from R_v in LIST$_B$, and senses CH_v. CH_v is free if V_v is not currently within the coverage area of BS_B, and the tracking algorithm at BS_B deletes R_v from LIST$_B$. Otherwise, the tracking algorithm at BS_B encrypts D_Sig$_v$ using the key B_2 to generate ED_Sig$_v$ and encrypts B_2 to generate K_2 using FUNC_12. It generates a signature packet containing (ED_Sig$_v$, K_2) pair of size 172 bits, and sends the signature packet to V_v through CH_v. It adds ED_Sig$_v$ as attribute in R_v of LIST$_B$. It also switches on a timer, and initializes it to τ_v. The timer τ_v indicates the time span during which ED_Sig$_v$ is assigned to V_v. It senses CH_v when τ_v expires. If CH_v is free, the tracking algorithm removes τ_v and deletes R_v from LIST$_B$. Otherwise it reassigns ED_Sig$_v$ to V_v through CH_v, and switches on τ_v again.

FUNC_12

Input is D_Sig$_v$ and B_2. Output is ED_Sig$_v$ and K_2. It selects a random number (B_2) of size 4 bits, rotates D_Sig$_v$ by B_2 bits towards left to generate ED_Sig$_v$, pads B_2 with 8 zero bits from most significant bits, and encrypts B_2 as $K_2 = B_2$ XOR $T_{v\_present}$.

V_v reads (ED_Sig$_v$, K_2) pair from the signature packet, decrypts ED_Sig$_v$ to generate D_Sig$_v$ using FUNC_13, and stores D_Sig$_v$.

FUNC_13

Input is ED_Sig$_v$ and K_2. Output is D_Sig$_v$. It computes B_2 as K_2 XOR $T_{v\_present}$ and rotates ED_Sig$_v$ towards right by B_2 bits to generate D_Sig$_v$.

10.4.5 Algorithmic Complexity

In this section, the algorithmic complexity of all the three phases is elaborated.

10.4.5.1 Algorithmic Complexity of Root-BS Registration Phase

The complexity of the Root-BS registration algorithm for the registration of all the BSs under CA in M_1 and under R_CA in M_2 is O(NO_OF_BS) and O (NO_OF_BS × NO_OF_FCA) respectively.

Proof The computation complexity for the registration of BS_B under root is due to encryption of ID$_R$, decryption of EID$_R''$, encryption of ID$_B$ and decryption of EID$_B''$. As the size of ID$_R$, $T_{R\_present}$, EID$_R$, EID$_R'$, EID$_R''$ is fixed to 2 bits, 12 bits,

12 bits, 12 bits, 14 bits respectively, the computation complexity of encrypting ID_R at root and decrypting EID_R'' at BS_B is $O(1)$. Again, as the size of ID_B, $T_{R\_present}$, EID_B, EID_B', EID_B'' is fixed to 2 bits, 12 bits, 12 bits, 12 bits, 14 bits respectively, the computation complexity of encrypting ID_B at BS_B and decrypting EID_B'' at root is $O(1)$. From the above discussion, it can be concluded that the complexity of the Root-BS registration algorithm for the registration of all the BSs under root is O (NO_OF_BS) in M_1 and $O(NO\_OF\_BS \times NO\_OF\_FCA)$ in M_2.

10.4.5.2   Algorithmic Complexity of BS-Vehicle Registration Phase

In the worst case, V_v visits all the BS in VANET. The complexity of the BS-Vehicle registration algorithm for the registration of V_v within the coverage area of all the BSs in VANET is $O(NO\_OF\_BS)$ in M_1 and $O(NO\_OF\_BS \times NO\_OF\_FCA)$ in M_2, respectively.

Proof The computation complexity for the registration of V_v under BS_B is due to the encryption of VIN_v, decryption of E_VIN_v, encryption of ID_B, decryption of EID_B'', searching for CH_v and insertion of R_v in $LIST_B$. As the size of VIN_v, VIN_v', VIN_v'', VIN_v''', E_VIN_v, $T_{v\_present}$ is fixed to 136 bits, 136 bits, 136 bits, 272 bits, 272 bits, 12 bits respectively, the computation complexity of encrypting VIN_v at V_v and decrypting E_VIN_v at BS_B is $O(1)$. Again, as the size of ID_B, ID_B', ID_B'', ID_B''', EID_B'', $T_{v\_present}$ is fixed to 2 bits, 12 bits, 12 bits, 14 bits, 14 bits, 12 bits respectively the computation complexity of encrypting ID_B at BS_B and decrypting EID_B'' at V_v is $O(1)$. The number of channels at BS_B is fixed as 48 [1, 2] and so the computation complexity of searching CH_v is $O(1)$. The size of R_v is fixed as 154 bits and the computation complexity of inserting R_v in $LIST_B$ is also $O(1)$. From the above discussion, it can be concluded that the complexity of the BS-Vehicle registration algorithm for V_v is $O(NO\_OF\_BS)$ in M_1 and $O(NO\_OF\_BS \times NO\_OF\_FCA)$ in M_2, respectively.

10.4.5.3   Algorithmic Complexity of Authentication Phase

The complexity of executing identification, authentication and tracking algorithm for V_v in VANET is $O(TOT\_V \times NO\_OF\_BS) + O(NO\_OF\_BS)$ in M_1 and O $(TOT\_V \times NO\_OF\_BS \times NO\_OF\_FCA) + O(NO\_OF\_BS \times NO\_OF\_FCA)$ in M_2, respectively.

Proof In the worst case, though V_v visits all the BS in VANET, its authentication is verified only once in M_1 but NO_OF_FCA times in M_2. The identification and tracking algorithm for V_v are executed by all BS in VANET, in the worst case.

Identification Algorithm

The computation complexity of the identification algorithm for V_v at BS_B is due to the searching of VIN_BS_B for VIN_v and reading of D_Sig_v from the record of V_v in VIN_BS_B. The size of VIN_v is fixed as 136 bits, the maximum number of records in VIN_BS_B is TOT_V, and the maximum size of each record is fixed as 296 bits. Hence the computation complexity of searching VIN_BS_B for VIN_v is O(TOT_V). The computation complexity of reading D_Sig_v of size 160 bits [18] from the record corresponding to VIN_v in VIN_BS_B is O(1). From the above discussion, it can be concluded that the computation complexity for the identification algorithm of V_v at all the BSs in VANET is O(NO_OF_BS × TOT_V) in M_1 and O (NO_OF_BS × NO_OF_FCA × TOT_V) in M_2.

Authentication Algorithm

The computation complexity for verifying the authentication of V_v is the sum of the computation complexity of verification function at root of the hierarchy and insertion function at all the BS under root. The computation complexity of the verification function is due to the reading of E_VIN_v from the authentication packet, insertion of E_VIN_v in queue at root, decryption of E_VIN_v, verification of the validity of VIN_v, generation of D_Sig_v from VIN_v, generation of B_v and encryption of B_v.

The size of E_VIN_v is fixed as 272 bits. So the computation complexity of reading E_VIN_v from the authentication packet, inserting E_VIN_v in CA_VIN_Queue in M_1 and R_CA_VIN_Queue in M_2 by the verification function is O(1). As the size of E_VIN_v, VIN_v''', VIN_v, VIN_v'' is fixed to 272, 272, 136, 136 bits respectively, the computation complexity of decrypting E_VIN_v at root is O(1).

The size of VIN_v is fixed as 136 bits and hence the computation complexity of verifying the validity of VIN_v is O(1). The computation complexity of generating D_Sig_v from VIN_v using SHA-1 algorithm is O(1) due to its fixed 80 rounds of operations. The size of B_v is fixed as 296 bits and hence the computation complexity of generating B_v is O(1). As the size of EB_v, B_v, B_2, K_2, $T_{R\_present}$ is fixed to 296, 296, 4, 12, 12 bits respectively, the computation complexity of encrypting B_v to generate EB_v is O(1). So the computation complexity of the verification function for V_v is O(1) in M_1 and O(NO_OF_FCA) in M_2.

The computation complexity of the insertion function at BS_B is due to the reading of (EB_v, K_1) pair from the insertion packet, decryption of EB_v and insertion of (VIN_v, D_Sig_v) pair in VIN_BS_B. The size of (EB_v, K_1) pair is fixed as 308 bits and hence the computation complexity of reading (EB_v, K_1) pair from the insertion packet is O(1). As the size of EB_v, B_v, B_2, K_2, $T_{R\_present}$ is fixed as 296, 296, 4, 12, 12 bits respectively, the computation complexity of decrypting EB_v and inserting B_v in VIN_BS_B by the insertion function at BS_B is O(1). The size of (VIN_v, D_Sig_v) pair is fixed as 296 bits and hence the computation complexity of inserting (VIN_v, D_Sig_v) pair in VIN_BS_B is O(1). So the computation complexity of executing the insertion function at all the BSs in VANET is O(NO_OF_BS) in M_1 and O(NO_OF_BS × NO_OF_FCA) in M_2.

Hence, the computation complexity of executing the authentication algorithm for V_v in VANET is O(NO_OF_BS) in M_1 and O(NO_OF_BS × NO_OF_FCA) + O(NO_OF_FCA) in M_2.

Tracking Algorithm

The computation complexity of the tracking algorithm for V_v at BS_B is due to the searching of $LIST_B$ for VIN_v, reading of $(T_{v\_present}, CH_v)$ pair from R_v, encrypting D_Sig_v to generate ED_Sig_v, generation of signature packet and addition of ED_Sig_v as attribute in R_v at $LIST_B$.

The maximum size of $LIST_B$ is 48 and hence the computation complexity of searching $LIST_B$ for VIN_v is $O(1)$. The computation complexity of reading $T_{v\_present}$ of size 12 bits and CH_v of size 6 bits from R_v in $LIST_B$ is $O(1)$. As the size of B_2, K_2, D_Sig_v, $T_{v\_present}$ is fixed at 4, 12, 160, 12 bits respectively, the computation complexity of encrypting D_Sig_v by the tracking algorithm at BS_B and decrypting ED_Sig_v by V_v is $O(1)$. As the size of signature packet is fixed as 172 bits its generation complexity is $O(1)$. The complexity of adding ED_Sig_v of size 160 bits in R_v of $LIST_B$ is $O(1)$. Hence the computation complexity of executing tracking algorithm at all the BSs in VANET for V_v is $O(NO\_OF\_BS)$ in M_1 and O $(NO\_OF\_BS \times NO\_OF\_FCA)$ in M_2.

From the above discussion, it can be concluded that the complexity of executing identification, authentication and tracking algorithm for V_v in the authentication phase is $O(TOT\_V \times NO\_OF\_BS) + O(NO\_OF\_BS)$ in M_1 and $O(TOT\_Vx$ $NO\_OF\_BS \times NO\_OF\_FCA) + O(NO\_OF\_BS \times NO\_OF\_FCA) + O$ (NO_OF_FCA) in M_2.

Correctness Proof of the Algorithms

In this section the correctness proof of the algorithms in the three phases are elaborated.

Correctness Proof of Algorithms in Root-BS Registration Phase

The correctness proof of the Root-BS registration algorithm is elaborated for the registration of BS_B under root. It is not possible to decrypt EID_R'' at BS_B and EID_B'' at root without knowing the key $T_{R\_present}$. BS_B knows $T_{R\_present}$ at the time of its registration under root. The attacker can't decrypt EID_R'' and EID_B'' without knowing the key $(T_{R\_present})$ and decryption algorithm. Hence only the intended recipient can decrypt the registration packet and registration response packet which ensures confidentiality of the algorithm.

The registration packet and the registration response packet are exchanged between the root and BS_B in encrypted form. BS_B knows the key $T_{R\_present}$ by decrypting the registration packet of root and uses the same key in the registration response phase. The root verifies the key after extracting it from the registration response packet of BS_B. The registration of BS_B is successful if the key is verified as valid at root which ensures that BS_B receives the exact registration packet and the root receives the exact registration response packet which in turn ensures integrity of the algorithm.

$T_{R\_present}$ is generated at the root during encryption of the registration packet and calculated at BS_B during decryption of the registration packet. An attacker cannot get this key, cannot perform any encryption/decryption and hence cannot pretend to be root or BS_B.

An attacker cannot send a registration packet to BS_B on behalf of root or registration response packet to root on behalf of BS_B as it does not know the packet format as well as the key. So root and BS_B cannot decline the transmission of registration packet and registration response packet respectively.

Correctness Proof of Algorithms in BS-Vehicle Registration Phase

The correctness proof of the BS-Vehicle registration algorithm is elaborated for the registration of V_v under BS_B. It is not possible to decrypt E_VIN_v at BS_B and EID_B'' at V_v without knowing $T_{v\_present}$. BS_B knows $T_{v\_present}$ at the time of registration of V_v. The attacker can't decrypt E_VIN_v and EID_B'' without knowing the key and decryption algorithm. Hence only the intended recipient can decrypt the registration packet and registration response packet which ensures confidentiality of the algorithm.

The registration packet and registration response packet are exchanged among BS_B and V_v in encrypted form. BS_B knows the key $T_{v\_present}$ by decrypting the registration packet of V_v and uses the same key in the registration response phase. V_v verifies the key after extracting it from the registration response packet of BS_B. The registration of V_v is successful if the key is verified as valid at V_v which ensures that BS_B receives the exact registration packet and V_v receives the exact registration response packet which in turn ensures integrity of the algorithm.

The key $T_{v\_present}$ is generated at V_v during encryption of the registration packet and calculated at BS_B during decryption of the registration packet in the BS-Vehicle registration phase. An attacker cannot get this key, cannot perform any encryption/decryption and hence cannot pretend to be V_v or BS_B.

An attacker cannot send a registration packet to BS_B on behalf of V_v without knowing the key ($T_{v\_present}$) and packet format. BS_B sends registration response packet to V_v through CH_v, so that it is not accessible by an attacker. Hence V_v and BS_B cannot decline the transmission of registration packet and registration response packet respectively.

Correctness Proof of Algorithms in Authentication Phase

The correctness proof of the identification, authentication and tracking algorithm is elaborated for V_v under BS_B. The identification algorithm for V_v is executed inside BS_B after the successful registration of V_v whereas the authentication algorithm for V_v is executed inside BS_B if V_v is an authentic vehicle. Otherwise the authentication algorithm at BS_B triggers the verification function at root by sending the authentication packet. The verification function at root verifies the authentication of V_v and calls the insertion function at all the BSs under CA in M_1 and under F_CA_C in M_2 by sending the insertion packet if V_v is authentic. The tracking algorithm at BS_B sends the signature packet to V_v through CH_v. Hence confidentiality, integrity, masquerade protection and non-repudiation protection are required to ensure the security of the authentication packet, insertion packet and signature packet during their transmission.

The authentication packet is generated by the identification algorithm at BS_B. This packet contains E_VIN_v. FUNC_9 at root decrypts this packet to retrieve VIN_v. An attacker cannot decrypt E_VIN_v without knowing the decryption algorithm. Hence only the intended recipient can decrypt the authentication packet which ensures confidentiality of this packet.

The insertion packet is generated by the root and decrypted by BS_B. Both root and BS_B use the same key B_1 which is generated at root and computed at BS_B. Moreover, the insertion packet contains the key B_1 in encrypted form (K_1). Hence, it is not possible to decrypt the insertion packet without knowing the key and decryption algorithm by the attacker. Hence, only the intended recipient can decrypt the insertion packet which ensures confidentiality of this packet.

The signature packet is generated by BS_B and decrypted by V_v. Both BS_B and V_v use the same key B_2 which is generated at BS_B and computed at V_v. Moreover, the signature packet contains the key B_2 in encrypted form (K_2). Hence it is not possible to decrypt the signature packet without knowing the key and decryption algorithm by the attacker which ensures confidentiality of the signature packet.

The authentication packet, insertion packet and signature packet contain data in encrypted form. An attacker cannot retrieve data from these packets without knowing the packet format, decryption algorithm, and key. Moreover, BS_B sends the signature packet to V_v through CH_v which is not accessible by the attacker. It ensures that the root receives the exact authentication packet, BS_B receives the exact insertion packet and V_v receives the exact signature packet which in turn ensures integrity.

The authentication packet contains E_VIN_v which is generated by V_v using the key $T_{v\_present}$. The insertion packet is generated at root and signature packet is generated at BS_B using the key B_1 and B_2 respectively. An attacker cannot get these keys, cannot perform any encryption/decryption and hence cannot pretend to be V_v or BS_B.

An attacker cannot send the authentication packet to the root on behalf of BS_B without knowing the packet format and encryption algorithm. It cannot send the insertion packet to BS_B on behalf of the root without knowing B_1 and packet format. BS_B sends the signature packet to V_v through CH_v which is not accessible to an attacker. Hence the root cannot decline the transmission of insertion packet and BS_B cannot decline the transmission of authentication packet and signature packet.

10.5 LoWVR

In this section, the proposed LoWVR scheme is elaborated by considering v^{th} vehicle (V_v) as sender and wth vehicle (V_w) as receiver. uid_v and uid_w are the unique identification of V_v and V_w respectively. key_v is the random number generated by V_v after observing an event.

10.5.1 Network Description

System Model

The system model considers hierarchical VANET [1] as in LIAT. Each vehicle generates message after observing an event within its coverage area. The coverage area of a BS is divided into a number of lanes, each of which is again divided into areas which helps a vehicle to identify the exact location of an event. In this architecture, the mode of communication among vehicles is broadcast. LoWVR verifies the vehicle authentication to identify unauthentic vehicles and message integrity to identify alteration attackers. Each vehicle maintains a certificate revocation list (CRL) for storing the identification of unauthentic vehicles and vehicles which are identified as alteration attacker for future protection from such vehicles.

Attack Model and Design Goal

The unique characteristics of VANET, such as mobility constraints, infrastructure less framework, short duration of links between vehicles and highly dynamic change in network topology, makes it susceptible to the attacks. An attacker can easily mix up with other vehicles in VANET by occupying same resources like authentic vehicles. Hence the secure and reliable message communication need to be achieved by ensuring the authentication of the vehicle and integrity of the message during its dissemination among vehicles. Hence, the attack model of LoWVR considers the attacks which are based on the principle of security such as authentication, integrity, modification detection, availability, privacy and accuracy.

10.5.2 Algorithms

In this section, the format of M and generation of M' from M are discussed. The algorithms that are executed by the sender and receiver vehicles are also elaborated.

Format of M and Generation of M'

M refers to the form including: Type, MC, Lane_id, Location_id. The type field in M indicates the type of the message. The size of this field is assumed as 3 bits so that LoWVR is able to support 8 different types of messages. The MC field indicates the message content, i.e. the events or infotainments for which M is generated. The size of this field is assumed as 6 bits to accommodate maximum number of events or infotainments in VANET during simulation.

V_v generates M after observing an event at a location (Location_id) in a lane (Lane_id). The size of Lane_id is assumed as 11 bits by considering the transport system of Kolkata [20] as a benchmark during simulation. The size of Location_id is assumed as 12 bits. Hence the size of M is 32 bits. V_v generates a timestamp

(T) corresponding to the time of generation of M of size 32 bits [21] and pads it with M to generate M' of size 64 bits.

ALGO_1

It is a deformed message generation algorithm. V_v uses this algorithm to generate DVOM of size 64 bits from M'. Input is M'. Output is DVOM. Algorithm steps are as follows:

1. Divide M' into 8 blocks each of size 8 bits; $[(B_7), (B_6), (B_5), (B_4), (B_3), (B_2), (B_1), (B_0)]$
2. Complement each block to generate $[(B_7')(B_6')(B_5')(B_4')(B_3')(B_2')(B_1')(B_0')]$
3. Compute $[(O_7)(O_6)(O_5)(O_4)(O_3)(O_2)(O_1)(O_0)]$; $[(O_7 = B_7')(O_6 = B_6' \text{ XOR } B_5')(O_5 = B_5')(O_4 = B_4' \text{ XOR } B_3')(O_3 = B_3')(O_2 = B_2' \text{ XOR } B_1')(O_1 = B_1')(O_0 = B_0')]$
4. Compute $[(O_7')(O_6')(O_5')(O_4')(O_3')(O_2')(O_1')(O_0')]$; $[(O_7' = O_7 \text{ XOR } O_6)(O_6' = O_6)(O_5' = O_5 \text{ XOR } O_4)(O_4' = O_4)(O_3' = O_3 \text{ XOR } O_2)(O_2' = O_2)(O_1' = O_1 \text{ XOR } O_0)(O_0' = O_0)]$
5. Divide i^{th} $(0 \leq i \leq 7)$ block into two parts, each of size 4 bits to generate

    $[(O_{7L}' O_{7R}'), (O_{6L}' O_{6R}'), (O_{5L}' O_{5R}'), (O_{4L}' O_{4R}'), (O_{3L}'O_{3R}'), (O_{2L}' O_{2R}'), (O_{1L}' O_{1R}'), (O_{0L}' O_{0R}')]$
6. Swap O_{7R}' with O_{0R}', O_{6R}' with O_{1R}', O_{5R}' with O_{2R}', O_{4R}' with O_{3R}' to generate DVOM as

    $[(O_{7L}'O_{0R}'), (O_{6L}' O_{1R}'), (O_{5L}' O_{2R}'), (O_{4L}' O_{3R}'), (O_{3L}' O_{4R}'), (O_{2L}' O_{5R}'), (O_{1L}' O_{6R}'), (O_{0L}'O_{7R}')]$
7. Return $[(O_{7L}'O_{0R}'), (O_{6L}' O_{1R}'), (O_{5L}' O_{2R}'), (O_{4L}' O_{3R}'), (O_{3L}' O_{4R}'), (O_{2L}' O_{5R}'), (O_{1L}' O_{6R}'), (O_{0L}' O_{7R}')]$ //64 bits DVOM

ALGO_2

It is MD and MD_DVOM generation algorithm. V_v uses this algorithm to generate MD from DVOM and to generate MD_DVOM by concatenating MD and DVOM. It makes a copy of DVOM (Copy_DVOM), generates MD of size 20 bits from Copy_DVOM and concatenates MD with DVOM to generate MD_DVOM of size 84 bits.

Input is Copy_DVOM in the form $[(O_{7L}'O_{0R}'), (O_{6L}' O_{1R}'), (O_{5L}' O_{2R}'), (O_{4L}' O_{3R}'), (O_{3L}' O_{4R}'), (O_{2L}' O_{5R}'), (O_{1L}' O_{6R}'), (O_{0L}'O_{7R}')]$. Output is MD_DVOM. Steps are as follows:

1. Generate a key K of size 8 bits by taking the ith bit of each ith block of Copy_DVOM; $K = K_7K_6K_5K_4K_3K_2K_1K_0$
2. Count the number of 1's in K
3. if (number of 1's in step 2 is odd) then
4. Replace each odd block of Copy_DVOM by its complement
5. else

6. Replace each even block of Copy_DVOM by its complement
7. end if
8. Compute K' as complement of K; $K' = K_7'K_6'K_5'K_4'K_3'K_2'K_1'K_0'$
9. Compute K'' by performing XOR operation among consecutive 2 bits of K'; $K'' = ((K_7'' \text{ XOR } K_6'), (K_5' \text{ XOR } K_4'), (K_3' \text{ XOR } K_2'), (K_1' \text{ XOR } K_0'))$
10. Perform XOR operation among the two parts of each block of Copy_DVOM and replace the same block by the result of XOR operation of size 4 bits to obtain $[(P_7)(P_6)(P_5)(P_4)(P_3)(P_2)(P_1)(P_0)]$ where $P_7 = O_{7L}' \text{ XOR } O_{0R}'$, $P_6 = O_{6L}' \text{ XOR } O_{1R}'$, $P_5 = O_{5L}' \text{ XOR } O_{2R}'$, $P_4 = O_{4L}' \text{ XOR } O_{3R}'$, $P_3 = O_{3L}' \text{ XOR } O_{4R}'$, $P_2 = O_{2L}' \text{ XOR } O_{5R}'$, $P_1 = O_{1L}' \text{ XOR } O_{6R}'$, $P_0 = O_{0L}' \text{ XOR } O_{7R}'$
11. Compute (Q_4'), (Q_3'), (Q_2'), $(Q_1')(Q_0')$ as $Q_4' = P_4$, $Q_3' = P_7 \text{ XOR } P_6$, $Q_2' = P_5 \text{ XOR } P_4$, $Q_1' = P_3 \text{ XOR } P_2$, $Q_0' = P_1 \text{ XOR } P_0$
12. Compute 20 bits MD as $[(MD_4)(MD_3)(MD_2)(MD_1)(MD_0)]$ where $MD_4 = Q_4'$, $MD_3 = Q_3' \text{ XOR } K''$, $MD_2 = Q_2' \text{ XOR } K''$, $MD_1 = Q_1' \text{ XOR } K''$, $MD_0 = Q_0' \text{ XOR } K''$
13. Generate MD_DVOM by concatenating DVOM and MD as (DVOM)(MD)
14. Return $[(O_{7L}'O_{0R}'), (O_{6L}' O_{1R}'), (O_{5L}' O_{2R}'), (O_{4L}' O_{3R}'), (O_{3L}' O_{4R}'), (O_{2L}' O_{5R}'), (O_{1L}' O_{6R}'), (O_{0L}' O_{7R}') (MD_4)(MD_3)(MD_2)(MD_1)(MD_0)]$
    //84 bits MD_DVOM

ALGO_3

It is watermark generation algorithm. V_v uses this algorithm to generate n bits watermark bits (WM_v) from n bits uid_v and n bits key_v.

Input is $uid_v = u_{n-1}u_{n-2}.......u_1u_0$ and $key_v = k_{n-1}k_{n-2}......k_1k_0$. Output is $WM_v = w_{n-1}w_{n-2}......w_1w_0$. Steps are as follows:

1. Perform XOR operation among all the bits of uid_v $(u_{n-1}u_{n-2}....u_1)$ with u_0 to generate uid_v' keeping u_0 unchanged
2. Perform XOR operation among all the bits of key_v $(k_{n-2}....k_1k_0)$ with k_{n-1} to generate key_v' keeping k_{n-1} unchanged
3. Generate WM_v by performing bitwise XOR operation among uid_v' and key_v'
   $w_{n-1} = u_{n-1} \text{ XOR } k_0$, $w_{n-2} = u_{n-2} \text{ XOR } k_1$, ... $w_1 = u_1 \text{ XOR } k_{n-2}$, $w_0 = u_0 \text{ XOR } k_{n-1}$
4. Return WM_v

ALGO_4

It is embedding algorithm. V_v uses this algorithm to generate WM_MD_DVOM by embedding WM_v, uid_v, key_v in MD_DVOM. The size of MD_DVOM is 84 bits. The size of WM_v, uid_v and key_v are n bits. If (84 mod n) = 0 then embedding algorithm stores MD_DVOM in Pad_MD_DVOM of size 84 bits. Otherwise the embedding algorithm pads $[n - (84 \text{ mod } n)]$ number of 0's to the MSB position of MD_DVOM to generate Pad_MD_DVOM of size $[84 + n - (84 \text{ mod } n)]$ bits so that Pad_MD_DVOM is divisible by n. The size of Pad_MD_DVOM is Size_Pad_MD_DVOM. The size of WM_MD_DVOM (Size_WM_MD_DVOM) is (Size_Pad_MD_DVOM + 3n) bits.

Input is Pad_MD_DVOM, WM_v, uid_v and key_v. Output is WM_MD_DVOM. Steps are as follows:

1. Divide Pad_MD_DVOM into n blocks each of size $\frac{Size\_Pad\_MD\_DVOM}{n}$ bits

$$[(BL_{n-1})(BL_{n-2})\ldots\ldots\ldots(BL_1)(BL_0)]$$

2. for (i = 0; i $\leq$ n − 1; i ++)
3. Count the number of 1's (N_i) in i^{th} block (BL_i) of Pad_MD_DVOM
4. if (N_i is odd)
5. Append complement of (w_i, u_i, k_i) to the right of BL_i to generate i^{th} block of WM_MD_DVOM ($BL_i{}'$) of size $\frac{Size\_Pad\_MD\_DVOM}{n}$ +3 bits //w_i, u_i, k_i are i^{th} bit of WM_v, uid_v, key_v respectively
6. else
7. Append (w_i, u_i, k_i) to the right of BL_i to generate i^{th} block of WM_MD_DVOM ($BL_i{}'$) of size $\frac{Size\_Pad\_MD\_DVOM}{n}$ +3 bits
8. end if
9. end for
10. Return $\left[(BL'_{n-1})(BL'_{n-2})\ldots\ldots\ldots(BL'_1)(BL'_0)\right]$ //Size_WM_MD_DVOM bits WM_MD_DVOM

ALGO_5

It is detection algorithm. V_w uses this algorithm to detect WM_v, uid_v, key_v and Pad_MD_DVOM from WM_MD_DVOM. Input is WM_MD_DVOM. Output is WM_v, uid_v, key_v, Pad_MD_DVOM. Steps are as follows:

1. for (i = 0; i $\leq$ n − 1; i ++)
2. Read the most significant $\frac{Size\_Pad\_MD\_DVOM}{n}$ bits from $BL_i{}'$//$BL_i{}'$ is i^{th} block of WM_MD_DVOM
3. Count the number of 1's (N_i) in $\frac{Size\_Pad\_MD\_DVOM}{n}$ bits data as obtained in step 2
4. if (N_i is odd)
5. Compute (w_i, u_i, k_i) as complement of the least significant 3 bits of $BL_i{}'$and delete the least significant 3 bits of $BL_i{}'$ to generate BL_i//BL_i is the ith block of Pad_MD_DVOM
6. else
7. Compute (w_i, u_i, k_i) as the least significant 3 bits of $BL_i{}'$ and delete the least significant 3 bits of $BL_i{}'$ to generate BL_i
8. end if
9. end for
10. Search CRL_w for uid_v
11. if found

//V_v is already identified as an unauthentic vehicle or an alteration attacker
12. Discard Pad_MD_DVOM
13. else
14. Trigger ALGO_6 by sending WM_v, uid_v, key_v, Pad_MD_DVOM

ALGO_6

ALGO_6 consists of three functions. V_w uses FUNC_1 of ALGO_6 to verify the authentication of V_v, FUNC_2 of ALGO_6 to verify whether V_v is an alteration attacker and FUNC_3 of ALGO_6 to compute M from DVOM if V_v is authentic and not an alteration attacker.

- **FUNC_1**: It reads WM_v, uid_v, key_v and Pad_MD_DVOM from the received trigger. It generates WM_v' from uid_v and key_v using ALGO_3. WM_v' is compared with WM_v. In case of mismatch, V_v is not authentic, CRL_w is updated by inserting uid_v and Pad_MD_DVOM is discarded. Otherwise V_v is authentic and FUNC_1 triggers FUNC_2 by sending Pad_MD_DVOM.
- **FUNC_2**: It reads the least significant 84 bits from Pad_MD_DVOM as MD_DVOM, reads the most significant 64 bits of MD_DVOM as DVOM and reads the least significant 20 bits of MD_DVOM as MD. It makes a copy of DVOM (Copy_DVOM), calculates MD' from Copy_DVOM following the steps 1 to 12 of ALGO_2 and compares MD' with MD. In case of mismatch V_v is identified as an alteration attacker, CRL_w is updated by inserting uid_v and Pad_MD_DVOM is discarded. Otherwise V_v is not an alteration attacker and FUNC_2 triggers FUNC_3 by sending DVOM.
- **FUNC_3**: Input is DVOM in the form [$(O_{7L}'O_{0R}')$, $(O_{6L}'\ O_{1R}')$, $(O_{5L}'\ O_{2R}')$, $(O_{4L}'\ O_{3R}')$, $(O_{3L}'\ O_{4R}')$, $(O_{2L}'\ O_{5R}')$, $(O_{1L}'\ O_{6R}')$, $(O_{0L}'\ O_{7R}')$]. Output is M. Steps are as follows:

1. Swap O_{7R}' with O_{0R}', O_{6R}' with O_{1R}', O_{5R}' with O_{2R}', O_{4R}' with O_{3R}' to generate [$(O_{7L}'\ O_{7R}')$, $(O_{6L}'\ O_{6R}')$, $(O_{5L}'\ O_{5R}')$, $(O_{4L}'\ O_{4R}')$, $(O_{3L}'O_{3R}')$, $(O_{2L}'\ O_{2R}')$, $(O_{1L}'\ O_{1R}')$, $(O_{0L}'\ O_{0R}')$]
2. Merge two parts of each block to generate [(O_7'), (O_6'), (O_5'), (O_4'), (O_3'), (O_2'), (O_1'), (O_0')]
3. Generate [$(O_7)(O_6)(O_5)(O_4)(O_3)(O_2)(O_1)(O_0)$] as $O_0 = O_0'$, $O_1 = O_1'$ XOR O_0, $O_2 = O_2'$, $O_3 = O_3'$ XOR O_2, $O_4 = O_4'$, $O_5 = O_5'$ XOR O_4, $O_6 = O_6'$, $O_7 = O_7'$ XOR O_6
4. Generate [$(B_7')(B_6')(B_5')(B_4')(B_3')(B_2')(B_1')(B_0')$] as $B_0' = O_0$, $B_1' = O_1$, $B_2' = O_2$ XOR B_1', $B_3' = O_3$, $B_4' = O_4$ XOR B_3', $B_5' = O_5$, $B_6' = O_6$ XOR B_5', $B_7' = O_7$
5. Generate M' in the form [(B_7), (B_6), (B_5), (B_4), (B_3), (B_2), (B_1), (B_0)] as complement of [$(B_7')(B_6')(B_5')(B_4')(B_3')(B_2')(B_1')(B_0')$]
6. Read the most significant 32 bits of M' as M in the form [(B_7), (B_6), (B_5), (B_4)]
7. Return [(B_7), (B_6), (B_5), (B_4)]//32 bits M

10.5.3 Correctness Proof of Basic Properties

Property 1 (Authentication)

ALGO_6 at V_w generates WM_v' and compares it with WM_v to identify V_v as authentic vehicle successfully.

Proof: WM_MD_DVOM is generated by using ALGO_1 to ALGO_4 at V_v. WM_MD_DVOM contains WM_v. An attacker cannot retrieve WM_v without knowing the exact format of WM_MD_DVOM and ALGO_5. Hence only the authentic vehicles like V_w can retrieve WM_v and compare WM_v with WM_v' using FUNC_1 of ALGO_6. The successful comparison ensures the authentication of V_v.

Property 2 (Integrity)

V_w receives exactly the same M that was sent by V_v.

Proof: WM_MD_DVOM contains deformed version of M. An attacker cannot retrieve M from WM_MD_DVOM without knowing the format of WM_MD_DVOM, ALGO_5 and ALGO_6 in time. Hence only the authentic intended recipient like V_w retrieves M from the received WM_MD_DVOM. V_w also verifies the integrity of M by comparing MD and MD' using FUNC_2 of ALGO_6 to make sure that V_w receives the exact WM_MD_DVOM.

Property 3 (Modification detection)

V_w can detect if WM_MD_DVOM is modified during dissemination.

Proof: It is not possible by an attacker to retrieve M from WM_MD_DVOM and hence to alter M in time. So only the authentic intended recipient like V_w retrieves WM_v, uid_v, key_v and Pad_MD_DVOM from WM_MD_DVOM using ALGO_5. Then it triggers ALGO_6 where WM_v' is compared with WM_v and MD' is compared with MD. Hence any modification in WM_MD_DVOM by V_v is identified at V_w by comparing MD' with MD which ensures modification detection.

Property 4 (Availability)

V_w can receive M which is sent by V_v in the form of WM_MD_DVOM.

Proof: V_w receives M in the form of WM_MD_DVOM from V_v. V_w can retrieve M from WM_MD_DVOM by using ALGO_5 and ALGO_6. V_w discards WM_MD_DVOM if V_v is not authentic or is identified as alteration attacker. Hence M is available to the authentic vehicle like V_w if V_v is authentic and the content of WM_MD_DVOM is not modified.

Property 5 (Privacy)

V_v can generate uid_v without taking help from any other trusted third party.

Proof: V_v generates n bits uid_v itself without taking help from any other trusted third party before executing ALGO_3. Hence it ensures that V_v can control its personal information, i.e. uid_v without allowing any interference from outside world.

Property 6 (Accuracy)

V_w accepts M after verifying the authentication of V_v and accuracy of M.

Proof: V_w receives WM_MD_DVOM through wireless communication from V_v. V_w accepts M only after verifying the authentication of V_v. V_w also verifies the accuracy of M by checking its integrity using ALGO_6. It ensures the authenticity as well as accuracy of M. Again, V_w receives WM_MD_DVOM through wireless communication which is based on IEEE 802.11a. It ensures the fidelity of M.

10.6 Performance Analysis of LIAT

The qualitative and quantitative performance of LIAT are elaborated in this section.

10.6.1 Qualitative Performance

The qualitative performance of LIAT is studied in terms of security analysis and overhead analysis, as discussed below.

10.6.1.1 Security Analysis

Security analysis of LIAT is done for V_v under BS_B using two parameters viz. cracking probability and cracking time. An attacker may crack the data during its transmission in the form of packet (Mode_1). It may also capture the data which is stored at different nodes in the network (Mode_2).

All the communication between BS_B and V_v is through a dedicated channel (CH_v) after the successful registration of V_v under BS_B. So, an attacker cannot capture the registration response packet and signature packet that are transmitted by BS_B to V_v through CH_v.

Also, an attacker cannot capture the stored data at BS_B as it has no dedicated channel for communication with BS_B. Besides, an attacker cannot capture the stored data at V_v as there is no direct communication among neighbor vehicles in LIAT. However, an attacker can capture the stored data at root. The root sends the registration packet and insertion packet to BS_B immediately after generating them, decrypts the registration response packet and authentication packet immediately after receiving them from BS_B. But the root maintains a copy of $T_{R\_present}$ for future communication with BS_B. Hence an attacker can capture $T_{R\_present}$ which is stored at root.

Cracking Probability and Cracking Time

The attacker captures the data and performs brute-force operations to crack the captured data. In LIAT, the cracking probability and cracking time are calculated both for Mode_1 and Mode_2. The cracking probability is the probability of retrieving data after capturing it. It is measured as the reciprocal of the number of brute-force operations to crack the data. In the worst case, an attacker may have to perform all the brute-force operations to crack the captured data. The cracking time is calculated in terms of the number of clock cycles which is required to crack the captured data in the worst-case scenario. Number of clock cycle is

$$\frac{\text{Total number of computations} \times \text{Time which is required to perform each computation}}{\text{Time for a single clock cycle}}$$

where the time which is required to perform each computation (T_C) and the time for single clock cycle (T_{CC}) are determined during simulation. If the cracking time exceeds the time when the data is actually being used (either during transmission or during its stay in a node), it can be claimed that obtaining the data is not of any use and therefore is not a matter of concern. In this section, both cracking probability and cracking time are evaluated for V_v within the coverage area of BS_B under CA in M_1, and under F_CA$_C$ within the coverage area of R_CA in M_2.

Cracking Probability and Time for Data in Mode_1

- The registration packet from root to BS_B in M_1 and root to BS_B through F_CA$_C$ in M_2 contains EID$_R''$ of size 14 bits. The registration response packet from BS_B to root in M_1 and BS_B to root through F_CA$_C$ in M_2 contains EID$_B''$ of size 14 bits. Hence the number of brute-force operations to crack the registration packet and registration response packet is $\sum_{j=0}^{13} 2^j = 16383$ in M_1 and $2 \times \sum_{j=0}^{13} 2^j = 32766$ in M_2 (where j indicates the bit position). The cracking probability is 6.10388×10^{-5} in M_1 and 3.05194×10^{-5} in M_2. The cracking time is $\frac{16383 \times T_C}{T_{CC}}$ clock cycles in M_1 and $\frac{32766 \times T_C}{T_{CC}}$ clock cycles in M_2. The time which is required for the transmission of the registration packet from the root to BS_B and for the transmission of the registration response packet from BS_B to the root in M_1 is evaluated as 6.44 ms whereas the time which is required for the transmission of the registration packet from root to BS_B through F_CA$_C$ and for the transmission of the registration response packet from BS_B to root through F_CA$_C$ in M_2 is evaluated as 11.62 ms during simulation. The cracking time of the registration packet and registration response packet is evaluated as 20749.03 ms in M_1 and 41491.63 ms in M_2 during simulation which is much higher than 6.44 ms and 11.62 ms respectively. Hence, the cracking of the registration packet and registration response packet during Root-BS_B registration phase by the attacker is of no use.
- The registration packet from V_v to BS_B, authentication packet from BS_B to root in M_1 and BS_B to root through F_CA$_C$ in M_2 contain E_VIN$_v$ of size 272 bits. Hence the number of brute-force operations to crack the registration packet is $\sum_{j=0}^{271} 2^j = 7.5885 \times 10^{81}$, authentication packet in M_1 is 7.5885×10^{81}

and authentication packet in M_2 is $2 \times \sum_{j=0}^{271} 2^j = 15.1770 \times 10^{81}$. The cracking probability for the registration packet is 0.131778×10^{-81}, authentication packet in M_1 is 0.131778×10^{-81} and authentication packet in M_2 is 0.065889×10^{-81}. The cracking time for the registration packet is $\frac{7.5885 \times 10^{81} \times T_C}{T_{CC}}$ clock cycles, for the authentication packet in M_1 is $\frac{7.5885 \times 10^{81} \times T_C}{T_{CC}}$ clock cycles and for the authentication packet in M_2 is $\frac{15.1770 \times 10^{81} \times T_C}{T_{CC}}$ clock cycles. The time which is required for the transmission of the registration packet from V_v to BS_B is evaluated as 8.39 ms whereas its cracking time is evaluated as 9.6078×10^{84} ms both for M_1 and M_2 during simulation. Hence, the cracking of the registration packet during BS_B–V_v registration phase by the attacker is of no use. The time which is required for the transmission of the authentication packet from BS_B to root in M_1 and BS_B to root through F_CA$_C$ in M_2 is evaluated as 7.56 ms and 12.84 ms respectively during simulation. The cracking time of the authentication packet is evaluated as 9.6078×10^{84} ms in M_1 and 1.925×10^{83}ms in M_2 during simulation, which are much higher than 7.56 ms and 12.84 ms respectively. Hence the cracking of the authentication packet during the authentication phase of V_v by the attacker is of no use.

- The insertion packet from root to BS_B in M_1 and from root to BS_B through F_CA$_C$ in M_2 contains (EB_v, K_1) pair of size 308 bits. Hence the number of brute-force operations to crack the insertion packet is $\sum_{j=0}^{307} 2^j = 5.21481$ 10^{92} in M_1 and $2 \times \sum_{j=0}^{307} 2^j = 10.42962 \times 10^{92}$ in M_2. The cracking probability is 0.19176×10^{-92} in M_1 and 0.09588×10^{-92} in M_2. The cracking time is $\frac{5.21481 \times 10^{92} \times T_C}{T_{CC}}$ clock cycles in M_1 and $\frac{10.42962 \times 10^{92} \times T_C}{T_{CC}}$ clock cycles in M_2. The time which is required for the transmission of the insertion packet from root to BS_B in M_1 and root to BS_B through F_CA$_C$ in M_2 is evaluated as 8.14 ms and 13.23 ms respectively during simulation. The cracking time of the insertion packet is evaluated as 1.3876×10^{94} ms in M_1 and 1.3204×10^{94} ms in M_2 during simulation which is much higher than 8.14 ms and 13.23 ms respectively. Hence, the cracking of the insertion packet during the authentication phase of V_v by the attacker is of no use.

Cracking Probability and Time for Data in Mode_2

The storage of $T_{R\_present}$ at root contains 12 bits. Hence the number of brute-force operations to crack $T_{R\_present}$ is $\sum_{j=0}^{11} 2^j = 4095$, cracking probability is 2.44200×10^{-4} and cracking time is $\frac{4095 \times T_C}{T_{CC}}$ clock cycles. The time which is required to complete the registration of all the BSs under root is evaluated as 30.664 ms for M_1 and 34.49 ms for M_2 during simulation. The cracking time of $T_{R\_present}$ at root for generating the registration response packet is evaluated as 5184.69 ms during simulation which is much higher than 30.664 ms and 34.49 ms. The root knows the number of BSs under it after 30.664 ms for M_1 and 34.49 ms for M_2. Hence the registration response packet of attacker is not accepted by root after 5184.89 ms, attacker cannot pretend to be a BS under root and so cracking of $T_{R\_present}$ is of no use.

10.6.1.2   Overhead Analysis

The overheads of LIAT are measured for V_v in VANET. STO_OH$_v$, COMM_OH$_v$ and COMP_OH$_v$ are the storage, communication and computation overhead of V_v respectively. In the worst case, V_v visits all the BSs in VANET. Hence STO_OH$_v$, COMM_OH$_v$ and COMP_OH$_v$ include: (i) Communication (COMM$_1$), storage (STO$_1$) and computation (COMP$_1$) overhead for the registration of all the BSs in VANET; (ii) Communication (COMM$_2$), storage (STO$_2$) and computation (COMP$_2$) overhead for the registration of V_v at all the BSs in VANET; and (iii) Communication (COMM$_3$), storage (STO$_3$) and computation (COMP$_3$) overhead for the identification, authentication and tracking of V_v in VANET.

STO_OH$_v$

VIN_CA is used by CA and R_CA to gather knowledge about the available pattern of VINs of the vehicles that are already manufactured. As VIN_CA is not required during the registration phase and authentication phase of V_v, STO_OH$_v$ is assumed as independent on the size of VIN_CA. Hence STO_OH$_v$ is the sum of STO$_1$, STO$_2$ and STO$_3$.

The root and all the BSs under root store $T_{R\_present}$ of size 12 bits. Hence STO$_1$ is of size $12 \times (1 + $ NO_OF_BS$)$ bits in M_1 and $12 \times (1 + $ NO_OF_BS $\times$ NO_OF_FCA$)$ bits in M_2.

V_v stores $T_{v\_present}$ of size 12 bits till it resides within the coverage area of BS$_B$. BS$_B$ stores E_VIN$_v$ of size 272 bits in BS$_B$_VIN_Queue till its processing is over and (VIN$_v$, $T_{v\_present}$, CH$_v$, ED_Sig$_v$) of size 314 bits as R$_v$ in LIST$_B$ till V_v resides within its coverage area. Moreover, V_v resides within the coverage area of a particular BS at a time. Hence STO$_2$ is 598 bits.

The verification function stores E_VIN$_v$ of size 272 bits in CA_VIN_Queue in M_1 and R_CA_VIN_Queue in M_2 till its processing is over. The insertion function at all the BSs in VANET insert (VIN$_v$, D_Sig$_v$) pair of size 296 bits in their VIN_BS. V_v stores D_Sig$_v$ of size 160 bits and FUNC_13 of size 244 bytes. Hence STO$_3$ is $272 + (296 \times$ NO_OF_BS$) + 160 + 1952$ bits in M_1 and $272 + (296 \times$ NO_OF_BS $\times$ NO_OF_FCA$) + 160 + 1952$ bits in M_2.

STO_OH$_v$ is of size $[2994 + (308 \times$ NO_OF_BS$)]$ bits in M_1 and $[2994 + (308 \times$ NO_OF_BS $\times$ NO_OF_FCA$)]$ bits in M_2.

COMM_OH$_v$

It is the sum of COMM$_1$, COMM$_2$ and COMM$_3$. The root of the hierarchy sends registration packet of size 14 bits to all the BSs under it. BS$_B$ sends registration response packet of size 14 bits to CA in M_1 and to R_CA through F_CA$_C$ in M_2.

Hence COMM$_1$ is $28 \times$ NO_OF_BS bits in M_1 and $(14 + 42 \times$ NO_OF_BS) $\times$ NO_OF_FCA bits in M_2.

V_v sends registration packet of size 272 bits for registration under BS$_B$ and BS$_B$ sends registration response packet of size 14 bits to V_v through CH$_v$. Hence COMM$_2$ is $286 \times$ NO_OF_BS bits in M_1 and $286 \times$ (NO_OF_BS $\times$ NO_OF_FCA) bits in M_2.

The authentication algorithm at BS$_B$ calls the verification function at CA in M_1 and at R_CA through F_CA$_C$ in M_2 by sending the authentication packet of size 272 bits. The verification function at CA in M_1 calls the insertion function at all the BSs under it by sending the insertion packet of size 308 bits. The verification function at R_CA in M_2 calls the insertion function at all the BSs under F_CA$_C$ by sending the insertion packet of size 308 bits. The insertion function at BS$_B$ sends the executable version of FUNC_13 of size 244 bytes to V_v through CH$_v$. The tracking algorithm at BS$_B$ sends signature packet of size 172 bits to V_v through CH$_v$.

Hence, COMM$_3$ is $272 + (308 \times$ NO_OF_BS$) + 1952 + (172 \times$ NO_OF_BS$)$ bits in M_1 and NO_OF_FCA $\times$ [$(2 \times 272) + (308 \times (1 + $ NO_OF_BS$))$ $+(172 \times$ NO_OF_BS$) +1952$] bits in M_2. COMM_OH$_v = 2224 + 794 \times$ NO_OF_BS bits in M_1 and NO_OF_FCA $\times$ ($2818 + 808 \times$ NO_OF_BS) bits in M_2.

COMP_OH$_v$

It is the sum of COMP$_1$, COMP$_2$ and COMP$_3$. COMP$_1$ is $O($NO_OF_BS$)$ in M_1 and $O($NO_OF_BS $\times$ NO_OF_FCA$)$ in M_2 as discussed in Sect. 10.4.5.1. COMP$_2$ is $O($NO_OF_BS$)$ in M_1 and $O($NO_OF_BS $\times$ NO_OF_FCA$)$ in M_2 as discussed in Sect. 10.4.5.2. COMP$_3$ is $O($TOT_V $\times$ NO_OF_BS$) + O($NO_OF_BS$)$ in M_1 and $O($TOT_V $\times$ NO_OF_BS $\times$ NO_OF_FCA$) + O($NO_OF_BS $\times$ NO_OF_FCA$) + O($NO_OF_FCA$)$ in M_2 as discussed in Sect.10.4.5.3. COMP_OH$_v = O($TOT_V $\times$ NO_OF_BS$) + O($NO_OF_BS$)$ in M_1 and O (TOT_V $\times$ NO_OF_BS $\times$ NO_OF_FCA$) + O($NO_OF_BS $\times$ NO_OF_FCA$) +$ $O($NO_OF_FCA$)$ in M_2.

Experimental Analysis

COMM_OH$_v$ (A), STO_OH$_v$ (B) and COMP_OH$_v$ (C) of [1] and M_1 are compared with the existing schemes in [22, 23], that are simulated in a distributed environment to compare their performance with the distributed schemes [2] and M_2. COMM_OH$_v$ (A), STO_OH$_v$ (B) and COMP_OH$_v$ (C) of [2] and M_2 are compared with distributed [22] (identified as Dist_22 in the rest of the chapter) and

Table 10.2 Overheads

	[1]	M_1	[22]	[23]	[2]	M_2	Dist_22	Dist_23
A	888	4606	6144	23760	2664	26366	55296	192336
B	888	3918	16170	67152	1440	11310	96042	444624
C	84362	84407	87235	86214	84592	84677	87470	86449

distributed [23] (identified as Dist_23 in the rest of the chapter). The result of comparison is shown in Table 10.2.

COMM_OH$_v$, STO_OH$_v$ and COMP_OH$_v$ are evaluated in bits, bits, and clock cycles (cc) respectively during simulation. The number and size of the different packets that are exchanged among the different levels of the VANET hierarchy for verifying vehicle authentication in [22, 23], Dist_22, Dist_23 are much higher than in [1, 2], M_1, M_2. All the intermediate nodes in [22, 23], Dist_22, Dist_23 store these packets till their processing is over. Moreover, the computations in [22, 23], Dist_22, Dist_23 are much more complex than the computations in [1, 2], M_1, M_2.

Hence the communication, storage and computation overhead per vehicle are higher in [22, 23], Dist_22, Dist_23 than those in [1, 2], M_1, M_2.

The required energy for transmission of 1 byte of data is 0.534×10^{-15} µJ and per clock cycle computation is $0.3524788443 \times 10^{-16}$ µJ [21]. Table 10.3 shows the result of comparison on the basis of COMM_OH$_v$ and COMP_OH$_v$ in joule with all the competing schemes.

10.6.2 Quantitative Performance

The simulation experiment is conducted to evaluate the effectiveness of LIAT as reported in earlier sections. The simulation is conducted by considering the transportation system in Kolkata [20]. The roads used have four lanes each as in [20]. The size of VANET is varied from 20 to 30 km^2; the number of vehicles is varied from 5 to 250 and the velocity of vehicles is varied from 72 km/h to 120 km/h as in [20]. The lane length, lane width and distance between the two vehicles are assumed as 10 km, 30 m and 2 m respectively as in [20]. The data transmission rate is considered as 6 Mbps and τ_v is assumed as 20 s.

10.6.2.1 Simulation Metrics

The objective of LIAT is to maximize the number of vehicles whose identification and authentication can be verified within the stipulated simulation time. Hence the time to verify the identification and authentication of each vehicle under a BS and the loss of registration packet of vehicles at a BS should be as small as possible to maximize the number of vehicles whose identification and authentication can be verified within the stipulated simulation time. Hence the simulation is conducted to (1) determine the time which is required for verifying the identification and authentication of V_v at BS$_B$ (VIN_P_Time$_{B\_v}$), (2) determine the number of vehicles whose identification and authentication can be verified within the stipulated simulation time (Per_AV$_B$), (3) measure the loss of registration packet of the

Table 10.3 Overheads in μJ

	[1]	[2]	M_1	M_2	[22]	[23]	Dist_22	Dist_23
A	19.758×10^{-15}	177.82×10^{-15}	307.58×10^{-15}	1760.064×10^{-15}	273.408×10^{-15}	4806×10^{-15}	820.22×10^{-15}	14387×10^{-15}
C	2.9735×10^{-12}	2.9816×10^{-12}	2.9751×10^{-12}	2.9846×10^{-12}	3.07484×10^{-12}	3.0388×10^{-12}	3.0831×10^{-12}	3.0471×10^{-12}

vehicles at BS_B (Pac_RPV_B), (4) determine the number of block of service at BS_B due to its limited number of channels (SE_BL_B), and (5) determine the rate of tracking the vehicles within the coverage area of BS_B ($Rate\_Track_B$). In the worst case, V_v is a new vehicle in VANET. Hence it is required to verify the identification and authentication of V_v after its registration under BS_B. In the worst case, the number of vehicles within the coverage area of BS_B is 48 and hence $LIST_B$ contains 48 records. The time which is required by the tracking algorithm to track V_v within the coverage area of BS_B ($Track_v$) includes the searching time of $LIST_B$ for VIN_v, reading of R_v from $LIST_B$ and sensing CH_v. Hence the time which is required by the tracking algorithm to complete the tracking of 48 vehicles within the coverage area of BS_B is $\sum_{v=1}^{48}(Track_v)$ sec. The delay in tracking V_v within the coverage area of BS_B occurs if τ_v is less than the time which is required by the tracking algorithm to complete the tracking of all the vehicles within the coverage area of BS_B. Such delay causes a reduction of $Rate\_Track_B$ with the increase in number of vehicles within the coverage area of BS_B. Hence, it is required to study the performance of the schemes [1], [2], M_1, M_2 in terms of $Rate\_Track_B$. The five metrics are formally defined below.

$VIN\_P\_Time_{B\_v}$

It is the time which is required to assign D_Sig_v by BS_B to V_v after receiving E_VIN_v in [1, 2] and registration packet in M_1 and M_2 from V_v . $VIN\_P\_Time_{B\_v}$ in [1, 2] is the sum of the following: (i) Transmission time of E_VIN_v from V_v to BS_B both in [1, 2], (ii) Time to add E_VIN_v in $BS_B\_VIN\_Queue$ both in [1, 2], (iii) Waiting time of E_VIN_v in $BS_B\_VIN\_Queue$ both in [1, 2], (iv) Searching time of E_VIN_v in VIN_BS_B both in [1, 2], (v) Transmission time of E_VIN_v from BS_B to CA in [1] and BS_B to R_CA through F_CA_C in [2], (vi) Time to add E_VIN_v in CA_VIN_Queue in [1] and $R\_CA\_VIN\_Queue$ in [2], (vii) Waiting time of E_VIN_v in CA_VIN_Queue in [1] and $R\_CA\_VIN\_Queue$ in [2], (viii) Decryption time of E_VIN_v at CA in [1] and at R_CA in [2], (ix) Validity checking time of decrypted VIN_v at CA in [1] and at R_CA in [2], (x) Generation time of D_Sig_v from decrypted VIN_v at CA in [1] and at R_CA in [2], (xi) Transmission time of (E_VIN_v, D_Sig_v) pair from CA to BS_B in [1] and R_CA to BS_B through F_CA_C in [2], (xii) Insertion time of (E_VIN_v, D_Sig_v) pair by BS_B in VIN_BS_B both in [1, 2], (xiii) Allocation time of CH_v to V_v both in [1, 2], and (xiv) Transmission time of D_Sig_v to V_v through CH_v both in [1, 2].

$VIN\_P\_Time_{B\_v}$ in M_1 and M_2 is the sum of the time which is required for the registration of V_v under BS_B, for executing the identification algorithm, authentication algorithm, verification function, insertion function and tracking algorithm for V_v at BS_B.

- Time for the registration of V_v under BS includes the following: (i) Time to generate the registration packet at V_v, (ii) Time to transmit the registration packet from V_v to BS_B, (iii) Time to read E_VIN_v from the registration packet, (iv) Time to add E_VIN_v in $BS_B\_VIN\_Queue$, (v) Waiting time of E_VIN_v in

BS_B_VIN_Queue, (vi) Time to decrypt the E_VIN_v at BS_B, (vii) Time to search for a free channel, (viii) Time to insert R_v for V_v in $LIST_B$, (ix) Time to generate the registration response packet at BS_B, (x) Time to transmit the registration response packet from BS_B to V_v through CH_v, (xi) Time to read EID_B'' from registration response packet by V_v, (xii) Time to decrypt EID_B'' at V_v

- Time for executing the identification algorithm includes: (i) Time to search VIN_BS_B for VIN_v, (ii) Time to generate the authentication packet
- Time for executing the authentication algorithm includes the time to transmit the authentication packet from BS_B to CA in M_1 and to R_CA through F_CA_C in M_2
- Time for executing the verification function includes the following: (i) Time for reading E_VIN_v from the authentication packet, (ii) Time for inserting E_VIN_v in CA_VIN_Queue in M_1 and R_CA_VIN_Queue in M_2, (iii) Waiting time of E_VIN_v in CA_VIN_Queue in M_1 and R_CA_VIN_Queue in M_2, (iv) Time for decrypting E_VIN_v to generate VIN_v at CA in [1] and R_CA in [2], (v) Time to check the validity of VIN_v at CA in [1] and at R_CA in [2], (vi) Time to generate D_Sig_v from VIN_v, (vii) Time to generate B_v, (viii) Time to encrypt B_v, (ix) Time to generate the insertion packet, (x) Time to transmit the insertion packet from CA to all the BSs under CA in M_1 and R_CA to all the BSs under F_CA_C in M_2
- Time for executing the insertion function includes: (i) Time to read (EB_v, K_1) pair from the insertion packet, (ii) Time to decrypt EB_v, (iii) Time to insert (VIN_v, D_Sig_v) pair in VIN_BS_B
- Time for executing the tracking algorithm includes: (i) Time to search $LIST_B$ for VIN_v, (ii) Time to read $T_{v\_present}$ and CH_v from R_v in $LIST_B$, (iii) Time to encrypt D_Sig_v, (iv) Time to generate the signature packet, (v) Time to transmit the signature packet to V_v through CH_v, (vi) Time to insert ED_Sig_v in R_v of $LIST_B$, (vii) Time to track V_v under BS_B

Per_AV_B

It depends upon the percentage of vehicles whose authentication can be verified within the stipulated simulation time at BS_B. The stipulated simulation time may not be sufficient to verify the authentication of all the vehicles who's E_VINs are waiting in BS_B_VIN_Queue due to the VIN processing time. Per_AV_B is defined as

$$\left(\frac{Total\ number\ of\ vehicles\ being\ authenticated}{Total\ number\ of\ vehicles\ seeking\ authentication} \times 100 \right) per\ simulation\ time.$$

Pac_RPV_B

This is the difference of the total number of E_VINs received and processed by BS_B within the stipulated simulation time in [1, 2]. It is the difference of the total number of registration packets that are received and processed by BS_B within the stipulated simulation time in M_1 and M_2.

SE_BL$_B$

The number of channels available at BS_B is 48 [1, 2]. So BS_B can provide service to 48 vehicles only. The service block occurs at BS_B when all 48 channels are already assigned to 48 vehicles and hence not possible to assign a free channel to a new vehicle. But a vehicle resides within the coverage area of a particular BS for a very short time interval due to its high velocity. Hence the possibility of block of service at a BS is less as observed during simulation.

Rate_Track$_B$

It is the percentage of vehicles which are tracked by the tracking algorithm within the coverage area of BS_B during a stipulated simulation time. Rate_Track$_B$ is defined as

$$\left(\frac{Total\ number\ of\ vehicles\ being\ tracked}{Total\ number\ of\ vehicles\ present\ under\ BS_B} \times 100 \right) \text{per simulation time.}$$

10.6.2.2   Simulation Results

In this section, the quantitative performance of LIAT is elaborated by considering V_v as a new vehicle in VANET. Five sets of experiments are conducted to study the performance of LIAT on the basis of the above-mentioned performance metrics. The quantitative performance of M_1 and M_2 is compared with [1, 2] and with the existing schemes as mentioned in [10, 22, 24].

VIN_P_Time$_{B\_v}$

The first set of experiment is conducted to study the performance of LIAT on the basis of VIN_P_Time$_{B\_v}$. The plots of VIN_P_Time$_{B\_v}$ versus the number of vehicles for [1, 22], M_1 and [2], Dist_22, M_2 are shown in Fig. 10.1(i) and 10.1 (ii) respectively. The maximum number of vehicles per BS is assumed as 40 during

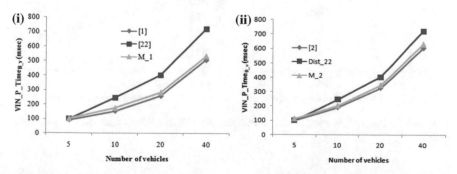

Fig. 10.1 (i) VIN_P_Times$_{B\_v}$ versus Number of vehicles (ii) VIN_P_Times$_{B\_v}$ versus Number of vehicles

simulation as assumed in [22]. Hence, in the worst case, there are 39 vehicles in front of V_v in the $BS_{B\_}VIN\_Queue$. The vehicle velocity and packet transmission range are assumed as 120 km/h and 300 m respectively during simulation.

Experimental Analysis

The number of vehicles within the coverage area of BS_B increases with the number of vehicles in VANET which in turn increases the length of queues at the different levels of VANET hierarchy. Hence VIN_P_Time per vehicle depends upon the number of vehicles in VANET and increases with the number of vehicles in VANET as observed from Fig. 10.1(i), (ii).

In [22] and Dist_22, each vehicle generates an encrypted request packet by using the identification of RSU, public key of RSU and a 32 bits timestamp value. The size of the encrypted packet is 864 bits. The vehicle sends the encrypted request packet to the root. The root generates an encrypted authentication report packet of size 1184 bits and sends it to the vehicle. The size of these packets is much higher than the size of the packets that are transmitted among the different levels of VANET hierarchy in [1, 2], M_1, M_2 during the execution of BS-Vehicle registration algorithm, identification algorithm, authentication algorithm and tracking algorithm for V_v at BS_B. Moreover, the root verifies the authentication of each vehicle and its parent RSU after receiving each new packet from that vehicle in [22] and Dist_22. Hence VIN processing time per vehicle per RSU of [22] and Dist_22 is higher than $VIN\_P\_Time_{B\_v}$ of [1, 2], M_1 and M_2 which can also be observed from graphs as shown in Fig. 10.1.

It can also be observed from Fig. 10.1 that $VIN\_P\_Time_{B\_v}$ of M_1 is higher than [1] and $VIN\_P\_Time_{B\_v}$ of M_2 is higher than [2] due to the extra computation overhead of $VIN\_P\_Time_{B\_v}$ in M_1 and M_2 than [1, 2] respectively as discussed earlier in Sect. 10.6.2.1.

The average value of $VIN\_P\_Time_{B\_v}$ increases by 1175% in [1], 1267% in M_1 and 2027% in [22] as observed from Fig. 10.1(i) whereas it is 1527% in [2], 1565% in M_2 and 2027% in Dist_22 as observed from Fig. 10.1(ii) with the

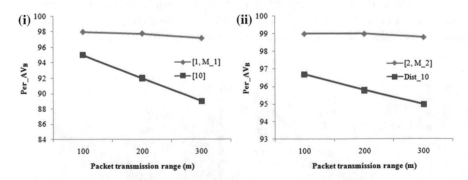

Fig. 10.2 (i) Per_AV$_B$ versus Packet transmission range (ii) Per_AV$_B$ versus Packet transmission range

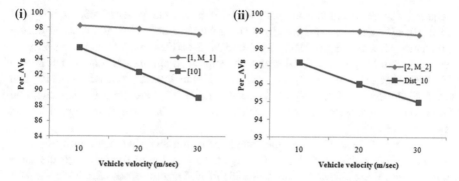

Fig. 10.3 (i) Per_AV$_B$ versus Vehicle velocity (ii) Per_AV$_B$ versus Vehicle velocity

increase in number of vehicles from 5 to 40. Hence it can be concluded that [1, 2], M_1, M_2 have more acceptable VIN_P_Time$_{B\_v}$ compared to [22] and Dist_22.

Per_AV$_B$

The second set of experiment is conducted to study the performance of LIAT on the basis of Per_AV$_B$. The existing scheme in [10] is simulated in a distributed environment (identified as Dist_10 in the rest of the chapter) to compare the performance of [2] and M_2 with Dist_10 on the basis of Per_AV$_B$.

The plots of Per_AV$_B$ versus packet transmission range for [1, 10], M_1 and [2], Dist_10, M_2 are shown in Fig. 10.2(i) and 10.2(ii) respectively when the vehicle velocity is 30 m/sec as assumed in [10].

The plots of Per_AV$_B$ versus vehicle velocity for [1, 10], M_1 and [2], Dist_10, M_2 is shown in Fig. 10.3(i) and 10.3(ii) respectively when the packet transmission range is 300 m as assumed in [10]. The simulation experiment is conducted for 40 vehicles in VANET.

Experimental Analysis

The coverage area of RSU in [10], Dist_10 and BS$_B$ in [1, 2], M_1 and M_2 is assumed as 300 m during simulation. It is identical to the packet transmission range. Hence a vehicle at a distance of 300 m from RSU or BS$_B$ can send a packet for its authentication verification. But such a vehicle or a vehicle having high velocity may go out of the coverage area of RSU in [10], Dist_10 and BS$_B$ in [1, 2], M_1 and M_2, may enter into the coverage area of a new RSU in [10], Dist_10 and new BS in [1, 2], M_1 and M_2 before the completion of its authentication verification. Hence Per_AV per RSU in [10], Dist_10 and Per_AV$_B$ in [1, 2], M_1, M_2 depends upon the packet transmission range and vehicle velocity. Moreover, they reduce with both of these two parameters as observed from Figs. 10.2 and 10.3.

In [1, 2], M_1 and M_2, a vehicle receives its digital signature even if it goes out from the coverage area of BS$_B$ and enters into the coverage area of a new BS during its authentication verification. But in [10] and Dist_10, the new RSU again initiates the authentication verification of the same vehicle. Moreover, the time which is required for verifying the authentication of a vehicle is 66.82 ms in [10] and Dist_10, 60.23 ms in [1, 2], and 62.41 ms in M_1 and M_2 as observed during simulation. Hence BS_B in [1, 2], M_1 and M_2 verifies the authentication of more vehicles than [10] and Dist_10 within the stipulated simulation time which in turn increases Per_AV_B of [1, 2], M_1 and M_2 than Per_AV per RSU of [10] and Dist_10. Moreover, Per_AV_B of [1], M_1 and [2], M_2 are almost independent on the packet transmission range and vehicle velocity as observed from Figs. 10.2 and 10.3 respectively.

Per_AV_B of [1] and M_1 are identical as equal number of vehicles are waiting for authentication verification during the stipulated simulation time. Due to the same reason, Per_AV_B of [2] and M_2 are also identical. The number of vehicles in VANET is higher in [2] and M_2 than [1] and M_1 due to its increased coverage area which in turn increases the number of vehicles within the coverage area of BS_B in [2] and M_2. Hence Per_AV_B of [2] and M_2 is higher than [1] and M_1 respectively.

The average value of Per_AV_B decreases by 0.4% in [1] and M_1, and 3% in [10] as observed from Fig. 10.2(i) whereas 0.1% in [2] and M_2, and 0.85% in Dist_10 as observed from Fig. 10.2(ii) with the increase in packet transmission range from 100 m to 300 m. The average value of Per_AV_B decreases by 5.5% in [1] and M_1, and 32% in [10] as observed from Fig. 10.3(i) whereas 1% in [2] and M_2, and 11% in Dist_10 as observed from Fig. 10.3(ii) with the increase in vehicle velocity from 10 m/sec to 30 m/sec. Hence it can be concluded that Per_AV_B is more acceptable in [1, 2], M_1, M_2 than [10], Dist_10.

Pac_RPV_B

The third set of experiment is conducted to study the performance of LIAT on the basis of Pac_RPV_B. The existing scheme [24] is simulated in a distributed environment (identified as Dist_24 in the rest of the chapter) to compare its performance on the basis of Pac_RPV_B with [2] and M_2. The plot of Pac_RPV_B versus the number of vehicles for [1, 24], M_1 and [2], Dist_24, M_2 is shown in Fig. 10.4(i) and 10.4(ii) respectively. The simulation experiment is conducted for 500 secs as in [24]. The vehicle velocity, packet transmission range and maximum number of vehicles per BS are considered as 20 m/sec, 250 m and 30 respectively during simulation as considered in [24].

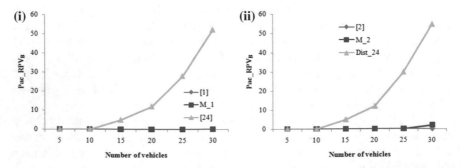

Fig. 10.4 (i) Pac_RPV_B versus Number of vehicles (ii) Pac_RPV_B versus Number of vehicles

Experimental Analysis

In [24] and Dist_24, a vehicle fragments data packet into multiple segments of different sizes to utilize the available bandwidth of service channel efficiently and sends the fragments to RSU. RSU assembles the data packet from the fragments, generates an authentication packet from the assembled data packet and sends it to the vehicle. The number of vehicles within the coverage area of RSU increases with the number of vehicles in VANET which in turn increases the number of fragments of data packets in [24] and Dist_24. So RSU takes a considerable amount of time for assembling the data packet from its fragments and generating the authentication packet. During this span of time, the vehicle may go out of the coverage area of RSU and hence may not be able to receive the authentication packet which causes packet loss. Hence packet loss at RSU increases with the number of vehicles for [24] and Dist_24 as observed from Fig. 10.4. The number of vehicles within the coverage area of BS_B increases with the number of vehicles in VANET. It increases the number of E_VINs in [1, 2] and registration packets in M_1 and M_2 at the BS_B_VIN_Queue which in turn increases Pac_RPV$_B$ with the number of vehicles for [1, 2], M_1 and M_2 as observed from Fig. 10.4. But in [1], M_1, [2] and M_2, VIN_P_Time$_{B\_v}$ is of the order of 1 s as observed during simulation which reduces Pac_RPV$_B$ in these schemes than [24] and Dist_24.

The average value of Pac_RPV$_B$ increases by 0% in [1] and M_1, and 208% in [24] as observed from Fig. 10.4(i) whereas 0% in [2], 4% in M_2 and 220% in Dist_24 as observed from Fig. 10.4(ii) with the increase in number of vehicles from 5 to 30. Hence it can be concluded that [1, 2], M_1 and M_2 have more acceptable Pac_RPV$_B$ compared to [24], Dist_24.

SE_BL$_B$

The fourth set of experiment is conducted to study the performance of LIAT on the basis of SE_BL$_B$. The plot of SE_BL$_B$ versus number of vehicles for [1], M_1, [10, 22] and [2], M_2, Dist_10, Dist_22 is shown in Fig. 10.5(i) and 10.5(ii) respectively. BS_B in [1, 2], M_1, M_2 can provide service to 48 number of vehicles whereas RSU in [10, 22], Dist_10, Dist_22 can provide service to 40 number of

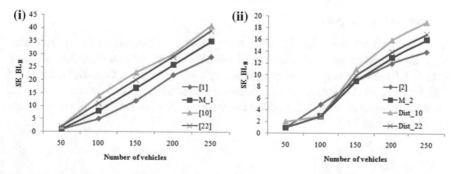

Fig. 10.5 (i) SE_BL$_B$ versus Number of vehicles (ii) SE_BL$_B$ versus Number of vehicles

Fig. 10.6 Per_TRV$_B$ versus
Number of vehicles

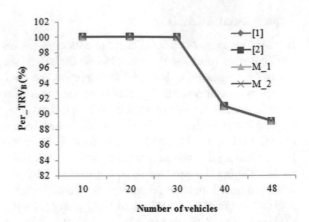

vehicles. The vehicle velocity and packet transmission range are considered as
72 km/h and 300 m respectively during simulation.

Experimental Analysis

The coverage area of VANET is higher in the distributed schemes [2] and M_2. It
increases the possibility that the vehicles remain distributed among the different
BSs in VANET which in turn reduces the service block per BS in [2], M_2 than [1],
M_1 as observed from Fig. 10.5.

The average SE_BL$_B$ increases by 14% in [1], 17% in M_1, 19.5% in [10] and
18.5% in [22] as observed from Fig. 10.5(i) whereas 6.5% in [2], 7.5% in M_2,
8.5% in Dist_10 and 8% in Dist_22 as observed from Fig. 10.5(ii) with the increase
in number of vehicles from 50 to 250. Hence, it can be concluded that [1], M_1 and
[2], M_2 have more acceptable SE_BL$_B$ compared to [10, 22] and Dist_10, Dist_22
respectively.

Rate_Track$_B$

The fifth set of experiment is conducted to study the performance of LIAT on the
basis of Rate_Track$_B$. The plot of Rate_Track$_B$ versus the number of vehicles for
[1, 2], M_1, M_2 is shown in Fig. 10.6.

Experimental Analysis

It can be observed from Fig. 10.6 that Rate_Track$_B$ is identical for [1, 2], M_1 and
M_2 as they have identical methods of tracking the vehicles within the coverage
area of BS$_B$. Moreover, the delay in tracking a vehicle within the coverage area of
BS$_B$ increases with the number of vehicles and hence Rate_Track$_B$ reduces with the
increase in number of vehicles within the coverage area of BS$_B$ for all the schemes
[1, 2], M_1 and M_2.

10.7 Performance Analysis of LoWVR

The qualitative and quantitative performances of LoWVR (Low Overhead Watermark-based Vehicle Revocation) are elaborated in this section.

10.7.1 Qualitative Performance

The performance of LoWVR is evaluated qualitatively in terms of security analysis followed by overhead analysis.

10.7.1.1 Security Analysis

Security analysis of LoWVR is done for the message M of V_v under BS_B using two parameters viz. cracking probability and cracking time.

Cracking Probability

This is the probability by which the attacker retrieves M from WM_MD_DVOM which is disseminated by V_v. The attacker may change the content of M and re-disseminate it. But it is not possible by the attacker to crack WM_MD_DVOM without knowing ALGO_5 and ALGO_6. The actual format of WM_MD_DVOM is also not known to the attacker. Hence the attacker tries to guess the content of WM_MD_DVOM of size Size_WM_MD_DVOM bits by performing brute-force operations. The number of brute-force operation ($\sum_{j=0}^{Size\_WM\_MD\_DVOM-1} 2^j$, where j indicates the bit position) is $2^{Size\_WM\_MD\_DVOM} - 1$ and the cracking probability

$$\left(\frac{1}{\sum_{j=0}^{Size\_WM\_MD\_DVOM-1} 2^j} \right) \text{ is } \frac{1}{2^{Size\_WM\_MD\_DVOM}-1}.$$

For example, the cracking probability is 1.97215×10^{-31} when the size of n is 6 bits and Size_WM_MD_DVOM is 102 bits.

Cracking Time

It is computed as the time duration between the time of attack and time of obtaining M by the attacker. In the worst case, an attacker cracks WM_MD_DVOM by performing all the brute-force operations. If the cracking time exceeds the time when M is actually being used by V_w, it can be claimed that obtaining M by the attacker is not of any use and therefore is not a matter of concern. This time is determined during simulation using 100 MHz processor [25] to prove that the cracking of WM_MD_DVOM by the attacker during its transmission from V_v to V_w is of no use.

Proof: During simulation, the required time to retrieve M from WM_MD_DVOM by V_w is evaluated as 9.8 ms. It has also been observed during simulation that the required time to execute each brute-force operation by the attacker over WM_MD_DVOM varies in between 3.56 and 3.92 ms. The required time to

Table 10.4 Cracking probability and cracking time for different values of Size_WM_MD_DVOM

n (bit)	Size_WM_MD_DVOM (bit)	Cracking probability	Cracking time (year)
6	102	1.97×10^{-31}	5.72×10^{20}
10	120	7.5231×10^{-37}	1.50×10^{26}
14	126	1.175×10^{-38}	9.60×10^{27}
18	144	4.484×10^{-44}	2.517×10^{33}

Fig. 10.7 Cracking
probability versus
Size_WM_MD_DVOM

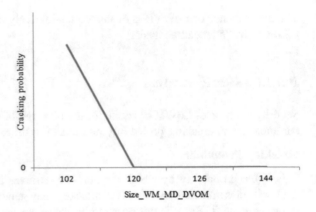

execute each brute-force operation is considered as 3.56 ms to calculate the minimum cracking time. Hence the cracking time to complete all the brute-force
operations by the attacker is $\left(2^{Size\_WM\_MD\_DVOM} - 1\right) \times 3.56$ ms which is
equivalent to $\dfrac{\left(2^{Size\_WM\_MD\_DVOM} - 1\right) \times 3.56}{31536 \times 10^6}$-year.

For example, the cracking time is 5.7240×10^{20} years when n is 6 bits and
Size_WM_MD_DVOM is 102 bits. Hence the message M of V_v is not available to
the attacker in time.

The variation of cracking probability and cracking time with
Size_WM_MD_DVOM is shown in Table 10.4. Size_WM_MD_DVOM depends
upon the size of M', DVOM, MD and the value of n. Now the size of M', DVOM
and MD are fixed to 64 bits, 64 bits and 20 bits, respectively. So
Size_WM_MD_DVOM depends on the size of n.

Figures 10.7 and 10.8 show the variation of cracking probability and cracking
time with Size_WM_MD_DVOM respectively. It can be observed from Figs. 10.7
and 10.8 that cracking probability decreases and cracking time increases respectively with Size_WM_MD_DVOM. Moreover, cracking probability is
7.5231×10^{-37} for 120 bits Size_WM_MD_DVOM as observed from Table 10.4
and Fig. 10.7 which is more than acceptable. But it can be observed from Fig. 10.8
that the most desirable cracking time starts when Size_WM_MD_DVOM is 126
bits. The cracking time is 9.60×10^{27} years when Size_WM_MD_DVOM is 126
bits. Hence, to satisfy the consideration of both, cracking probability and cracking

Fig. 10.8 Cracking time versus Size_WM_MD_DVOM

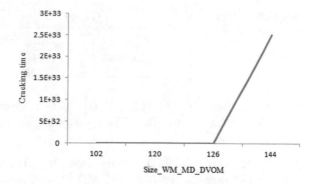

time, Size_WM_MD_DVOM is chosen as 126 bits and n is chosen as 14 bits in LoWVR.

10.7.1.2   Overhead Analysis

The overhead analysis of LoWVR is performed on the basis of STO_OH$_v$, COMM_OH$_v$ and COMP_OH$_v$ of V$_v$.

STO_OH$_v$

It is due to the storage of uid$_v$ and CRL$_v$. In the worst case, all the (NO_OF_V$_B$-1) number of vehicles are neighbor of V$_v$ and are identified as unauthentic or alteration attacker. So, the size of CRL$_v$ is [n*(NO_OF_V$_B$-1)] bits in the worst case and STO_OH$_v$ is [n + n*(NO_OF_V$_B$-1)] bits.

COMM_OH$_v$

It is due to the transmission and reception of messages by V$_v$. Let V$_v$ broadcasts GM$_v$ number of messages and receives RM$_v$ number of messages. The size of each message is Size_WM_MD_DVOM bits. So, COMM_OH$_v$ is [Size_WM_MD_DVOM*(GM$_v$ + RM$_v$)] bits.

COMP_OH$_v$

It includes the computation overhead of generating WM_MD_DVOM from GM$_v$ number of messages (COMP_OH_GM$_v$) and retrieving the original message (M) from RM$_v$ number of received WM_MD_DVOM (COMP_OH_RM$_v$). COMP_OH_GM$_v$ for generating WM_MD_DVOM from M includes: generation of M and M' at V$_v$, execution of ALGO_1 at V$_v$, execution of ALGO_2 at V$_v$, execution of ALGO_3 at V$_v$, execution of ALGO_4 at V$_v$.

Now the computation overhead of generating M and M' is O(1) due to their fixed size. The computation overhead of executing ALGO_1 and ALGO_2 is O(1) as these algorithms are executed over fixed size block. The computation overhead of executing ALGO_3 is O(n) where n is the number of watermark bits. ALGO_4

Table 10.5 Comparison of overhead analysis

	LoWVR	[22]	[23]
A	3780	353813	4608×10^5
B	2100	3096371	17664×10^3
C	1585	3486	3444

inserts 3n bits in Pad_MD_DVOM with computation overhead O(Size_ Pad_MD_DVOM * n). So COMP_OH_GM$_v$ is GM$_v$*[O(n* Size_Pad_MD _DVOM) + O(n)].

Let V$_v$ receives WM_MD_DVOM from V$_w$. The computation overhead of retrieving M from the received WM_MD_DVOM includes execution of ALGO_5 and ALGO_6 at V$_v$

ALGO_5 detects 3n bits from WM_MD_DVOM with computation overhead O (Size_WM_MD_DVOM*n). In the worst case all the (NO_OF_V$_B$-1) number of vehicles are neighbor of V$_v$ and (NO_OF_V$_B$-1) number of vehicles are already identified at V$_v$ as unauthentic or alteration attackers. So CRL$_v$ contains (NO_OF_V$_B$-1) number of uids and the computation overhead of searching CRL$_v$ for uid$_w$ using ALGO_5 is O(NO_OF_V$_B$).

ALGO_6 generates WM$_v$′ using ALGO_3 with computation overhead O(n), compares WM$_v$′ and WM$_v$ with computation overhead O(1), updates CRL$_v$ by inserting uid$_w$ with computation overhead O(1) if V$_w$ is identified as unauthentic, reads the least significant 84 bits from Pad_MD_DVOM as MD_DVOM, most significant 64 bits from MD_DVOM as DVOM and least significant 20 bits from MD_DVOM as MD with computation overhead O(1), computes MD′ from DVOM using ALGO_2 with computation overhead O(1), compares MD′ and MD with computation overhead O(1), updates CRL$_v$ by inserting uid$_w$ with computation overhead O(1) if V$_w$ is identified as an alteration attacker, generates M′ from DVOM and M from M′ with computation overhead O(1) due to the fixed size of DVOM, M and M′.

So, COMP_OH_RM$_v$ is RM$_v$*[O(n) + O(Size_WM_MD_DVOM*n) +O (NO_OF_V$_B$)]. Thus, COMP_OH$_v$ = COMP_OH_GM$_v$ + COMP_OH_RM$_v$.

Experimental Analysis

COMM_OH$_v$ (A), STO_OH$_v$ (B), and COMP_OH$_v$ (C) of LoWVR are compared with results in [22, 23]. The comparison is shown in Table 10.5. COMM_OH$_v$ (A), STO_OH$_v$ (B), and COMP_OH$_v$ (C) are evaluated in bits, bits, and clock cycles respectively during simulation. The overheads are measured by considering the value of n, GM$_v$, RM$_v$ and NO_OF_V$_B$ as 14, 15, 15 and 150 respectively for LoWVR and [22, 23].

In [22], the VANET hierarchy consists of Trusted Authority, State Level Trusted Authority, City Level Trusted Authority, RSU and the vehicles. V$_v$ is equipped with a pair of key (private key, public key), a shared key and also a short certificate. V$_v$ has to store also the identity of its immediate higher authority for its own authentication. As a result, STO_OH$_v$ is high in [22] than LoWVR.

In [22], V_v encrypts the identity of its immediate higher authority and sends it to the Trusted Authority through the VANET hierarchy. Finally, the Trusted Authority generates short certificate from the encrypted identity and sends the short certificate to V_v through VANET hierarchy. Such communication among the different levels of the hierarchy increases COMM_OH$_v$ and the complex computation for identity encryption increases COMP_OH$_v$ of [22] than LoWVR.

In [23], the VANET hierarchy consists of Trusted Authority, RSU and vehicles. The trusted authority assigns a certificate, a private signature key, a group-key and a group signature key to V_v. V_v has to store all these keys and the size of all these keys is almost 100 bytes which increases STO_OH$_v$ of [23] than LoWVR.

V_v belongs to a group of vehicles under the coverage area of RSU. The private signature key, group-key and group signature key are used by V_v during communication with the other members of the group as well as with immediate higher authority to send and receive messages of almost 100 bytes which in turn increases the COMM_OH$_v$ of [23] than LoWVR.

The certificate indicates the authenticity of V_v—it is then assigned to V_v for a session. Hence re-authentication verification of V_v is required after the expiry of a session. V_v has to participate in the group-key generation procedure which is controlled by Trusted Authority. This group-key generation procedure is a complex encryption-decryption process. The group-key is also required to be renewed during the re-authentication of V_v. The renewal process of certificate and group key is also a complex one which increases COMP_OH$_v$ of [23] than LoWVR.

It can be observed from Table 10.5, that the net gain of LoWVR in terms of STO_OH$_v$ is 99.9% as compared to that in [22, 23]. The net gain of LoWVR in terms of COMM_OH$_v$ is 98.9% over [22] and 99.9% over [23]. The net gain of LoWVR in terms of COMP_OH$_v$ is 54.5% over [22] and 53.97% over [23]. Hence LoWVR is a low overhead scheme due to the significant improvement of its net overhead gain over [22, 23].

The variation of overheads with Size_WM_MD_DVOM is shown in Table 10.6. 0.534×10^{-15} µJ [21], 0.4005×10^{-15} µJ [21] and 0.3524 788443×10^{-16} µJ [21, 25] are the required energy for 1byte data transmission, 1byte data reception and 1 cc computation. It can be observed from Table 10.6, that STO_OH$_v$, COMP_OH$_v$ and COMM_OH$_v$ increase with Size_WM_MD_DVOM.

10.7.2   Quantitative Performance

The performance of LoWVR is evaluated quantitatively by conducting the simulation experiment in Lane$_k$. The simulation experiment is conducted by varying the VANET size from 20 to 30 km^2 and vehicle velocity from 72 to 120 km/h, by assuming the lane length as 10 km, lane width as 30 m and inter vehicle distance as 2 m [20].

Table 10.6 Overheads for different values of Size_WM_MD_DVOM

Size_WM_MD_DVOM (bit)	n (bit)	STO_OH$_v$ (bits)	COMP_OH$_v$ (μJ)	COMM_OH$_v$ (μJ)	
				Transmitting	Receiving
102	6	900	5.5621×10^{-14}	102.12×10^{-15}	76.595×10^{-15}
120	10	1500	5.572×10^{-14}	120.15×10^{-15}	90.112×10^{-15}
126	14	2100	5.586×10^{-14}	126.15×10^{-15}	94.618×10^{-15}
144	18	2700	5.6×10^{-14}	144.18×10^{-15}	108.13×10^{-15}

10.7.2.1   Simulation Metrics

The quantitative performance of LoWVR is evaluated in terms of the time of verifying authentication of V_v by V_w (Time_A), time of disseminating M by V_v (Time_D) and delay in message reception by V_w (Time_Delay). The required time to verify the authentication of a vehicle should be as small as possible so that the authentication of a maximum number of vehicles can be verified within a stipulated simulation time. Hence, Time_A which is the required time of verifying the authentication of V_v by V_w is an important simulation metric. Any delay in message dissemination may cause injuries even death. Hence the time taken by V_v to disseminate M among neighbors (Time_D) and the time taken by V_w to get the original message (Time_Delay) are also important parameters to study the performance of LoWVR.

Time_A

It includes time to extract uid_v, key_v, WM_v from WM_MD_DVOM at V_w, time to search CRL_w for uid_v, time to compute WM_v' from uid_v and key_v and time to compare WM_v' with WM_v.

Time_D

It includes the time to generate M after observing an event, time to generate M', time to generate DVOM, time to generate MD from DVOM, time to generate MD_DVOM, time to generate WM_v, time to generate WM_MD_DVOM and time to broadcast WM_MD_DVOM.

Time_Delay

It includes time to verify authentication of V_v (Time_A), time to extract MD_DVOM from Pad_MD_DVOM, time to extract DVOM and MD from MD_DVOM, time to compute MD' from DVOM, time to compare MD' with MD, time to generate M' from DVOM and time to generate M from M'.

10.7.2.2   Simulation Results

Three sets of experiments are conducted based on the above-mentioned simulation metrics for comparing the performance of LoWVR with [22, 23]. The simulation experiment is conducted at $Lane_k$ under BS_B. The maximum number of vehicles and maximum number of events at $Lane_k$ are assumed as 150 and 15 respectively. The maximum data transmission range and transmission rate are considered as 300 m and 6 Mbps respectively. Time_A, Time_D and Time_Delay vary with the number of vehicles, rate of occurrence of events and rate of generation of messages. Again, the rate of generation of messages increases with the rate of occurrence of

Fig. 10.9 Time_A versus
Number of messages/sec

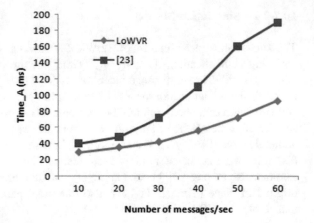

Fig. 10.10 Time_A versus
Number of vehicles

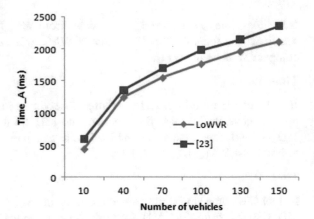

Fig. 10.11 Time_D versus
Number of messages/sec

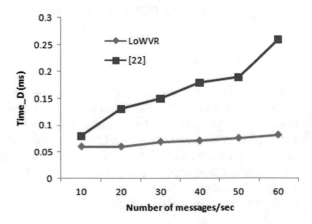

Fig. 10.12 Time_D versus
Number of vehicles

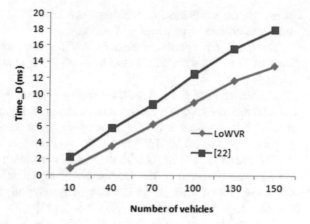

Fig. 10.13 Time_Delay
versus Number of messages/
sec

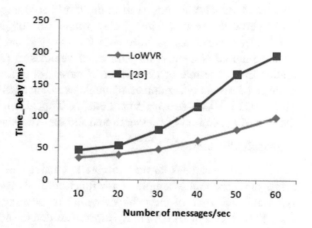

Fig. 10.14 Time_Delay
versus Number of vehicles

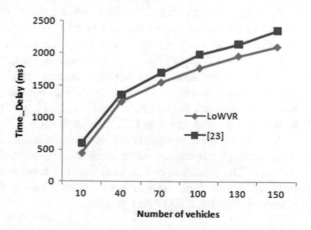

event. Hence simulation experiments are conducted by varying the number of vehicles and rate of generation of messages.

The quantitative performance of LoWVR is compared with results in [23] on the basis of Time_A, with [22] on the basis of Time_D and with [23] on the basis of Time_Delay.

The variation of Time_A with the rate of generation of messages and number of vehicles are observed in the first set of experiment. Figures 10.9 and 10.10 show the plot of Time_A versus rate of generation of messages and number of vehicles respectively both for LoWVR and [23].

The variation of Time_D with the rate of generation of messages and number of vehicles are observed in the second set of experiment. Figures 10.11 and 10.12 show the plot of Time_D versus rate of generation of messages and number of vehicles respectively both for LoWVR and [22].

The variation of Time_Delay with the rate of generation of messages and number of vehicles are observed in the third set of experiment. Figures 10.13 and 10.14 show the plot of Time_Delay versus rate of generation of messages and number of vehicles respectively both for LoWVR and [23].

The number of events and number of vehicles in $Lane_k$ increase with the simulation time. The rate of generation of messages at $Lane_k$ also increases with the number of events and number of vehicles. Hence, as observed from Figs. 10.9, 10.10, 10.11, 10.12, 10.13, 10.14, Time_A, Time_D and Time_Delay increase with the rate of generation of messages and number of vehicles in $Lane_k$.

Experimental Analysis

In [23], RSU generates its own private key using one random number, two large prime numbers, and a complex hash function. RSU uses this private key for verifying authentication of the vehicles within its coverage area. As can be observed from Figs. 10.9 and 10.10, Time_A increases due to such extra computation in [23] than LoWVR.

In [22], a sender vehicle collects all segments of the master key from its neighbours, reassembles the master key from the received segments, and encrypts the message before its dissemination using the master key. Moreover, the sender vehicle verifies 1024 bits RSA signature and uses it to make the message authentic before its dissemination which helps the receiver vehicle to verify the authentication of the received message. All these operations consume a considerable amount of time. Hence, as can be observed from Figs. 10.11 and 10.12, Time_D increases due to such extra computation in [22] than LoWVR.

In [23], the receiver vehicle executes a complex hash function based on SHA-3 by using its private signature-key to retrieve the original message from the received message. The execution of this hash function consumes a considerable amount of time, and hence causes an increase of Time_Delay in [23], more than LoWVR as observed from Figs. 10.13 and 10.14.

10.8   Conclusion

A novel mechanism for revocation of unauthentic and misbehaving vehicles from VANET is presented in this chapter. The first part of the chapter elaborates the mechanism to identify and revoke unauthentic vehicles whereas the second part of the chapter discusses a mechanism to identify and revoke alteration attackers as misbehaving vehicles. Both the mechanisms are implemented in traditional VANET.

The traditional VANET has several technical challenges in deployment and management. It is non-collaborative and non-scalable; and has limited network awareness. It supports poor connectivity among vehicles due to their high mobility.

The traditional VANET cannot disseminate emergency messages timely and reliably due to the limited transmission range of DSRC. Moreover, the future vehicles are smart objects equipped with multi-sensor platform, a set of communication technologies, robust computational units, IP-based connectivity to the Internet and to other vehicles. Such vehicles need to communicate with other vehicles for sharing traffic data like road accident reports, traffic jam notification, road construction reports, emergency vehicle warnings, etc. to enhance traffic management and road safety. The traditional VANET is not able to analyze, process and evaluate such massive amount of traffic data efficiently. Such technical challenges of traditional VANET along with the requirement of resources like computing, storage, and networking are satisfied by merging VANET with cloud computing provision. The cloud satisfies the requirements like to store huge data, to handle complex computation, and to process huge data in a distributed manner.

The increase in number of vehicles in urban areas consumes time and energy in uploading to and downloading from the cloud. It provides service in close proximity that is within a single hop to the vehicles. The increase in number of hops increases delay and reduces throughput in communication. Moreover, the traditional cloud is not able to satisfy the requirements related to mobility, low latency, and location awareness of smart vehicles in future VANET. The merging of fog computing with cloud computing is the solution of such issues in future VANET.

In a vehicular networked environment, the multi-layered fog computing architecture has vehicles in the first layer, fog platform including fog devices like road side units in the second layer, and conventional cloud in the third layer. The fog computing devices collect the raw data from the vehicles, examine the collected data to find the specific information, and enhance the data processing at the vehicles in real time. The selected data is then sent to the cloud for future use. Fog computing devices also ensure data privacy by examining sensitive data locally, instead of analysing in cloud. The storage of time-sensitive data in fog devices at the close proximity of the vehicles reduces latency, increases throughput, and also reduces the Internet dependency. The presence of fog devices in between the vehicles and cloud eliminates the exchange of information between vehicles and cloud and hence it becomes very difficult for the attackers to access such data. Fog computing also supports high mobility of vehicles in future VANET.

References

1. Mondal A, Mitra S (2012) Identification, authentication and tracking algorithm for vehicles using VIN in centralized VANET. In: International conference of communication, network and computing (CNC), proceedings. LNICST, vol 108. Springer, Berlin, pp 115–120. ISBN no. 978-3-642-35614-8
2. Mitra S, Mondal A (2012) Identification, authentication and tracking algorithm for vehicles using VIN in distributed VANET. In: International conference on advances in computing, communications and informatics (ICACCI), proceedings. ACM digital library, pp 279–286. ISBN no. 978-1-4503-1196-0
3. Mondal A, Mitra S (2018) Extending security horizon through identification authentication and tracking services in hierarchical vehicular networks. Int J Comput Electr Eng 69:479–496. Elsevier
4. Mondal A, Mitra S (2019) LoWVR: low overhead watermark based vehicle revocation scheme in VANET. Int J Sens Wirel Commun Control 9(1):2019
5. Fan CI, Sun WZ, Huang SW, Juang WS Huang JJ (2014) Strongly privacy preserving communication protocol for VANETs. In: 9th Asia joint conference on information security, pp 119–26
6. Moses SJ, Angelin PAC (2013) Enhancing the privacy through pseudonymous authentication and conditional communication in VANETs. J Eng Sci 2(7):45–49
7. Liu Y, Wang L, Chen H (2014) Message authentication using proxy vehicles in vehicular ad hoc networks. IEEE Trans Veh Technol 64(8):3697–3710
8. Fulewar VV, Dorle S (2014) A scheme for authentication with multiple levels of anonymity in vehicular communication. J ResComput CommunTechnol 3(4):550–555
9. Dolev S, Krzywiecki L, Panwar N, Segal M (2014) Dynamic attribute based vehicle authentication. In: IEEE international symposium on network computing and applications, pp 1–8
10. Ashritha M, Sridhar CS (2015) RSU based efficient vehicle authentication mechanism for VANETs. In: IEEE international conference on intelligent systems and control, pp 1–5
11. Lee KH, Kim SK (2015) Authentication scheme based on biometric key for VANET information system in M2 M application service. Appl Math Inf Sci 9:645–651
12. Bharat M, Sree KS, Kumar TM (2014) Authentication solution for security attacks in VANETs. Int J Adv Res Comput Commun Eng 3(8):7661–7664
13. Caballero-Gil P, Fernandez FM, Caballero-Gil C (2014) Tree-based revocation for certificate less authentication in vehicular ad hoc network. Int J Comput Commun 2:14–21
14. Abdulhussain SH (2013) Enhanced management of certificate caching and revocation lists in VANET. Int J Comput Appl 83(12)
15. Kenney JB (2011) Dedicated short-range communications (DSRC) standards in the United States. IEEE SAE Stand Wirel Access Veh Environ (WAVE) 99(7):1162–1182
16. Hess KM, Orthmann CH, Cho HL (2013) Criminal investigation, 11 edn. Cengage Learning. ISBN: 978-1-285-86261-3, 2013
17. Xiao Y, Li J, Pan Y (2005) Security and routing in wireless networks. In: Wireless network mobile computing. Nova science publishers, New York; E-Publishing Inc
18. Kahate A. Cryptography and network security, 3rd edn. McGrill education (India) Pvt. Ltd, India. 013

19. Vehicle identification number (2016). https://en.wikipedia.org/wiki/Vehicle_identification_ number. Accessed 31 Mar 2016
20. https://www.kmcgov.in/KMCPortal/jsp/BasicStatistics.jsp (2017)
21. Mondal A, Mitra S (2016) TDHA: a timestamp defined hash algorithm for message authentication in VANET. In: International conference on computational modeling and security (CMS), proceedings. Elsevier (Proc Comput Sci J), pp 182–189. ISSN: 1877-0509
22. Chaurasia BK, Verma S (2011) Infrastructure based authentication in VANETs. J Multimed Ubiquitous Eng 6(2)
23. Kim SH, Lee IY (2014) A secure and efficient vehicle-to-vehicle communication scheme using bloom filter in VANETs. Int J Secur Appl 8(2):9–24
24. Hassnawi LA, Ahmad RB, Salih MH, Warip MNM,Elshaikh M (2014) Measurement study on packet size and packet effects over vehicular ad hoc network performance. J Theoret Appl Inf Technol 70(3):475–481
25. NI PXIe-4339, National Instrument. http://sine.ni.com/nips/cds/view/p/lang/en/nid/212632

Chapter 11
Cloud-Based Secured VANET with Advanced Resource Management and IoV Applications

Sachin Sharma and Seshadri Mohan

Abstract In this chapter, we propose to integrate cloud computing with Vehicular Ad Hoc NETworks (VANETs) so that the vehicles in the network can share network resources and avail of a variety of information collected by them to make useful decisions. We present an architecture that includes a cloud-based VANET. Then, we study the Internet of Vehicles (IoV) application management system in this cloud-based VANET. The proposed architecture facilitates the recognition of available resources in real time. In addition, it provides cloud-based IoV applications to cloud-based VANET enabled vehicles. The presented work also demonstrates a security algorithm suitable for a cloud-based VANET. We propose a distributed methodology for a secure vehicular network to communicate; and demonstrate the potential of our architecture for real-time access to IoV applications in cloud-based VANET environment.

Keywords Cloud computing · Internet of Vehicles · IoV · Vehicular adhoc network · VANET · Internet of Things · IoT · Autonomous vehicles · Security · Intelligent transport systems

11.1 Introduction

Recent advances in the Internet of Things (IoT) and Intelligent Transport Systems (ITS) have paved the way for the onset of intelligent connected vehicles. Aided by the various sensors and Internet of Things mounted on them, the connected vehicles are now capable of gathering enormous amount of data, as much as 1 Tbyte/hour, from the adhoc vehicular environment. The data collected is then processed and

S. Sharma (✉)
Graphic Era Deemed to be University, Dehradun, UK, India
e-mail: sachin.cse@geu.ac.in; sxsharma88@gmail.com

S. Mohan
University of Arkansas at Little Rock, Little Rock, AR, USA
e-mail: sxmohan@ualr.edu

© Springer Nature Switzerland AG 2020
Z. Mahmood (ed.), *Connected Vehicles in the Internet of Things*,
https://doi.org/10.1007/978-3-030-36167-9_11

useful information extracted using machine learning, artificial intelligence, or other such tools, that can then be shared with drivers, other connected vehicles in the Ad Hoc network, and other infrastructure as well as applications so as to engineer an environment for safe driving [1]. With autonomous vehicles, IoV will have the intelligence, and learning capabilities to anticipate the drivers' intentions and, with the help of the information available, can take smart and safe decisions so as to facilitate a safe driving environment. By integrating cloud computing with Vehicular Ad Hoc NETwork (VANET) [2], machine learning applications and other novel multimedia applications and services can be developed that can avail of the computing resources available to them. Autonomous and connected vehicles will then be able to offer to the drivers increased accessibility to multimedia services, novel cloud-based applications, and enormous computing power [3]. This chapter proposes such an architecture for cloud-based VANET that enhances resource management.

Many emerging IoV applications, such as real-time video sharing [4], require larger bandwidth, secure storage, complex computation, enhanced resource management, cloud-based dynamic bandwidth sharing [5], and social media sharing [6, 7]. VANET's connected vehicles continuously collect data related to real-time traffic, driver behavior, road condition, and bandwidth utilization [8], and upload the data to the cloud servers for further processing. The machine learning tools executing in the cloud serves may then mine useful information that can help in improving driver and vehicle safety, easing traffic congestion, and enhancing driving comfort. The cloud servers may then transmit the mined information to data centers for further processing and share the information with the connected vehicles of VANET for safe route optimization. Due to the availability of mined information, it is possible to reduce traffic congestion, and, consequently, traveling duration. The process of Big Data mining transforms a VANET into a smart VANET [9, 10].

The field of IoV stands at the confluence of many evolving disciplines, a few of which are: the Internet of Things (IoT) consisting of a multitude of sensors that are housed in a vehicle, on the roadside, and in the devices worn by pedestrians; the radio technology along with the protocols that can establish an adhoc vehicular network; the cloud technology; the field of Big Data; and machine intelligence tools. IoV is an emerging technology that provides secure vehicle-to-vehicle (V2V) communication and safety to drivers.

The vehicle sensors gather information like GPS location, vehicle health conditions, roadside businesses, driver behavior and road conditions and upload to the cloud-based so that the information mined from the data can be utilized to optimize the performance of vehicles and improve the safety and comfort of drivers, passengers, and pedestrians [11, 12]. The IoV applications in the cloud-based VANET provide virtual connectivity between vehicles and hence among drivers and passengers. They can also share their interests and manage their resources during trips [13, 14]. In cloud-based VANET, the autonomous vehicle and non-autonomous vehicles can efficiently participate to maintain smooth traffic flow in urban areas and highways. Network operators can manage the bandwidth requirements of the vehicles in an efficient manner as well.

The autonomous vehicle systems are expected to be complex that empower a new generation of reliable applications and services using cloud-based VANET architectures. Cloud-based VANET resource management systems rely on a large ecosystem of autonomous vehicles systems where collaboration at large scale will take place. However, this is only possible due to the distributed, autonomous, intelligent, proactive, fault tolerant, reusable systems, which offer their capabilities and functionalities as services located in the "IoV service cloud".

The rest of this chapter is organized as follows. Section 11.2 describes the cloud-based VANET architecture. Section 11.3 illustrates the VANET resource management system. Section 11.4 explains the IoV applications of cloud-based VANET. Section 11.5 discusses the VANET communication in distributive manner. Section 11.6 depicts the proposed distributive security algorithm for VANET communication modes and Sect. 11.7 delineates the conclusion.

11.2   Cloud-based VANET Architecture

Cloud based Vehicular Adhoc NETwork (VANET) architecture is a type of computing architecture that enables access to multiple sources on the Internet on a demand basis. There are many vendors across the world which offer cloud computing architecture-based technologies to access, install and update software on a demand basis. These cloud-based companies provide cost-effective services and server maintenance to their clients over the Internet [15]. Cloud-based VANET architecture enables multiple services, such as Software-as-a-Service (SaaS) [16, 17], Hardware-as-a-Service (HaaS), and Data-as-a-Service (DaaS), all of which together are known as 'Platform-as-a-Service (PaaS) [18, 19]. It is an innovative technology suitable for intelligent transport systems [20–22]. Vehicles equipped with numerous sensors communicate and share the information they collect about traffic, weather, driver behavior and street views with each other using multi-hop Ad Hoc networking [23, 24]. Cloud-based VANET provides a new and efficient approach to gathering and accessing real-time information by vehicles using V2V communication than acquiring by traditional methods of traffic information collection from roadside traffic cameras [25, 26]. This facilitates vehicles to acquire information about traffic conditions far in advance of actual encounter with them so that the vehicles can make intelligent decisions and reroute themselves to their destinations.

Figure 11.1 illustrates the cloud-based VANET architecture which consists of three parallel cloud networks: local VANET cloud network (LVCN), wide VANET cloud network (WVCN), and central VANET cloud network (CVCN). These are briefly described below:

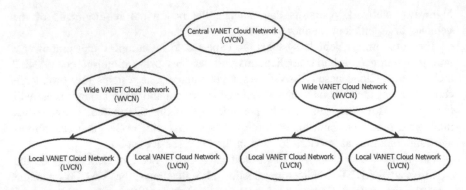

Fig. 11.1 Cloud-based VANET architecture

- Local VANET cloud network (LVCN): A LVCN may be established among a set of VANET enabled vehicles spanning a few miles in range. The vehicles in an LVCN share their computational, storage, and bandwidth resources thereby enhancing the resource utilization. Each vehicle can retrieve on a demand basis cloud services for its own benefit.
- Wide VANET cloud network (WVCN): This may be established among a set of LVCN vehicles through the Internet, Wi-Fi, or dedicated short range communications (DSRC). There are dedicated LVCN servers attached to other VANET cloud networks. A vehicle accesses a WVCN by V2V communications with other vehicles located within the coverage area of the cloud site. The WVCN is a trusted, resource-rich cloud network that offers cloud services to vehicles passing by. A vehicle can choose a WVCN option for their system configuration update, customization and share large files data.
- Central VANET cloud network (CVCN): A CVCN may be established among a group of WVCN servers via Internet. A vehicle retrieves a CVCN by cellular communications. A CVCN has more resources for intricate computations; and facilitates vehicles to make decisions guided complex global policies. There are various commercial software platforms available such as Amazon Web Services, Microsoft Azure, IBM, Google Cloud Platform, Salesforce.com, Adobe, Oracle cloud, SAP, Rackspace and Workday that provide cloud services.

The cloud-based VANET architecture enhances the utilization of physical resources of all the three levels of cloud networks. This architecture permits all the vehicles of the VANET to access all the three layers, as illustrated in Fig. 11.1, of cloud networks as needed.

The emergence of smart cities paradigm will transform the lifestyle of urbanites. The cloud-based VANET will play a key role in enabling access to Intelligent Transport System (ITS) applications as well as various types of other services in a smart city.

Cloud-based VANET proposed in this work may usefully facilitate a rapid implementation of smart cities. When the vehicles in a smart city are equipped with VANET-capable functionality, with an efficient, reliable and fast V2V, V2I, or simply V2X communication based on a certain optimal resource allocation and routing strategy, they will be able to play a major role in supporting the operations management system of the city. The V2I architecture includes key sites such as traffic signals, bus stops, parks, train stations, tourist attractions, and other landmarks, which function as access points to the networks and serve as local data collection platform. The architecture facilitates passengers to browse the Internet and access multimedia application services. In this scenario, some relevant systems are mentioned below:

- **Smart traffic management**: It is a system that regulates and controls all traffic signals in response to real-time operations. Efficient and smart traffic management is central to providing users information such as early warnings about traffic on heavily traveled roads and highways, reducing traffic congestion, and keeping travelers aware of locally available facilities, businesses, and attractions. The government issued policies, like odd-even day vehicle bans, may not be able to control traffic efficiently and municipal corporations and traffic authorities do not necessarily take into account the real-time traffic conditions. Cloud-based VANET is intended to enable smart traffic management that is dynamic and take into account traffic conditions, traffic congestion, road works, accidents, detours, road conditions, pedestrian traffic on local roads, and other factors that could impact traffic.
- **Smart accidents warning systems**: The safety of passengers and drivers is a key concern that should be taken into account in planning and designing a smart city. A smart warning system could considerably reduce and probably eliminate the loss of lives and infrastructure-related injury arising from accidents. Cloud-based VANET facilitates the development and deployment of a smart accidents warning system that can immediately notify all the vehicles in the vicinity in real-time with the precise location of accidents. The intelligent sensors deployed in a smart city may also facilitate the collection of traffic-related data and upload to the cloud-based VANET information such as any fire incidents, gas leakage and weather-related damages to buildings and infrastructure in the city, which may then be made available to all the vehicles in the vicinity of such incidents by VANET.
- **Cost-effective web services**: The cloud-based VANET helps to build a large high-speed Wi-Fi network system. It allows multiple communication devices to use cost-effective web services by reducing its dependence on fixed cellular base stations.
- **Privacy and security**: One of the key challenges in engineering a smart city is to install adequate mechanisms for protecting the privacy of people's primary information data and guarding it against any security threats. A cloud-based VANET offers a platform for building applications that offer such services. Each vehicle has its own unique ID which helps to connect vehicle's owner with

multiple services or facilities using a single sign-on (SSO) option. The proposed architecture has capabilities to recognize malicious users and block them immediately.

- **Smart city design**: The network design in a smart city facilitates multiple services to the whole city population. A cloud-based VANET architecture is capable of providing a dynamic vehicular network, albeit Ad Hoc that can supply information concerning real-time traffic and road conditions, and other related information to augment the smart city's network system. Significant information for planning purposes can then be extracted from the system and reported to city council officials. The real-time information can also help the city to manage and improve sustainable infrastructure design and development. Relevant information from the system may also help to monitor environmental health, waste management, the structural health of buildings, prevailing noise level, and energy consumption, and help implement smart lighting and automation in public buildings in the city. The augmented system is also capable of extracting statistics in real-time concerning hazards such as gas emissions and pollution, which may help regulatory authorities to formulate and implement effective policies to limit environmental damages.

The proposed architecture for real-time information gathering offers various levels of services for the benefits of individuals, people, society and authorities in smart cities. The process involves three steps: real-time data gathering, transmission of data, and analysis of huge volumes of data.

- In the real-time data gathering step, the system continuously gathers data from multiple sources including vehicle sensors and web APIs and uploads into the cloud server. The available web APIs provide detailed information and real-time statistics about the current weather such as temperature, humidity, wind speed, and precipitation, as well as road conditions with traffic. Though the on-board units (OBUs), continuously update the cloud-based VANET with real-time data useful to end user's services, the communication among vehicles in VANET is however opportunistic and not always guaranteed due to the speed of vehicles or where the vehicles may be sparsely populated and spread too far apart to be able to establish an Ad Hoc network.
- The real-time data transfer may potentially be helpful to end users in accessing the useful shared data by other vehicles. When the user requests multiple services, then an OBU sends a request to a cloud server associated with the VANET.
- After receiving a user request the cloud server searches for and retrieves the appropriate information and analyzes the search to generate an evaluation report and sends the results back to the user with very little delay.

The benefit of using a cloud server is that it carries out heavy computations in the cloud and analyzes big data in almost real-time. The cloud server uses various web APIs as well to solve complex calculations and retrieve useful information.

Cloud-based VANET may play a major role in a smart city development. The key technologies in this context are as follows:

- Bandwidth characteristics: Flexible bandwidth, low Doppler shift, and reduced building occlusions are the benefits of cloud-based VANET in a smart city.
- Dynamic network topology: Due to the mobility of the vehicle, the topology of vehicles is essentially dynamic, thereby necessitating an optimal routing strategy through the VANET that helps to reduce congestion.
- Optimal routing strategy: An optimal routing strategy is helpful for optimizing bandwidth allocation and avoiding traffic congestion.
- Congestion avoidance: A congestion avoidance technique enhances the throughput of the cloud-based VANET.
- Security and privacy: In the cloud-based VANET, privacy and security of users is to identify any malicious activity of an attacker and block it immediately and take actions to prevent future attacks.

11.3   VANET Resource Management System

The resource management system of the cloud-based VANET consists of three stages as illustrated in Fig. 11.2.

- *Client management system*: The client management system consists of a vehicle's sensors information and bandwidth resources. It provides global communication features of vehicles connectivity with IoV applications access.
- *Gateway management system*: This gateway management system consists of multiple ways of homogeneous and heterogeneous interconnection methods such as vehicle- to-vehicle, vehicle-to-roadside and vehicle-to-Internet. This stage facilitates global real-time network access to each vehicle in the cloud-based VANET.
- *Cloud management system*: This cloud management system facilitates multiple computational operations and various services such as logistics, transport, emergency, business, safety and environment. This stage includes virtualization, authentication, real-time data collection, operation, scheduling, monitoring, and controlling of applications management.

The following resulting potential directions may play a key role in the design of the VANET resource management system. More specifically, these are:

- *Resource monitoring*: The monitoring of resources is of key importance especially in a highly complex heterogeneous system. In the complex systems, it will be practically impossible to carry out effective information acquisition with the conventional approach. The promising approach is to have a task-driven infrastructure coupled with service-oriented architectures. Due to the close relationship between resources monitoring, control and processing the

Fig. 11.2 VANET resource
management system

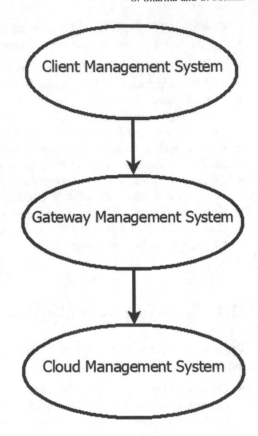

components required to build a large-scale eco-friendly task-driven system, it
also extends to personnel, material, energy, and to other aspects of the resources
that are used in processes.

- *Inter compatibility*: The future autonomous vehicles are expected to be constantly evolving. As such, it is important to be compatible in order to avoid
  disruption of the existing functionality and compatibility with existing
  interfaces.
- *Building joint applications, services and tools*: The future trends in software
  applications industry is rapid development of software by combining existing
  functionality in a complex way. It is anticipated that this trend will also
  empower next generation IoV applications. Since often the development of such
  system is very much task-oriented, new tools will need to be developed that
  ideally can be easily combined into building a larger system.
- *Real-time integration and operation*: The focus here is the optimization at
  architectural and process levels of the logical and physical system. To achieve
  this, several actions must be taken, for example: (i) identify task and information
  flows, and their operational requirements, (ii) investigate new novel

technologies that can meet the performance requirements, (iii) direct standards for Quality of Service (QoS), (iv) identify optimal system performance patterns.

- *Infrastructure development management*: Although, autonomous vehicle system infrastructures have up to now been designed for the long run, e.g., with 30–35 years lifetime in some cases, in the future they are expected to be more often modernized for increased reliability using latest technologies and provision new advanced functionality.

- *System management*: The next generation autonomous vehicle systems will be composed of millions of devices with different hardware and software configurations. There will be a need to automate the monitoring functions and also the soft-control of such systems such as dynamically identifying devices, systems and services offered by the infrastructures. It should be possible to perform software upgrades or re-configure existing systems. System management of a heterogeneous network of devices is critical for the feasibility of a cloud-based large-scale architecture. The need to use devices and systems from different manufacturers adds to system management functions, additional requirements such as flexibility and extensibility. The use of a feasible communication architecture will help mitigate some of these constraints. Scalability and robustness are also important factors when the number of managed (SOA enabled) devices increases.

- *Adaptability support*: As the autonomous vehicle systems evolve, the operators should not be bound to specialized control centers. Instead, they should be able to control and monitor the processes from anywhere. Adaptability support will need to be considered towards different perspectives of support for mobile devices, mobility of devices, mobile users and interaction with static and mobile infrastructure and mobility of services, where services can migrate between various systems and devices as per preferences set in a user's profile.

- *Real-time information processing*: It is a significant challenge to process information in real-time collected from data sources, such as sensors, with low latency and deliver the results to the end user, such as dashboard (or operator in front of dashboard), or database. Real-time information processing requires high performance set-ups by utilizing in-memory databases, efficient algorithms and even effective collaborative approaches for pre-filtering or pre-processing of information for a specific (business) objective and complex analysis of relevant (stream) events in-network and on-device.

- *Real-world business processes*: New and sustainable generation of business processes can rely on timely acquired information exchange and effective resource management. This has the potential to enhance and integrate real world enterprise services.

- *Scalability*: Scalability is a key feature required of complex systems such as autonomous vehicle system so that both vertical scalability (scale up) and horizontal scalability (scale out) are addressed. The service-oriented architecture paradigm must support, at all levels. Heterogeneous set of devices, networks and capabilities ranging from gigabit networks to low bandwidth, energy-constrained networked sensors and actuators connected over unreliable wireless links. This also implies that the overall network must be able to support cross-network interaction with devices that have completely different capabilities such as processing power, bandwidth and energy availability.

- *System simulation feasibility*: A holistic system analysis is needed in order to identify possible conflicts and side effects at an early stage before implementation of complex system. Traditionally, simulations of process systems are carried out at different levels with varying detail. It is expected that system-wide simulations will assist in designing, developing and operating future infrastructure and their interactions.

- *Unique resource identification*: For a complex system to be able to support service-oriented communication, it should incorporate standard universal resource identification and addressing a unique mechanism. This process should be flexible, scalable without additional overhead in configuration, performance and complexity.

11.4 IoV Applications of Cloud-based VANET

The cloud based VANET facilitates new potential real-time IoV applications via the local VANET cloud network, wide VANET cloud network, and central storage. Some such applications and scenarios are elaborated below

- *VANET cloud network with enhanced resource management*: such as Intelligent Disaster Management Reinforcement System (IDMRS) which is an efficient V2V communication and computational environment can be provided by the cloud-based VANET. For example, if a natural disaster like earthquake happens, then the Emergency Medical Services (EMS) vehicles should be able to coordinate with each other in order to process the evacuation. This process needs secure communication among multiple entities of resources such as nearby hospitals, ambulances, police vehicles, alternate escape routes, and real-time images of affected area. This IoV application provides multiple essential services such as routing techniques, dynamic bandwidth sharing methods, and various attack protection to autonomous or non-autonomous vehicles. The IDMRS is an application in an urban environment. The existing disaster management reinforcement system will be more effective by deploying IDMRS model. In the IDMRS model, real-time information of incidents along a road, such as an accident and its location, will be delivered to public safety vehicles through VANET. This model deployment may increase the total number of evacuated people from the roadside accident place in urban area. The proposed model can be used by the Department of Transportation (DoT) for efficient management of transport services and policy planning. The cloud-based VANET platform connected with a data source for real-time visual streaming will enhance the resource management capabilities for decision makers. The IDMRS supports a strategic operational process from decision making authorities of planning, protection, mitigation, response and recovery. IDMRS model performs real-time visual information collection, analysis, and sharing strategy from decision making authorities for fast efficient implementation among multiple entities such as Emergency Medical Services (EMS) vehicles, police vehicles, local and state government authorities and emergency services.

- *Bandwidth resource sharing*: This application provides unique feature to cloud-based VANET accessing vehicles to share their bandwidth resources as per their requirement. It is difficult to upload/download large volume files and access multimedia applications due to high-speed vehicles and limited wireless bandwidth. A cloud-based VANET may provide such application by upload or download files in distributive manner. A central storage VANET cloud network observes and analyzes the vehicles bandwidth requirements and share their bandwidth resources.

- *Real-time three-dimensional vehicle tracking, storing and sharing*: The CVCN resources utilized for traffic management analysis in a real-time vehicle tracking and sharing application. This application may provide other than traditional GPS, the real-time three-dimensional map, street views, intersection views with an adaptive optimize route to avoid traffic congestion. A vehicle in LVCN can request this application from WVCN and/or CVCN. Then the traffic analysis process starts in requested networks for the particular requested route. After the analysis the requested network suggests several routes based on the current traffic situations. Once a vehicle selects the route then requested cloud-based VANET keeps updating the driver with real-time traffic situations on the road during the entire trip. This application allows each vehicle to store the necessary map and share their trip maps with friends and family members, insurance partners and auto expert engineers on social networking sites.

- *High definition trip videos monitoring, storing and sharing*: A very large volume of hard disk is needed to store high definition trip video contents for an entire trip. Modern autonomous vehicles may be equipped with as many as ten cameras and possibly more so in the future. The information they generate may be on the order ten terabytes. In the case of accident, the police department is not able to make timely and proper decisions immediately after an accident. The cloud-based VANET architecture, a new distributed storage paradigm proposed here, can address this problem. This application enables vehicles to monitor, store, retrieve and share their high definition trip videos into their LVCN. A vehicle in a LVCN can request this application-based cloud services from WVCN and/or CVCN. The video contents will start upload or download process from CVCN in a real-time. The video contents will be separating and storing into multiple segments of the above-mentioned cloud networks along the entire trip. When an accident occurs, the police department can request high definition trip videos from different levels of the cloud-based VANET architecture to make the proper decision immediately, with due regard to the privacy of those involved in accidents. Besides being able to determine the exact events at the time of accidents, the users may derive other benefits such as incentives from their vehicle insurance providers.

A Unique Vehicle Identity (UVI) provides administrative authentication for IoV applications access. An UVI integrates a vehicle's smart information like on-board diagnostics and shares that information with the cloud. The vehicles of the cloud-based VANET instantly create a mobile Internet or an IoV in which the vehicles function as mobile routers. The IoV enables a multitude of features that are unique mobile Internet through moving vehicles with multi applications access. It introduces unique features to avoid identity infringement issues and enhances privacy in private data protection and trusted ID services. In the future, IoV cloud terminals be equipped with man-to-machine interfaces, be able to connect to in-vehicle screens, auto insurance, rescue operations, and vehicle remote inspections, and update the system with the vehicle's UVI.

11.5 VANET Communication Modes

The inter-vehicular communication can be categorized into two classes:

- Broadcast (1 × N): The information to exchange in VANET does not necessarily require an established connection between vehicles. Some information just needs to be broadcast over the VANET and, unlike connection oriented communication, it is not important to ensure if the broadcast is received by all the other cars of the VANET. For instance, the traffic and accident scenarios require an anonymous broadcast rather than full duplex communication between vehicles. The most common security threat to VANET in broadcast scenario is *impersonation attack* and *fabrication attack* as discussed in [21]. We further discuss the proposed solution to such attacks in a later section.
- Bilateral (N × M): This is a full duplex communication in VANET to establish connection between multiple (M) vehicles for information exchange. The sources and destinations are required to exchange certain information before exchanging and contents. The formal way to implement would be the exchange of a unique identification in VANET. It can impose the security threats such as wormhole attacks, fabrication attacks, impersonation attacks, reply attacks, denial of service attacks and physical attacks [27].

11.6 Proposed Distributive Security Algorithm for VANET Communication Modes

The security algorithm for the cloud-based VANET considers various types of internal and external attacks on resource management functions. The possible attacks on the proposed scheme include: message integrity, active level attack, acknowledgment message attack, message modeling attack, message mutation attack, message voiding attack, service message attack, bandwidth demand attack, bandwidth supply attack, Denial of service (DoS) attack and Sybil attack [2]. The fundamental requirement of the proposed algorithm is to have an authentication process in place that ensures the privacy of a VANET user. We have made some assumptions to implement our proposed security algorithm so that each vehicle will have access to the cloud-based VANET. Cloud-based VANET being a safety information sharing medium, needs secure and safe environment from attacks due to its highly dynamic nature, wireless communication medium and topology. Security issues of cloud-based VANET have very high impact on VANET system functions.

Unique vehicle identity (UVI) is a trusted entity in our proposed algorithm. The proposed algorithm also guarantees the message integrity in the cloud-based VANET by ensuring the delivery off the original unaltered message to the intended recipient. The proposed algorithm ensures the privacy and security of a VANET user in case of any malicious act and the true identity will be revealed only to the law enforcement agencies. The proposed algorithm includes a fast identity verification process, and security of management of resources required by the VANET users. In the beginning, the cloud-based VANET vehicles register themselves using a Certified Signature (CS), which they share via the cloud. The sender message consists of the following information.

$$CCV_S \longrightarrow CCV_R(UVI_S, CS_S, t, n, m)$$

where

CCV_S	Cloud-based VANET vehicle (sender)
CCV_R	Cloud-based VANET vehicle (receiver)
UVI_S	Unique vehicle identity (sender)
CS_S	Certified signature (sender)
t	time stamp
n	Certified signature expiration time
m	message

Then the identity verification process starts that first checks the vehicles UVI in the database in the given time n. In case the UVI does not exist in the database, then security alarms are generated and stored against that particular UVI. Those alarms need to be resolved subsequently by passing through the identity verification and enrollment process. The process generates a pseudo identity (P_{ID}) and stores in the database for that vehicle's credential and gives temporary access to the cloud applications with limited time stamp. This vehicle's credential will be used for communication among vehicle and network subsequently. Thus,

$$CCV_5 \longrightarrow CCV_R(UVI_S, CS_5, t, n, P_{ID}, m)$$

Each vehicle checks the authenticity of the message before broadcasting to others. The cloud-based VANET reports the malicious behavior of any vehicle to the law enforcement agency (LA). It contains the P_{ID} with UVI and report the malicious event:

$$CCV_R \longrightarrow LA(UVI_S, CS_S, t, n, P_{ID}, m)$$

Once the LA receives negative reports, it sends a cancel request (CR) with UVI_S and CS_S to the cloud which subsequently serves to block that CCV. The report is stored in a database for future use. Once the UVIs and CS_S are canceled by the LA. Until then CCV_S may continue to send messages:

$$LA \longrightarrow Cloud(CR, UVIs, CSs)$$

The cloud-based VANET divides the geographical region size based on the number of average vehicles in that region. The UVI consists of region information in the form of geo-coordinates data that helps to develop 3-D map for user.

We propose a distributive information approach to avoid the security threats mentioned in the previous section. Let B be the broadcast channel and F the full duplex channel, which are used to transmit or receive in the two modes discussed in the previous section. In a broadcast scenario, the broadcasting triggers due to several variables that could be from internal or external environment of vehicle. All the vehicles are registered with an authentication center and are authenticated using cloud connectivity. Various symbols and acronyms are defined below.

B   Broadcast channel
F   Full duplex channel
BM   Broadcast mode
FM   Full duplex mode
CC   Cross cloud
AUTH   Authentication
CCR   Cross VANET
TODS   Temporary On-board Record Database
COMP   Comparative
N   Number of vehicles in VANETVANET

Figure 11.3 illustrates COMP algorithm compare the broadcasted received packet with other packets received from same vicinity. As N increases, the comparative algorithm reliability increases. If some of the packets are malicious and being broadcast by an unauthorized user, they will be discarded and sender's information will be recorded in TODS. On the other hand, if the FM is enabled, the cross authentication is required from VANET and the cloud. In the FM mode, the communication is intended to be long term and connection oriented which is why

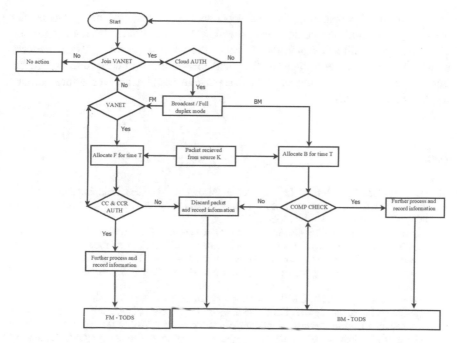

Fig. 11.3 Distributive security algorithm for VANET

the cross authentication is processed after receiving the packet. The bidirectional arrows show that the information is exchanged both ways between the entities. The distributive TODS (FM = BM) increases the speed for lookup and comparison.

11.7   Conclusion

This chapter provides an insight into the communication and security aspects of connected vehicles and proposes a cloud-based VANET with security-aware resource management and IoV applications. This chapter also provides a comparative survey of the previous work done in inter-vehicular communication with respect to security measures for information exchanged between over vehicular network. Future generation of IoV applications will need to be developed at a rapid pace in order to capture the agility required by modern businesses trends. We propose a distributive and comparative methodology for a secure vehicular network to communicate. The concept of cross authentication of users between VANET and cloud is introduced to block the penetration of unauthorized users for any malicious attack. Typical IoV-related software development approaches will need to be adjusted to the new paradigm of complex VANET systems with the collaboration and multi-layer interactions among systems. Significant work needs to be invested towards further investigating the interdependencies of resource management and needs of all targeted service domains as well as the technologies for realizing them. The proposed architecture attempts to cover the basic needs of VANET resource monitoring, management, data handling and integration by taking into consideration the advanced technologies and concepts that could empower future autonomous vehicle systems.

References

1. Gerla M, Lee EK, Pau G, Lee U (2014) Internet of vehicles: from intelligent grid to autonomous cars and vehicular clouds. In: Proceedings of the IEEE world forum internet things (WF-IoT), pp 241–246, Mar 2014
2. Sharma S, Mohan S (2016) Cognitive radio adhoc vehicular network (VANET): architecture, applications, security requirements and challenges. In: IEEE international conference on advanced networks and telecommunications systems (ANTS). IEEE, pp 1–6
3. Fangchun Y, Shangguang W, Jinglin L, Zhihan L, Qibo S (2014) An overview of internet of vehicles. China Commun 11(10):1–15
4. Jiang X, Cao X, Du DHC (2014) Multihop transmission and retransmission measurement of real-time video streaming over DSRC devices. In: Proceedings of the IEEE 15th international symposium on world wireless, mobile multimedia network (WoWMoM), pp 1–9, June 2014
5. Sharma S, Mohan S (2016) Dynamic spectrum leasing methodology (DSLM): a game theoretic approach. In: 2016 IEEE 37th Sarnoff symposium. IEEE, pp 43–46
6. Ma T et al (2015) Social network and tag sources based augmenting collaborative recommender system. IEICE Trans Inf Syst E98-D(4):902–910
7. Fu Z, Sun X, Liu Q, Zhou L, Shu J (2015) Achieving efficient cloud search services: multi-keyword ranked search over encrypted cloud data supporting parallel computing. IEICE Trans Commun 98(1):190–200
8. Chen J et al (2011) Measuring the performance of movement assisted certificate revocation list distribution in VANET. Wirel Commun Mob Comput 11(7):888–898
9. Su K, Li J, Fu H (2011) Smart city and the applications. In: Proceedings of the IEEE international conference on electronics, communications and control (ICECC), Zhejiang, China, pp 1028–1031, Sept 2011
10. Campolo C, Iera A, Molinaro A, Paratore SY, Ruggeri G (2012) SMARTCAR: an integrated smartphone-based platform to support traffic management applications. In: Proceedings of the IEEE international workshop vehicular traffic management for smart cities (VTM), Dublin, Ireland, pp 1–6, Nov 2012
11. Amadeo M, Campolo C, Molinaro A (2012) Enhancing IEEE 802.11p/WAVE to provide infotainment applications in VANETs. Ad Hoc Netw 10(2):253–269
12. Guerrero-Ibez JA, Flores-Corts C, Zeadally S (2013) Vehicular adhoc networks (VANETs): architecture, protocols and applications. In: Next-generation wireless technologies. Springer, London, U.K., pp 49–70
13. Aslam B, Wang P, Zou CC (2013) Extension of internet access to VANET via satellite receive-only terminals. Int J Ad Hoc Ubiquitous Comput 14(3):172–190
14. Toor Y, Muhlethaler P, Laouiti A (2008) Vehicle ad hoc networks: applications and related technical issues. IEEE Commun Surv Tutor 10(3):74–88 (3rd Quarter)
15. Murray P (2009) Enterprise grade cloud computing. In: Proceedings of the third workshop on dependable distributed data management, Nuremberg, Germany, pp 1–1
16. Böhm M, Stefanie L, Christoph R, Helmut K (2011) Cloud computing–outsourcing 2.0 or a new business model for IT provisioning? In: Application management. Gabler, pp 31–56

17. Youseff L, Butrico M, Da Silva D (2008) Toward a unified ontology of cloud computing. In: Grid computing environments workshop, 2008. GCE 08, pp 1–10
18. Bernstein D, Vidovic N, Modi S (2010) A cloud PAAS for high scale, function, and velocity mobile applications with reference application as the fully connected car. In: 5th international conference on systems and networks communications (ICSNC). IEEE, pp 117–123
19. Chee BJS, Curtis FJ (2010) Cloud computing: technologies and strategies of the ubiquitous data center. CRC Press, Inc.
20. Wang M, Liu D, Zhu L, Xu Y, Wang F (2014) LESPP: lightweight and efficient strong privacy preserving authentication scheme for secure VANET communication. Springer J Comput 1–24
21. Sharma S, Baig Awan M, Mohan S (2017) Cloud enabled cognitive radio adhoc vehicular networking (CRAVENET) with security aware resource management and internet of vehicles (IoV) applications. In: 2017 IEEE international conference on advanced networks and telecommunications systems (ANTS). IEEE, pp 1–6
22. Sharma S, Muhammad A, Mohan S (2018) Cloud enabled cognitive radio adhoc vehicular networking with security aware resource management and internet of vehicles applications. U. S. Patent Application 16/058,488, filed 6 Dec 2018
23. Sharma S, Muhammad A, Mohan S (2018) Smart vehicular hybrid network systems and applications of same. U.S. Patent Application 15/705,542, filed 29 Mar 2018
24. Ghanshala KK, Sharma S, Mohan S, Joshi RC (2018) Cloud-based cognitive radio adhoc vehicular network architecture: a next-generation smart city. In: 2018 IEEE 9th annual information technology, electronics and mobile communication conference (IEMCON). IEEE, pp 145–150
25. Ghanshala KK, Sharma S, Mohan S, Nautiyal L, Mishra P, Joshi RC (2018) Self-organizing sustainable spectrum management methodology in cognitive radio vehicular adhoc network (CRAVENET) environment: a reinforcement learning approach. In: 2018 first international conference on secure cyber computing and communication (ICSCCC). IEEE, pp 168–172
26. Sharma S, Mohan S (2017) Human bond communication using cognitive radio approach for efficient spectrum utilization. In: Human bond communication: the Holy Grail of Holistic communication and immersive experience, pp 97–113
27. Wei Y (2014) An anonymous routing protocol with authenticated key establishment in wireless ad hoc networks. Int J Distrib Sens Netw 2014(Article ID 212350)

Index

© Springer Nature Switzerland AG 2020
Z. Mahmood (ed.), *Connected Vehicles in the Internet of Things*,
https://doi.org/10.1007/978-3-030-36167-9